Histocompatibility Testing

The Practical Approach Series

SERIES EDITORS

D. RICKWOOD
Department of Biology, University of Essex
Wivenhoe Park, Colchester, Essex CO4 3SQ, UK

B. D. HAMES
Department of Biochemistry and Molecular Biology,
University of Leeds, Leeds LS2 9JT, UK

Affinity Chromatography
Anaerobic Microbiology
Animal Cell Culture (2nd Edition)
Animal Virus Pathogenesis
Antibodies I and II
Biochemical Toxicology
Biological Data Analysis
Biological Membranes
Biomechanics—Materials
Biomechanics—Structures and Systems
Biosensors
Carbohydrate Analysis
Cell–Cell Interactions
Cell Growth and Division
Cellular Calcium
Cellular Neurobiology
Centrifugation (2nd Edition)
Clinical Immunology
Computers in Microbiology
Crystallization of Nucleic Acids and Proteins
Cytokines
The Cytoskeleton
Diagnostic Molecular Pathology I and II
Directed Mutagenesis
DNA Cloning I, II, and III
Drosophila
Electron Microscopy in Biology
Electron Microscopy in Molecular Biology
Electrophysiology
Enzyme Assays
Essential Molecular Biology I and II
Eukaryotic Gene Transcription
Experimental Neuroanatomy
Fermentation
Flow Cytometry
Gel Electrophoresis of Nucleic Acids (2nd Edition)
Gel Electrophoresis of Proteins (2nd Edition)
Genome Analysis
Growth Factors
Haemopoiesis
Histocompatibility Testing
HPLC of Macromolecules
HPLC of Small Molecules
Human Cytogenetics I and II (2nd Edition)
Human Genetic Diseases
Immobilised Cells and Enzymes

Immunocytochemistry
In Situ Hybridization
Iodinated Density Gradient Media
Light Microscopy in Biology
Lipid Analysis
Lipid Modification of Proteins
Lipoprotein Analysis
Liposomes
Lymphocytes
Mammalian Cell Biotechnology
Mammalian Development
Medical Bacteriology
Medical Mycology
Microcomputers in Biochemistry
Microcomputers in Biology
Microcomputers in Physiology
Mitochondria
Molecular Genetic Analysis of Populations
Molecular Neurobiology
Molecular Plant Pathology I and II
Monitoring Neuronal Activity
Mutagenicity Testing
Neural Transplantation
Neurochemistry
Neuronal Cell Lines
Nucleic Acid and Protein Sequence Analysis
Nucleic Acid Hybridisation
Nucleic Acids Sequencing
Oligonucleotides and Analogues
Oligonucleotide Synthesis
PCR
Peptide Hormone Action
Peptide Hormone Secretion
Photosynthesis: Energy Transduction
Plant Cell Culture
Plant Molecular Biology
Plasmids
Pollination Ecology
Postimplantation Mammalian Embryos
Preparative Centrifugation
Prostaglandins and Related Substances
Protein Architecture
Protein Engineering
Protein Function
Protein Phosphorylation
Protein Purification Applications
Protein Purification Methods
Protein Sequencing
Protein Structure
Protein Targeting
Proteolytic Enzymes
Radioisotopes in Biology
Receptor Biochemistry
Receptor–Effector Coupling
Receptor–Ligand Interactions
Ribosomes and Protein Synthesis
RNA Processing
Signal Transduction
Solid Phase Peptide Synthesis
Spectrophotometry and Spectrofluorimetry
Steroid Hormones
Teratocarcinomas and Embryonic Stem Cells
Transcription Factors
Transcription and Translation
Tumour Immunobiology
Virology
Yeast

Histocompatibility Testing

A Practical Approach

PHILIP DYER

Northwest Regional Tissue Typing Laboratory, Manchester

and

DEREK MIDDLETON

Northern Ireland Tissue Typing Laboratory, City Hospital, Belfast

at
OXFORD UNIVERSITY PRESS
Oxford New York Tokyo

Oxford University Press, Walton Street, Oxford OX2 6DP
Oxford New York Toronto
Delhi Bombay Calcutta Madras Karachi
Kuala Lumpur Singapore Hong Kong Tokyo
Nairobi Dar es Salaam Cape Town
Melbourne Auckland Madrid
and associated companies in
Berlin Ibadan

Oxford is a trade mark of Oxford University Press

A Practical Approach is a registered trade mark
of the Chancellor, Masters, and Scholars of the University of Oxford
trading as Oxford University Press

Published in the United States
by Oxford University Press Inc., New York

A catalogue record for this book is available from the British Library

Library of Congress Cataloging in Publication Data
Histocompatibility testing : a practical approach / edited by P. Dyer and D. Middleton.
p. cm.—(The Practical approach series)
Includes bibliographical references and index.
1. Histocompatibility testing. I. Dyer, P. A. II. Middleton, D.
III. Series.
QR187.H57H578 1993 616.07'92—dc20 92–38629

ISBN 0–19–963364–9 (h/b)
ISBN 0–19–963363–0 (p/b)

Typeset by Footnote Graphics, Warminster, Wilts
Printed in Great Britain by Information Press Ltd, Eynsham, Oxon

Foreword

The HLA system developed out of the definition of antigens that came from the application of some simple statistical ideas to fairly crude serological data using cumbersome leukocyte agglutination assays starting in the early 1960s. From this has grown our extraordinary depth of information on the HLA system and the genetic region on the short arm of chromosome 6 that surrounds it. Many HLA genes are now defined at the DNA level and alongside them, between the genes of the HLA-A, -B, -C, and HLA-D regions, are a whole host of other genes, some with functions related to the immune system but many others not. The DNA sequence data has also confirmed and enormously extended the extraordinary range of individual variation that has been observed for the genes of the HLA system.

In parallel with the serological definition of HLA antigens, the development of the genetics, and our knowledge at the DNA level, has come the functional analysis of the HLA genes. Now we know the essential role they play in presenting antigens, in the form of peptides derived from processed protein, to the T cell receptor. A whole new technology has recently developed for the study of peptides bound to HLA molecules, and now we can understand immune response variation controlled by the HLA region in terms of differential affinity of bound peptides. Thus, we are now in a position of extraordinary power in understanding the role of the HLA system in determining susceptibility to a variety of diseases with an immune aetiology, as well as in controlling transplantation rejection.

Underlying the development of all of this knowledge has been the development of a wide variety of techniques for studying the genes of the HLA system and their expressed products. These techniques range from serological assays, now often carried out using monoclonal antibodies, the more cumbersome techniques for the cellular definition of antigens, which have, however, played a major role in functional analysis, to the techniques of DNA analysis that define variation at the level of the nucleotide sequence. This very useful volume is an up-to-date review of the variety of techniques used in clinical histocompatibility testing, together with essential background information on the HLA system itself and its pattern of variation in different populations. This collection of chapters should be a most valuable source book for anyone concerned with the major human histocompatibility system, HLA, especially its clinical applications.

London WALTER F. BODMER
September 1992

Contents

List of contributors xvii

Abbreviations xix

1. Introduction 1

Philip Dyer and Derek Middleton

1. Objectives of this book 1

2. History of HLA 1

3. Structure of the Mhc and HLA genes and molecules 3
- Genetic map of the human Mhc 3
- Structure of HLA class I molecules 4
- Structure of HLA class I genes 5
- Structure of HLA class II molecules 5
- Structure of HLA class II genes 5

4. Clinical applications 6
- Transplantation 6
- Disease studies 9

Acknowledgements 10

References 10

2. HLA typing reagents: alloantisera and monoclonal antibodies 13

Derek Middleton, Julia Bodmer, Judith Heyes, and Steven Marsh

1. Introduction 13

2. HLA alloantibodies 13
- Source of alloantibodies 13

3. Screening for HLA alloantibodies 14
- Lymphocytes used for class I antibody screening 15
- Lymphocytes used for class II antibody screening 18
- Screening in small laboratories 21

4. Procurement of large volumes of sera for use as reagents 21
- Storage of serum donation 22
- Testing of serum donation 23
- Manipulation of sera to produce useful reagent 23

5. Computer analysis 25

6. Records of sera 26

7. Exchange of sera 26
- With other laboratories 26
- With national organizations 27
- With commercial companies 27

8. Monoclonal antibodies 27
- Monoclonal antibodies versus alloantisera 28
- Production of monoclonal antibodies 29
- Immunization schedules 29
- Immunization protocols 30
- Example of a fusion schedule 34
- Screening of hybridoma supernatants following fusion 38
- The cloned, screened, monoclonal antibody 40

9. The use of monoclonal antibodies in cytotoxicity assays 41
- Form of monoclonal antibody 41
- Bulk storage 41
- Absorption 42
- Dilution of antibody 42
- Carryover 42
- Rabbit complement 43

10. Use of monoclonal antibodies for definition of sequence epitopes 43
- Initial monoclonal antibody screening 44
- Further screening 44
- Correlation of antibody reaction pattern with sequence 45
- Further reading 50

Acknowledgement 50

References 50

3. Clinical HLA typing by cytotoxicity 51

Christopher Darke and Philip Dyer

1. Introduction 51

2. Target cells, storage, and transport 52
- Preparation of lymphocyte suspensions for HLA-A, -B, -C typing 53
- Preparation of B lymphocytes for HLA-DR/DQ typing 56
- Storage 58
- Transport 59

3. Cytotoxicity assays 60
- Technique 60
- Complement 63

4. Automation, reading, and interpretation 65
- Automation 65

Reading of tissue typing trays 67
Interpretation 68

5. Family studies, antigen and allele frequencies, allelic association, ethnic groups 72
Family studies 72
HLA antigen and allele frequencies 75
HLA antigens and alleles in different ethnic populations 76

6. Statistical tests 76

7. Studies of HLA and disease 77

8. Reporting results 78
Clinical testing 79
Journal publication 79

References 79

4. Antibodies and crossmatching for transplantation 81

Susan Martin and Frans Claas

1. Screening for HLA alloantibodies 81
Introduction 81
Screening for alloantibodies in relation to organ transplantation 82

2. Antibody classes 88

3. Screening for non-HLA antibodies 90
Autoantibodies 90
Monocyte antibodies 91
Epithelial cell antibodies 91
Anti-idiotype antibodies 93

4. Crossmatch procedures 94
Introduction 94
Technical aspects 96

5. HLA matching and recipient selection 101
Renal transplantation 101
The need for quality controls in HLA typing and matching 102
Matching in highly sensitized patients 102
Matching for other solid organs and tissues 104

Acknowledgements 104

References 104

5. Molecular methods 107

David Savage, Lee Ann Baxter-Lowe, Jack Gorski, and Derek Middleton

1. Introduction 107

2. Application of restriction fragment length polymorphism typing 107

3. HLA class II cDNA probes 108
   - Source of cDNA probes 108
   - Preparation of plasmid DNA 109
   - Removal of probe DNA from plasmid 111

4. Extraction, assay, and storage of genomic DNA 113

5. Restriction endonuclease digestion of genomic DNA 114

6. Agarose gel electrophoresis 115

7. Southern blotting 117

8. Radiolabelling of probe DNA 119

9. Pre-hybridization and hybridization of membrane 120

10. Post-hybridization procedures 122
   - Stringency washes 122
   - Autoradiography 122
   - Dehybridization 122

11. Interpretation of RFLP data and trouble-shooting 123
   - Regime for interpretation 123
   - *Taq*I/DRB RFLPs 124
   - *Taq*I/DQB RFLPs 124
   - *Taq*I/DQA RFLPs 124
   - Typical DR/DQ associations 124
   - Novel RFLPs and atypical DR/DQ associations 125
   - Trouble-shooting 125

12. Theory and application of oligotyping 125

13. Amplification using the PCR 127

14. Sequence-specific oligonucleotide probe hybridization 131
   - Controls 131
   - Preparation of dot blots 132
   - SSOPH 134

References 141

6. Cellular methods 143

Nancy L. Reinsmoen

1. Mixed lymphocyte culture 143
   - Introduction 143
   - Results and interpretation 146

2. HLA-Dw typing using homozygous typing cells (HTC) 147
   - Results and interpretation 148

3. Primed lymphocyte test (PLT) 148
Results and interpretation 151

4. T cell precursors 151

5. Epstein–Barr virus (EBV) transformed lymphoblastoid cell lines 152

6. Propagation of graft infiltrating cells 154

7. Clinical applications 155

References 157

7. HLA class I gene and protein sequence polymorphisms 159

Ann-Margaret Little and Peter Parham

1. Introduction 159
Application 160
Non-serological techniques which define polymorphisms 161

2. Immunoprecipitation of HLA-A, -B molecules and electrophoretic characterization 161
Background 161
Methodology 161
Variations in methodology 165
Analysis of HLA class I IEF banding patterns 168

3. Nucleotide sequencing of HLA class I genes 169
Amplification of cDNA by the polymerase chain reaction (PCR) 169
Cloning amplified HLA genes into M13 172
Annealing and sequencing reactions 175
Sequencing analysis 182

4. Class I typing—the future 188

Acknowledgements 189

References 189

8. HLA antigens in tissues 191

Marlene L. Rose

1. Introduction 191

2. Methods 192
Preparation of tissue 192
Antibodies 193
Staining methods 194

3. Distribution of HLA antigens in normal and diseased tissue 199
Heart 199
Lungs 204
Liver 205
Kidney 205
Alterations in Mhc expression in non-transplant disease 206

4. Clinical applications 208

Acknowledgements 209

References 209

9. Paternity testing 211

Ernette du Toit

1. Introduction 211

2. Collection of samples 212

3. Techniques to determine the relevant genetic systems 213
Serological testing for HLA-A and -B antigens 213
Red cell antigens 214
Red cell enzyme and plasma protein systems 216
DNA typing 217

4. Analysis of data 218
HLA in paternity testing 219
Statistical analysis 220

5. Reporting 225

Acknowledgements 225

References 226

10. Mapping techniques used to isolate novel major histocompatibility complex genes 227

R. Duncan Campbell, John Trowsdale, and Ian Dunham

1. Introduction 227

2. Pulsed-field gel electrophoresis: a method for long-range mapping 228
Application of PFGE to mapping of the Mhc 235

3. Characterization of the Mhc genomic structure 239

4. Novel genes in the class II region 244

5. Novel genes in the class III region 245

6. Implication of novel genes in HLA disease associations 247
Acknowledgements 248
References 248

11. Data tables 251

1. Genes in the HLA region 251
2. Complete listing of recognized HLA specificities 252
3. Designations of HLA-A, -B, -C, and -E alleles 254
4. Designations of HLA-DR, -DQ, and -DP alleles 256
5. HLA-Bw4 and -Bw6 associated specificities 259
6. HLA-A, -B, -C, -DR, and -DQ allele frequencies in different ethnic groups 260
7. Frequencies and linkage disequilibrium for five-locus haplotypes in different ethnic groups (HLA-A, -C, -B, -DR, -DQ) 269
8. Frequencies of DRB1–DPB1 haplotypes in different ethnic groups 283
9. List of ethnic groups in Table 6 292
References 294

Appendix 295

Index 299

Contributors

LEE ANN BAXTER-LOWE
DNA Typing Laboratory, The Blood Centre of Southeastern Wisconsin, 1701 Wisconsin Avenue, Milwaukee, WI 53235, USA.

JULIA G. BODMER
Tissue Antigen Laboratory, Imperial Cancer Research Fund, PO Box 123, Lincoln's Inn Fields, London WC2A 3PX, UK.

WALTER F. BODMER
Imperial Cancer Research Fund, PO Box 123, Lincoln's Inn Fields, London WC2A 3PX, UK.

R. DUNCAN CAMPBELL
MRC Immunochemistry Unit, University of Oxford, South Parks Road, Oxford OX1 3QU, UK.

FRANS CLAAS
Department of Immunohaematology and Blood Bank, University Hospital, Building 1 E3-62, PO Box 9600, 2300 RC Leiden, The Netherlands.

CHRISTOPHER DARKE
Tissue Typing Laboratory, Regional Blood Transfusion Centre, Rhyd Lafar, St. Fagans, Cardiff CF5 6XF, UK.

IAN DUNHAM
Paediatric Research Unit, 8th Floor Guy's Tower, Guy's Hospital, London Bridge, London SE1 9RT, UK.

ERNETTE DU TOIT
Private Bag 4, Observatory 7935, Falmouth Road, Cape Town, South Africa.

PHILIP DYER
Tissue Typing Laboratory, St. Mary's Hospital, Hathersage Road, Manchester M13 0JH, UK.

JACK GORSKI
DNA Typing Laboratory, The Blood Centre of Southeastern Wisconsin, 1701 Wisconsin Avenue, Milwaukee, WI 53235, USA.

JUDITH HEYES
Tissue Antigen Laboratory, Imperial Cancer Research Fund, PO Box 123, Lincoln's Inn Fields, London WC2A 3PX, UK.

ANN-MARGARET LITTLE
Department of Cell Biology, Stanford University School of Medicine, Fairchild Centre, Stanford, CA 940305-5400, USA.

STEVEN MARSH
Tissue Antigen Laboratory, Imperial Cancer Research Fund, PO Box 123, Lincoln's Inn Fields, London WC2A 3PX, UK.

SUSAN MARTIN
Tissue Typing Laboratory, St. Mary's Hospital, Hathersage Road, Manchester M13 0JH, UK.

DEREK MIDDLETON
Tissue Typing Laboratory, City Hospital, Lisburn Road, Belfast BT9 7AD, UK.

PETER PARHAM
Department of Cell Biology, Stanford University School of Medicine, Fairchild Centre, Stanford, CA 940305-5400, USA.

NANCY L. REINSMOEN
Department of Surgery, University of Minnesota, Phillips-Wangensteen Building, 516 Delaware Street, S.E., Minneapolis, MN 55455, USA.

MARLENE L. ROSE
Immunology Laboratories, Harefield Hospital, Harefield, Uxbridge, Middlesex UB9 6JH, UK.

DAVID SAVAGE
Tissue Typing Laboratory, City Hospital, Lisburn Road, Belfast BT9 7AD, UK.

JOHN TROWSDALE
Human Immunogenetics Laboratory, Imperial Cancer Research Fund, PO Box 123, Lincoln's Inn Fields, London WC2A 3PX, UK.

Abbreviations

ABC-HRP	avidin–biotin complex with horseradish peroxidase
ACP1	acid phosphatase
ADA	adenosine deaminase
ADCC	antibody dependent cellular cytotoxicity
AHG	anti-human globulin
AK	adenylate kinase
AMPPD	3-(2′-spiroadamantane)-4-methoxy-4-(3″-phosphoryloxy)-phenyl-1,2-dioxetane
AMV	avian myeloblastosis virus
APAAP	alkaline phosphatase anti-alkaline phosphatase
APS	ammonium persulphate
AS	ankylosing spondylitis
ASHI	American Society of Histocompatibility and Immunogenetics
ATCC	American Type Culture Collection
BAL	bronchoalveolar lavages
BAT	HLA-B associated transcript
β_2-m	beta-2-microglobulin
Bf	properdin factor B
B-LCL	B cell lymphoid cell line
BSA	bovine serum albumin
C2	complement factor 2
C4	complement factor 4
CA2	carbonic anhydrase
CD8	cell differentiation antigen 8
cDNA	complementary DNA
CFB	complement-fixing buffer
CFDA	6-carboxyfluoroscein diacetate
CHO	carbohydrate
CLL	chronic lymphocytic leukaemia
cM	centi Morgan unit of genetic recombination
CO-11A2	collagen 11 gene
CPA UK	Clinical Pathology Accreditation United Kingdom
CPE	cumulative power of exclusion
CpG	cytidine-phosphate-guanosine
CREB	cyclic AMP response element binding
CREG	cross reacting group
CSA	cyclosporin A medium
CTL	cytotoxic T lymphocyte
CTLp	cytotoxic T cell precursor

CTS	Collaborative Transplant Study
CYP21	cytochrome P450 21-hydroxylase gene
DAB	3,3′-diaminobenzidine tetrahydrochloride
ddATP	2′,3′-dideoxyadenosine-5′-triphosphate
ddCTP	2′,3′-dideoxycytidine-5′-triphosphate
ddH_2O	deionized distilled water
ddGTP	2′,3′-dideoxyguanosine-5′-triphosphate
ddNTP	2′,3′-dideoxynucleoside-5′-triphosphate
ddTTP	2′,3′-dideoxythymidine-5′-triphosphate
DigMAAHS	digoxigenin-3-O-methylcarbonyl-ε-aminocaproic acid-N-hydroxy-succinimide
dITP	2′-deoxyinosine-5′-triphosphate
DMF	dimethylformamide
DMSO	dimethyl sulphoxide
DNase	deoxyribonuclease
dNTP	2′-deoxynucleotide-5′-triphosphate
DNV	double normalized values
DTT	dithiothreitol
4E	HLA class I MoAb reacts with HLA-A29–33 and all HLA-B antigens
EBV	Epstein–Barr virus
EDTA	ethylenediamine tetracetic acid
ELISA	enzyme-linked immunosorbent assay
ESD	esterase-D
f	antigen frequency
Fab	antibody binding fragment
FACS	fluorescent activated cell sorter
FCS	fetal calf serum
G11	non-HLA gene of human Mhc
GAG	β-galactosidase/anti-β-galactosidase
GLB	gel loading buffer
GLO	glyoxylase
G6PD	glucose-6-phosphate dehydrogenase
GVHD	graft-versus-host disease
H-2	the Mhc of the mouse
HAT	hypoxanthine, aminopterin, thymidine
HBSS	Hanks' balanced salt solution
HC	HLA class I heavy chain
HGPRT	hypoxanthine-guanine-phosphoribosyl transferase
HMT	hypoxanthine, methotrexate, thymidine
HP	haptoglobin
HPLC	high performance liquid chromatography
HSP70	heat shock protein 70
HTC	homozygous typing cell

HTF	*Hpa*II tiny fragment
HVR	hyper-variable regions
IDDM	insulin dependent diabetes
IEF	isoelectric focusing
IHW	International Histocompatibility Workshop
I.P.	intraperitoneal
IPTG	isopropyl-β-D-thiogalactopyranoside
I.V.	intravenous
KE3/5	non-H-2 genes in the mouse Mhc
LCL	lymphoblastoid cell line
LGT	low gelling temperature
LMP	low molecular weight polypeptide or large multifunctional protease
LSM	lymphocyte separation medium
M13	*E. coli* bacteriophage
M13 mp18	derivative of M13
M13 mp19	derivative of M13
Mhc	major histocompatibility complex
MLC	mixed lymphocyte culture
MoAb	monoclonal antibodies
MPC	magnetic particle concentrator
NBT	4-nitroblue tetrazolium chloride
NDS	0.05 M EDTA/10 mM Tris pH 9.5/1% Sarkosyl
NIH	National Institutes of Health
NLB	nuclei lysis buffer
OB	obliterative bronchiolitis
OFAGE	orthogonal field alteration gel electrophoresis
OLB	oligolabelling buffer
p	gene frequency
PAP	peroxidase anti-peroxidase
PBL	peripheral blood lymphocytes
PBMC	peripheral blood mononuclear cells
PBS	phosphate-buffered saline
PCR	polymerase chain reaction
PE	probability (or power) of exclusion
PEG	polyethylene glycol
PEY	phycoerythrin
PFGE	pulsed-field gel electrophoresis
PGD	phosphogluconate dehydrogenase
PGM1	phosphoglucomutase
PHA	phytohaemagglutinin
pI	isoelectric point
PI	paternity index
PLT	primed lymphocyte test

PMSF	phenylmethylsulphonyl fluoride
PVP	polyvinyl-pyrrolidone
RAM	rabbit anti-mouse
RCL	red cell lysis buffer
RD	non-HLA gene of the human Mhc
RFLP	restriction fragment length polymorphism
$r\mathrm{IL}_2$	recombinant interleukin 2
RING	novel genes of the HLA class II region
RNV	responder normalized values
RR	relative response
SA	*Staphylococcus aureus*
S.C.	subcutaneous
SD	standard deviation
SDS	sodium dodecyl sulphate
SET	sucrosc, EDTA, Tris
SPF	specific pathogen free
SSC	saline, sodium citrate
SSOP	sequence-specific oligonucleotide probes
SSOPH	sequence-specific oligonucleotide probe hybridization
SSPE	saline, sodium phosphate, EDTA
TAE	Tris-acetate, EDTA
TAP	transporter associated with antigen processing
TBE	Tris, boric acid, EDTA
TBS	Tris-buffered saline
TCR	T cell receptor
TE	Tris, EDTA
TEMED	N,N,N′,N′-tetramethylethylenediamine
TF	transferrin
6TG	6-thioguanine
TMAC	tetramethylammonium chloride
TNEN	Tris-HCl, NaCl, EDTA, Nonidet P-40
TNF	tumour necrosis factor
Tris	Tris(hydroxymethyl)aminomethane
TTBS	Tris-buffered saline with Tween 20
3UT	3′ antisense amplification primer
5UT	5′ universal amplification primer
VNTR	variable number of tandem repeats
W	likelihood of paternity
W6/32	HLA class I specific monoclonal antibody
WHO	World Health Organization
X-gal	5-bromo-4-chloro-3-indolyl-β-D-galactoside
YAC	yeast artificial chromosomes
YT	YT medium

1

Introduction

PHILIP DYER and DEREK MIDDLETON

1. Objectives of this book

This book aims to set out tried and tested protocols that will be used in a histocompatibility laboratory in support of both clinical service and research work. The techniques have been established by the chapter authors in their own laboratories but the reader is encouraged to amend and develop as necessary. The early chapters set out the serological and molecular techniques which are common in many histocompatibility laboratories. Later chapters deal with more specialized techniques which those keen to develop their range of services will find useful. The debate over the future of serological testing with the advent of molecular techniques continues but it must be remembered that the donor/recipient crossmatch remains of fundamental importance.

The design of this book and others in the series is to facilitate use at the bench. The field of histocompatibility is rapidly moving and the chapter authors have provided their most up-to-date methods. A direct enquiry to authors will give access to further modifications.

Chapter 11 stands apart from the other chapters since its aim is to provide key reference data. The most recent World Health Organization (WHO) HLA Nomenclature Committee report should always be consulted.

The remainder of this chapter serves as a brief introduction to human histocompatibility and the most recent International Histocompatibility Workshop (IHW) report will give more detail.

2. History of HLA

Early work on human histocompatibility antigens, as has often been the case in the field of genetics, grew from pioneering studies on animals. In the 1930s and 40s Peter Gorer and George Snell worked on the genetics and immunology of tumour transplantation in mice. Soon afterwards Peter Medawar persued the cellular rejection mechanisms responsible for murine skin and organ transplants.

In the early 1960s Dausset and Payne independently showed, respectively, that multi-transfused patients and parous women, often had circulating antibodies to the alloantigens or HLA antigens now known to be products of the human major histocompatibility complex (Mhc) genes. The term 'HLA' originated in discussions between pioneers of the field and it probably refers to an early nomenclature term '*H*uman *L*eucocyte antigen series *A*' meaning the first histocompatibility locus. The debate over the origin of this term has been entertainingly documented (1). More often and strictly speaking incorrect but appropriate, *HLA* is taken to mean *human leucocyte antigens.*

The impressive knowledge of the human Mhc and its antigenic products has been gathered in only 30 years, reflecting the remarkable international cooperation between many workers in the field. A series of collaborative IHWs has brought together scientists at regular intervals to discuss results of testing of shared reagents, initially antisera but more recently DNA probes. The IHWs have also been an opportunity for the introduction of new techniques and for the testing of a common set of reagents on many different human populations throughout the world. Following each IHW a WHO Nomenclature Committee has met to update the universally accepted vocabulary of HLA. The WHO Nomenclature reports are published in relevant scientific journals and become the standard for use by all workers (see Chapter 11). The WHO Nomenclature Committee designates recognized HLA alleles and specificities according to standard genetic practice, and the current nomenclature is a step in an evolutionary process which will continue as our knowledge of gene and protein sequences and serological reaction patterns grow. It is informative to review past reports to see how the current nomenclature has arisen. This is particularly important when reading past publications since old nomenclature may not obviously relate to that in current usage. The chief aim of the WHO Nomenclature is to provide a universally accepted workable system which allows communication between centres, and at the same time is informative of relationships between alleles and specificities.

The need to comprehend the human Mhc has been driven by the expansion of clinical organ transplantation. An important historical record from those surgeons developing the operating techniques and immunosuppressive drugs is available (2). A further catalyst in the development was the use of simple but powerful statistical techniques (see Chapter 3) and computer handling of large data sets to interpret complex reaction patterns. Much of this work was pioneered by van Rood who also proposed that the HLA antigens were coded for by a multi-locus system. The discovery that some diseases were closely associated with particular HLA antigens induced others to join HLA research.

The fundamental role of the lymphocytoxicity crossmatch in preventing hyperacute rejection in kidney transplanation was established by Kissmeyer-Nielsen who showed that transplantation of a kidney following a positive crossmatch between recipient serum and donor lymphocytes led to immediate irreversible rejection (see Chapter 4).

Many individuals have contributed to current knowledge of histocompatibility and Benacerraf, Dausset, and Snell were awarded the Nobel Prize in 1980 for their work on the role of the Mhc in the immune response, the discovery of human Mhc antigens, and in mouse immunogenetics respectively.

Initially, leuco-agglutination techniques were used to define human Mhc antigens but lymphocytoxicity (see Chapter 3) gained dominance once Terasaki and McClelland developed the microlymphocytoxicity test. This assay is now the routine method for serological typing and donor/recipient crossmatching. No kidney transplant should proceed without a lymphocytotoxic crossmatch test being performed. Modifications of the microlymphocytotoxicity assay have been made, mainly by altering the dyes used to assess the end point of cell killing or by changing incubation times and incubation temperatures (see Chapter 3), but the principle of target cell killing by HLA antigen specific antibodies remains unchanged. The development of monoclonal antibody (MoAb) technology promised better quality monospecific antisera for the serologist, but alloantisera from gravid women are still most commonly used due to the lack of cytotoxic MoAb for most HLA antigens (see Chapter 2).

The development of techniques for the identification of gene sequence variations gave rise to the linked restriction fragment length polymorphism (RFLP) procedure which can identify HLA-DR and -DQ serologically defined specificities (see Chapter 5). Furthermore, HLA-DP RFLP's gave the first opportunity for many laboratories to type for the polymorphism of this gene. With the advent of DNA amplification by the polymerase chain reaction (PCR) and knowledge of HLA allele sequences for the synthesis of sequence specific oligonucleotide probes (SSOP), a more precise definition of HLA gene polymorphisms using small amounts of DNA from almost any type of tissue is now possible.

3. Structure of the Mhc and HLA genes and molecules

3.1 Genetic map of the human Mhc

With the rapid advance in genetic mapping and sequencing techniques the number of genes found in the area of the human Mhc on the short arm of chromosome 6 is rapidly increasing. A recent comprehensive map of the Mhc listed 15 class I, 18 class II, and 49 other genes (see Chapter 10). Inevitably, some of these genes have no immunological function but are found in this part of the genome by chance. Alternatively, the genes may code for proteins whose functions are not fully understood but they may turn out to be important true Mhc genes.

To simplify discussion, Klein (3) proposed a classification of Mhc coded proteins grouping those with similar structure and function; thus HLA-A, -B, -C molecules were assigned to class I, with HLA-DR molecules assigned to class II. The gene products coded for in the region between HLA-DR and HLA-B of chromosome 6 were originally grouped in class III. With time this classification has been extended to the Mhc genes in addition to their products. It has been argued that the term 'class III' to describe neither class I nor class II has become increasingly vague since these genes and their products are structurally and functionally diverse. Use of 'class III' as a nomenclature should be discouraged. The extent of the Mhc has never been defined.

3.2 Structure of HLA class I molecules

Class I molecules consist of a heavy chain of 340 amino acids (45 kd) encoded by genes within the Mhc, and a light chain of 12 kd, the invariant β_2-microglobulin (β_2-m), coded for on chromosome 15. The functional class I molecule is not expressed at the cell surface in the absence of β_2-m; for example the cell line *Daudi* is class I negative since it lacks the β_2-m gene.

The intracellular portion of class I molecules is about 30 amino acids long and terminates in a hydrophilic carboxyl region. The short (25 amino acids) hydrophobic transmembrane portion is typical of that seen in other membrane bound proteins. The three extracellular domains of the heavy chain are each about 90 amino acids long and extend from the membrane adjacent $\alpha3$ region to the most membrane distal regions $\alpha2$ and $\alpha1$ which have carbohydrate moieties attached. Most of the sequence variation in class I molecules is found in the $\alpha1$ and $\alpha2$ regions and occurs in hyper-variable regions (HVR) (see Chapter 7). The light chain β_2-m associates non-covalently with the $\alpha3$ domain.

The three dimensional crystallographic structure of HLA-A2 was resolved in 1987 (4) and this gave rise to many studies on the functional relevance of the sites of polymorphism and binding to the T cell receptor (TCR). The two alpha helices of the $\alpha1$ and $\alpha2$ regions lie adjacent on a beta pleated sheet; between these is a groove which is most open at the centre of the structure (see Chapter 7). Surprisingly the crystallographic studies revealed a small (approximately 18 amino acids) peptide to be bound in the groove and attached to the beta pleated sheet. This is now known to be endogenous peptide and there has been much speculation over its function (5).

The location of polymorphic antigenic sites (epitopes) in the class I HLA molecule groove, near to bound peptide and near to the sites of interaction with the TCR, has led to speculation on the mechanistic role of these polymorphisms in organ transplantation and disease susceptibility (6). It is interesting to note that serological specificities known to form cross reacting antigen groups (CREGs, see Chapter 3) are not always located in the same part of the molecule; they can be found on either of the $\alpha1$ or $\alpha2$ helices, or on the beta pleated sheet. Some parts of the molecule are likely to be more accessible to the TCR than others, which means some specificities could be of greater functional relevance than others.

3.3 Structure of HLA class I genes

There are three commonly recognized HLA class I genes, HLA-A, -B, and -C, coding for serologically defined antigens, and in addition three other class I genes, HLA-E, -F, and -G, are known to code for functional molecules. Also recognized are two pseudogenes, HLA-H and HLA-J. The typical HLA class I gene consists of eight exons, including a signal peptide, occupying 3.5 kb (7). Exons two to four correlate with the extracellular regions α1–α3 respectively, and most polymorphic variation is found in sequences coding for amino acids 60–100 of α1 and 150–200 of α2.

3.4 Structure of HLA class II molecules

HLA class II molecules consist of a heterodimer with a heavy chain (α) of 31–34 kd and a light chain (β) of 26–29 kd; the size difference is accounted for by carbohydrate moieties. The α chain of HLA-DR molecules is invariant. Both α and β chains have intracellular, transmembrane, and two extracellular regions. The three dimensional structure of class II molecules is presumed to resemble that of class I molecules, and a hypothetical model has been proposed (8). The peptide bound to HLA class II proteins is usually accepted to be of exogenous origin in contrast to class I bound endogenous peptide. Again the most polymorphic regions of the molecule are found in the terminal regions of the extracellular domains. There are HVRs between amino acids 25 to 40 and 65 to 80 of HLA-DRB1 and these are the target regions for typing methods used to define class II gene polymorphisms by PCR amplification of DNA (see Chapter 5).

HLA-DQ and -DP class II molecules are polymorphic in both α and β chains. Definition of HLA-DQ polymorphisms by serological means is restricted to HLA-DQB, however molecular techniques are used successfully to define HLA-DQ and -DP A and B gene polymorphisms.

There is evidence in the mouse (9) that hybrid class II αβ heterodimers exist and these can even be coded for by genes in *trans* combination on different haplotypes, contrasting with the usual *cis* conformation of the same haplotype. In man such hybrid molecules are possible and their expression may be relevant to susceptibility of an individual to disease.

3.5 Structure of HLA class II genes

Whilst there is only one HLA-DR α chain gene (HLA-DRA) there are five HLA-DR β chain genes (HLA-DRB1–5), and four pseudogenes (HLA-DRB6–9). Not all HLA-DRB genes occur on all haplotypes and some combinations are haplotype specific (see Chapter 11). Both HLA-DQ and HLA-DP regions have pseudogenes located close to the expressed genes (see Chapters 10 and 11). The typical α chain gene has five exons; the leader sequence contains the 5′ untranslated region, the signal sequence, and codes for the first few amino acids of protein. Two exons code for α1 and α2 domains while the connecting peptide, transmembrane and cytoplasmic domains are coded in a fourth exon. Most of the 3′ untranslated region is coded for in the fifth exon. The β chain gene is similar except that the cytoplasmic domain has two or three exons (10).

4. Clinical applications

4.1 Transplantation

4.1.1 Introduction

The main clinical application of HLA has been in the field of human transplantation. The various aspects of crossmatching are dealt with in Chapter 4. The application of matching in transplantation varies with the nature of the transplant.

4.1.2 Kidney

The role of HLA antigens in transplant survival can be best demonstrated by the excellent graft outcome of kidneys transplanted between HLA identical siblings. Although the benefit of matching in sibling to sibling transplants is generally agreed, some centres have not been able to show an effect of matching for HLA antigens in cadaveric kidney transplantation. This could be due to the low number of cadaveric transplants analysed, to the low number of well-matched transplants performed, or to the optimization at that centre of other important variables in a transplant, minimizing the benefits of matching. The positive effect of HLA-A and -B matching in cadaveric kidney transplantation were first reported in the late 1960s'. It was not until 1978, when the antigens could first be detected by serology, that the outcome of HLA-DR matching on graft survival could be analysed. In some reports it would appear that matching for HLA-DR improves graft survival more than matching for HLA-A and -B.

Most multi-centre registries have concluded that HLA matching improves cadaveric kidney graft survival. The Collaborative Transplant Study (CTS) has reported a stepwise decline in graft survival with an increase in the number of HLA mismatches (11). In the United Kingdom it has been shown that cadaver kidneys transplanted with a maximum of either one HLA-A or -B antigen mismatched and no HLA-DR antigens mismatched, referred to as a 'beneficial' matched graft, have a superior graft survival to other grafts (12). A scheme has been implemented in which more than 90% of transplant units in the UK have agreed to share at least one kidney from a cadaver donor if there is a *beneficial* matched recipient in another centre. In the USA a scheme to share kidneys when all six HLA antigens (-A, -B, -DR) are matched between recipient and donor has proved very successful not only in obtaining a one year graft survival of 87%, compared to 76% in the recipients of the contralateral kidney, but also in transplanting sensitized patients (13). It has been calculated that in a recipient pool size of 8000, transplants with six antigens matched could be provided for 9%, transplants with no HLA-A, -B, and -DR antigens mismatched could be provided for 27%, and transplants with no HLA-B and -DR antigens mismatched could be provided for 66%.

HLA matching has also been reported to be advantageous in the following specific instances;

- in patients treated with Cyclosporin (14)
- on the survival of patients (11)
- in centres with excellent graft survival (centres with an overall 87% one year success rate) (15)
- on the function of grafts. At one year 52% of grafts with no mismatch for HLA-A, -B, and -DR had a creatinine of <130 μmol/litre whereas only 37% of grafts with six mismatches had a creatinine of <130 μmol/litre (15).

One problem in the typing of recipients and donors is the lack of suitable reagents. It has been shown that HLA matching is more influential in recipients and donors who have been typed for HLA-A and -B antigen splits rather than broad antigens (16). The advent of DNA techniques has highlighted discrepancies in HLA-DR assignment between serological (see Chapter 3) and RFLP (see Chapter 5) methods (17). The percentage of discrepancies varies from laboratory to laboratory and may well be indicative of the effort that a laboratory inserts into screening and exchanging reagents (see Chapter 2). Antigen misassignment is also an additional explanation for the lack of a matching effect in single centre studies. A recent report (18) has shown that if antigen misassignments are eliminated from the determination of HLA-DR antigens by using RFLP, the success rate of HLA matched cadaveric transplants at one year (90%) approaches that of HLA identical sibling grafts (93%). The time constraints of RFLP and of oligonucleotide typing in its present form prevent the prospective typing of cadaver donors. However new procedures such as the use of allele specific primers in the PCR (19) should see, at least for HLA-DR, the application of molecular techniques to prospective typing of cadaver donors.

There has been some indication that matching for HLA-DQ may have some benefit in graft survival (20, 21) but this question remains difficult to answer because of the strong linkage disequlibrium between HLA-DR and -DQ. Investigations have been performed, using oligonucleotide typing, into the effect that HLA-DP matching may have on graft survival. No effect has been found either in transplants from a single centre or in transplants selected from multiple centres with no mismatches for HLA-A, -B, and -DR (21, 22).

4.1.3 Pancreas

Both a single centre study and an international study have shown a beneficial effect of matching for HLA-DR on pancreas allograft survival (23, 24). A scheme to implement HLA-DR matching for pancreas transplantation was initiated in 1991 by Eurotransplant. The case for matching for HLA-A and -B in pancreas transplantation remains to be proven.

4.1.4 Liver

Previous reports have found no beneficial effect of HLA compatibility in liver transplants, indeed it was reported that HLA compatibility was associated with diminished allograft survival. This was thought to be due to a dualistic effect, on the one hand matching reduced the rejection process, but on the other hand it enhanced other immunological mechanisms leading to allograft dysfunction (25, 26). However recently a report from the same Pittsburgh group has found that HLA-A, and -B matching significantly improved the graft survival in liver transplantation whereas HLA-DR matching had a minimal effect (27).

4.1.5 Heart

Although no significant improvement in graft outcome was found with a decreasing number of HLA-A, -B, and -DR mismatches, a significant improvement has been reported in graft outcome for <2 HLA-B and -DR mismatches versus ⩾2 HLA-B and -DR mismatches in the CTS analysis (28). In a single centre study no correlation was seen between HLA-A and -B mismatching and rejection or rejection episodes. However better HLA-DR matching led to a reduction in the number of rejection episodes and an increase in the percentage of patients with no or mild rejection in a series of 519 patients (29).

For matching to be operative for heart, liver, and pancreas transplants it is necessary for the donor to be tissue typed before the removal of the organs using peripheral blood lymphocytes. This has been greatly facilitated by the use of magnetic beads, with suitable attached antibodies to class I or class II antigens, to isolate the required lymphocytes for typing (see Chapter 3 and 4).

4.1.6 Cornea

A significant benefit of HLA-A and -B matching in patients with or without vascularization has been demonstrated (30). However contradictory results regarding the role of HLA-DR matching have been reported (31, 32).

4.1.7 Bone marrow

The majority of bone marrow transplants are performed between HLA identical siblings. Some transplants have taken place with one haplo-identical siblings with additional sharing of some of the HLA-A, -B, and -DR antigens on the other haplotype. The chances of having a HLA identical sibling increases with the number of available siblings according to the formula $1 - 0.75^n$ where n is the number of available siblings. In order to help those patients without an HLA identical sibling, registers of unrelated tissue typed bone marrow donors have been established.

4.1.8 Platelets

An additional advantage of the above registers, when they have been based on blood donors, is that they provide a supply of HLA matched platelet donors. These are used on those occasions when a recipient has been sensitized by previous platelet transfusions and requires further platelets which are HLA matched in order for the therapy to be beneficial.

4.1.9 Blood transfusions

A further use of a bone marrow register has been in obtaining HLA-DR matched blood for transfusion to potential kidney transplant recipients. This has been performed following the finding that matched blood was less likely to cause sensitization than blood from random donors (33). The benefit of giving HLA-DR matched blood to reduce sensitization has still to be confirmed.

4.2 Disease studies

4.2.1 Introduction

There are two different approaches to studying whether the Mhc is relevant to a disease.

(a) Family studies. To reveal if there is *genetic linkage* between a HLA locus and a gene controlling the disease.

(b) Population studies. To reveal if there is an *association* between HLA antigens and the disease by comparing the frequency of HLA antigens in patients to the corresponding frequencies in controls.

HLA typing has been applied to numerous diseases (see Chapter 3). Initially HLA-A and -B antigens were examined, but with the advent of HLA-DR typing it was shown that in the majority of diseases the association was stronger with HLA-DR than HLA-A or -B. In some diseases it is difficult to ascertain which HLA locus has the highest association due to linkage disequilibrium. This is particularly true of HLA-DR and -DQ.

4.2.2 Aid to diagnosis

To date of all the associations found, only that of HLA-B27 with ankylosing spondylitis is requested routinely to aid diagnosis. Even in this disease HLA typing is probably requested too frequently and should only be used selectively for particular clinical problems. HLA typing of siblings of the proband can be performed to determine those siblings who are at risk for haemachromatosis or 21-OH deficiency.

4.2.3 Impact of DNA techniques

Molecular biology DNA techniques have provided a new impetus in the study of HLA and disease. They have removed the imprecise nature of serology allowing associations to be more accurately defined. They have enabled other HLA loci such as HLA-DQ and -DP to be examined and will allow investigations into those genes [transporter associated with antigen processing (TAP) and large multifunctional proteases (LMP)] which map close to HLA-DP (see Chapter 10).

Acknowledgements

We thank our past and present colleagues who have performed HLA typing in our laboratories for the benefit of patients; many of them participated in the development of some of the techniques set out in this book. Adrian Ivinson, Jeanie Martin, Susan Martin, and Bill Ollier made helpful comments and we are grateful for their support. The secretarial help of Karen Mills is much appreciated.

References

1. Terasaki, P. I. (ed.) (1990). *History of HLA: Ten Recollections.* UCLA Press, Los Angeles.
2. Terasaki, P. I. (ed.) (1991). *History of Transplantation: Thirty Five Recollections.* UCLA Press, Los Angeles.
3. Klein, J. (1979). *Science,* **203,** 516.
4. Bjorkman, P. J. (1987). *Nature,* **329,** 506.
5. Neefjes, J. J., Stollorz, V., Peters, P. J., Geuze, H. J., and Ploegh, H. L. (1990) *Cell,* **61,** 171.
6. Lechler, R. I., Lombardi, G., Batchelor, J. R., Reinsmoen, N., and Bach, F. H. (1990). *Immunol. Today,* **11(3),** 83.
7. Strachan, T. (1987). *Brit. Med. Bull.,* **43,** 1.
8. Brown, J. H., Jardetzky, T., Saper, M. A., Samtaoui, B., Bjorkman, P. J., and Wiley, D. C. (1988). *Nature,* **332,** 845.
9. Charron, D. J., Lotteau, V., and Turmel, P. (1984). *Nature,* **312,** 157.
10. Trowsdale, J. (1987). *Brit. Med. Bull.,* **43,** 15.
11. Opelz, G. (1991). *Transplant. Proc.,* **23,** 46.
12. Gilks, W. R., Bradley, B. A., Gore, S. M., and Klouda, P. T. (1987). *Transplantation,* **43,** 669.
13. Takemoto, S., Carnahan, E., and Terasaki, P. I. (1991). *Transplant. Proc.,* **23,** 1318.
14. Opelz, G. (1988). *N. Engl. J. Med.,* **318,** 1289.
15. Opelz, G., Schwarz, V., Engelmann, A., Back, D., Wilk, M., and Keppel, E. (1991). *Transplant. Proc.,* **23,** 373.

16. Opelz, G. (1988). *Lancet,* **2,** 61.
17. Opelz, G., Mytilineos, J., Scherer, S., Dunckley, H., Trejaut, J., Chapman, J., Middleton, D., Savage, D., Fischer, G., Bignon, J. D., Bensa, J. C., Albert, E., and Noreen, H. (1992). *Transplant. Int.,* **5,** S580.
18. Opelz, G., Mytilineos, J., Scherer, S., Dunckley, H., Trejaut, J., Chapman, J., Middleton, D., Savage, D., Fischer, G., Bignon, J. D., Bensa, J. C., Albert, E., and Noreen, H. (1991). *Lancet,* **338,** 461.
19. Olerup, O. and Zetterquist, H. (1991). *Tissue Antigens,* **37,** 197.
20. Sengar, D. P. S., Couture, R. A., Raman, S., Jindal, S. L., and Lazarovits, A. I. (1987). *Transplant. Proc.,* **19,** 3422.
21. Middleton, D., Savage, D., Trainor, F., and Taylor, A. (1992). *Transplantation,* **53,** 1138.
22. Middleton, D., Mytilineos, Y., Savage, D., Ferrara, G. B., Angelini, G., Amoroso, A., Trainor, F., Gaweco, A., Mazzola, G., Delfino, L., Berrino, M., and Opelz, G. (1992). *Transplant. Proc.,* **24,** 2439.
23. So, S. K. S., Moudry-Munns, K. C., Gillingham, K., Minford, E. J., and Sutherland, D. E. R. (1991). *Transplant. Proc.,* **23,** 1634.
24. Sutherland, D. E. R., Moudry-Munns, E. C., and Gillingham, K. (1989). In *Clinical Transplants* (ed. P. I. Terasaki), p. 19. UCLA Tissue Typing Laboratory, Los Angeles.
25. Malatack, J. J., Zitelli, B. J., Gartner, J. C., Shaw, B. W., Iwatsuki, S., and Starzl, T. E. (1983). *Transplant. Proc.,* **15,** 1292.
26. Markus, B. H., Duquesnoy, R. J., Gordon, R. D., Fung, J. J., Vanek, M., Klintmalm, G., Bryan, C., van Thiel, D., and Starzl, T. E. (1988). *Transplantation,* **46,** 372.
27. Iwaki, Y. (1992). In *HLA 1991* (ed. K. Tsuji, M. Aizawa, and T. Sasazuki). Oxford University Press.
28. Opelz, G. (1989). *Transplant. Proc.,* **21,** 794.
29. Smith, J. D., Danskine, A. J., Rose, M. L., Pomerance, A., and Yacoub, M. H. (1992). *Transplantation,* (In press).
30. Volker-Dieben, H. J., D'Amaro, J., and Kok-van Alphen, C. C. (1987). *Aust. N. Z. J. Ophthamol.,* **15,** 11.
31. Hoffman, F., von Keyserlingk, H. J., and Wiederholt, G. (1986). *Cornea,* **5,** 139.
32. Volker-Dieben, H. J., D'Amaro, J., Kruit, P. J., and de Lange, P. (1989). *Transplant. Proc.,* **21,** 3135.
33. Lagaaij, E. L., Hennemann, P. H., Ruigrok, M., de Haan, M. W., Persijn, G. G., Termijtelen, A., Hendriks, G. F. J., Weimar, W., Claas, F. H. J., and van Rood, J. J. (1989). *N. Engl. J. Med.,* **321,** 701.

2

HLA typing reagents: alloantisera and monoclonal antibodies

DEREK MIDDLETON, JULIA BODMER, JUDITH HEYES, and STEVEN MARSH

1. Introduction

In order to determine an individual's HLA type two important ingredients are required. A suitable cell suspension, especially a B lymphocyte preparation when typing for class II antigens, and quality reagents containing specific antibodies against HLA antigens. The reagents are of two forms—alloantibodies produced by various sensitization procedures which are described in the first part of this chapter, and monoclonal antibodies which are described in the latter part of the chapter.

2. HLA alloantibodies

Regardless of the source of alloantibodies the normal policy is to screen sera from small amounts of blood (<1 ml) for suitable antibodies and then to obtain large volumes of suitable sera for use as reagents. Many laboratories screen and procure their own reagents and then exchange these with other laboratories. Other laboratories rely on obtaining their reagents from National Tissue Typing Reference Laboratories or by purchase from commercial companies.

2.1 Source of alloantibodies

2.1.1 Pregnancy

The majority of alloantisera used are obtained from parous women. In most cases the antibody is detectable for only a few months after birth and screening programmes are built around this fact. It is advisable to screen women in their second or subsequent pregnancy as distinct from primigravidae as they are a richer source of antibodies. The frequency with which HLA antibodies are found in sera increases with the number of pregnancies, from approximately 5% in primigravidae to around 40% in women who have had five or more pregnancies. Antibodies formed during a first pregnancy are usually of low titre, although often of narrow specificity. In subsequent pregnancies the antibodies may be of higher titre and are often directed against more than one antigen. On average we find that 20% of women who have had more than one pregnancy have HLA antibodies but less than 1% have antibodies useful as HLA typing reagents.

2.1.2 Blood donors

In some instances the antibody formed during pregnancy has a long life. Previous experience has shown that 220 (3.0%) of 7230 female blood donors have HLA antibody (1). These donors contained both females who had been pregnant and those who had not been pregnant. Obviously the percentage with HLA antibodies would have been improved if female parous blood donors had been selected. The blood donors tested yielded a much higher proportion of monospecific HLA antibodies than sera taken during or shortly after pregnancy. This was probably because weaker and cross reacting antibodies had declined in activity to undetectable levels. (A total of 48% of the female donors who had HLA antibodies had not been pregnant nor been transfused in the previous five years.) Thus the ease of taking blood from regular blood donors can make this a rewarding procedure in a Blood Transfusion Centre.

2.1.3 Transfused and/or transplanted recipients

HLA antibodies are also formed as a result of transfusion or transplantation (see Chapter 4). Although these antibodies are common in patients on dialysis or transplanted patients it is rare that a useful reagent can be obtained.

2.1.4 Immunization

Planned immunization of volunteers using whole blood or lymphocyte suspensions has been performed but this practice is not common at present. However patients with recurrent spontaneous abortions are sometimes immunized with lymphocytes of their spouse in an attempt to have a normal delivery. In this laboratory we screen a sample of serum three weeks after immunization and this serum sometimes contains a suitable HLA antibody. A larger donation of blood can then be obtained.

3. Screening for HLA alloantibodies

The techniques used to test for HLA alloantibodies are identical to those used in determining the HLA type. For screening, lymphocytes from individuals whose HLA types are known are used as target cells to determine the presence of antibodies. For class I antibodies, sera are incubated with lymphocytes in a Terasaki tray for 30 minutes, rabbit complement is added, and the incubation continued for a further 60 minutes. For class II antibodies the sera are incubated with B lymphocyte rich cell preparations for 60 minutes followed by the addition of rabbit complement for a further 120 minutes (see Chapter 3).

In this laboratory sera for screening is obtained from small amounts of blood (1 ml) taken at ante-natal clinics or from samples sent to a Blood Transfusion Centre for routine ante-natal screening. Sera are collected from Monday to Friday, separated into microtest tubes (Robbins 1009-04-0) on day of receipt, kept at 4°C, plated on Friday, and the plates kept at −20°C before testing. The number of plates prepared corresponds to the number of individuals in the panel of typing cells. Another option is to freeze sera immediately after separation and plate out when enough sera are collected to fill the plate.

The ante-natal sera for screening are obtained at approximately 32 weeks of pregnancy. This coincides with what is for many patients their first appointment at the ante-natal clinic. This allows us to perform testing on samples at an advanced stage of pregnancy and thus more likely to have antibodies and to complete the testing (approximately three months) not long after delivery.

3.1 Lymphocytes used for class I antibody screening

The sera are screened against panels of lymphocytes. For class I screening the lymphocytes are selected from healthy individuals who have been typed in this laboratory for various reasons, e.g. bone marrow donor registry, relatives for live kidney transplant. Some tissue typing laboratories use the husband's lymphocytes to test the sera from the pregnant spouse. This enables them to quickly determine whether antibodies are present.

Isolation of lymphocytes is performed according to *Protocol 1* (see also Chapter 3).

Protocol 1. Isolation of lymphocytes from citrated blood

1. Prepare a 80 ml plastic container (Medfor M4281) containing 0.6 ml 2.94% $CaCl_2$, two drops thrombin made up fresh daily (Ortho 731200), and approximately 15 glass beads (BDH 330554Q). One container for each 10 ml blood.
2. Add 10 ml of blood to the container, replace lid, and mix on rotashaker until fibrin clot is formed (at least 7 min).
3. Carefully layer blood on to 2 ml lymphocyte separation medium (LSM) (Lymphoprep, Nycomed 221397) in 5 ml plastic Kahn tubes.
4. Centrifuge at 400 *g* for 30 min.
5. Remove lymphocyte layer and add to RPMI 1640 (Gibco 04102400H) in a 10 ml centrifuge tube.
6. Centrifuge at 400 *g* for 5 min.
7. Remove supernatant, resuspend lymphocytes in 2 ml RPMI 1640, and count the lymphocytes in a haemocytometer.

3.1.1 Selection of panel

We use three panels of lymphocytes, each composed of approximately 15 individuals and each selected to cover, whenever possible, all known antigens. Thus individuals with uncommon antigens are selected in order to minimize as far as possible the number of individuals required for each panel. The panels are continually assessed with newly typed individuals replacing existing panel members when this leads to greater efficiency. If available a cell panel should include cells with genotypic blanks to detect new specificities and cells from various ethnic groups. Each serum is tested against one panel. Only if a serum has positive reactions is it tested against the other two panels. Possibly some suitable antibodies may be missed but this procedure has the great advantage of testing a high number of sera. The panels are rotated, one being used each month. The scheme is devised in order to facilitate the routine screening of sera from patients awaiting renal transplantation, which is performed using the same panels at the same time to maximize the use of the panels (see also Chapter 4). Other laboratories may prefer to use random cells for their screening, in which case a minimum of 40 individuals is recommended.

3.1.2 Storage of lymphocytes

i. Freezing of lymphocytes in cryotubes

The lymphocytes from suitable individuals are frozen in liquid nitrogen and thawed as required (*Protocol 2*).

Protocol 2. Freezing and thawing lymphocytes

- freezing mixture: 20% dimethyl sulphoxide (DMSO, BDH 28216), 25% fetal calf serum (FCS, Gibco 01306290H), 55% RPMI 1640

1. Centrifuge isolated lymphocytes (from *Protocol 1*) at 400 *g* for 5 min, remove supernatant, and resuspend in 25% FCS in RPMI 1640 to give a concentration of 4×10^6/ml.
2. Dispense 1 ml amounts into pre-labelled cryotubes (Nunc 3-68632).
3. Place cryotubes into fresh crushed ice. Cool freezing mixture in crushed ice.
4. After 30 min slowly add 1 ml of freezing mixture to each cryotube, mixing while adding.
5. Place the cryotubes in a vertical position in a polystyrene box previously cooled to 4°C. Place immediately in −70°C freezer.
6. On the next working day transfer cryotubes from −70°C freezer to liquid nitrogen. Whenever possible leave cells in liquid nitrogen for at least two weeks before use. This ensures a more stable viability as it appears that cells prone to dying do so in the first two weeks of storage.

7. To thaw lymphocytes add 4 ml of 50% FCS in RPMI 1640 to a centrifuge tube.
8. Place in fresh crushed ice at 4°C.
9. After 30 min, remove cryotube from liquid N_2, thaw by gently shaking in 37°C water-bath until a small ice crystal remains. Transfer contents of cryotube directly to medium in centrifuge tube.
10. Centrifuge at 400 *g* for 5 min.
11. Remove supernatant, allow to drain for 30–60 sec, and resuspend cells in 0.5 ml 20% FCS/complement-fixing buffer (CFB Oxoid BR16).
12. Add one drop of cells to one drop of trypan blue (BDH 34078), and count cell number and viability using a haemocytometer.
13. After counting, wash cells, and dilute in 20% FCS/CFB to give concentration of 2.4×10^6/ml.
14. Add cells to test serum in a Terasaki tray as soon as possible but keep cells in 22°C incubator if testing is not immediate.

If a laboratory has the facilities available cryotubes should be frozen in a controlled rate biological freezer at 1–1.5°C/min. By correct programming the machine can compensate for the heat of crystallization at the eutectic point. Beyond −30°C cooling can be accelerated to 7°C/min.

The viability of lymphocytes after thawing may be improved, if required, by treatment with deoxyribonuclease (DNase) (*Protocol 3*) or by incubation at 37°C for 30 min.

Protocol 3. DNase method for improving lymphocyte viability

1. Resuspend lymphocytes in RPMI 1640 in 2 ml aliquots at a concentration of 1.5×10^6/ml.
2. Add 200 μl DNase containing 1000 U, (Sigma D4527 reconstituted in 8 ml dH_2O and frozen at −20°C in 200 μl aliquots) mix, and incubate for 5 min in a 37°C water-bath.
3. Add 3 ml 20% FCS/RPMI 1640 and centrifuge at 400 *g* for 5 min.
4. Wash twice in 5 ml 20% FCS/RPMI 1640, centrifuging at 400 *g* for 5 min.
5. Resuspend in 20% FCS/CFB and check viability before use.

---

ii. Freezing of lymphocytes in microplates

In contrast to freezing lymphocytes in vials some laboratories use cell panels pre-loaded and frozen in microplates. This method is described in Chapter 4. This method is particularly advantageous when the result of the screening is required quickly. For example in Blood Transfusion Centres, where much of the donor blood is separated into packed red cells and plasma, this method offers the opportunity to hold the plasma units for 24 hours whilst the screening is complete. Therefore if suitable HLA antibodies are detected there is no requirement for the donor to return for blood donation. The frozen tray panels can also be obtained commercially from several suppliers, (e.g. Biotest Diagnostics, One Lambda Inc.).

3.2 Lymphocytes used for class II antibody screening

In order to screen for antibodies to class II antigens a suspension of B lymphocytes is required. Because of the high percentage of B lymphocytes, blood from patients with B cell chronic lymphocytic leukaemia (CLL) is suitable and can be obtained on a regular basis from Haematology Clinics. The blood is tested for the percentage of B lymphocytes present using the sheep cell rosetting method (*Protocol 4*). If available a Flow Cytometer could be used to determine the percentage of B lymphocytes.

Protocol 4. Determination of the percentage of B lymphocytes using sheep cell rosetting

1. Dissolve neuraminidase (10 U) (Sigma N2876) in 10 ml distilled water, split into 0.4 ml amounts, and store at −20°C.
2. Centrifuge 10 ml of sheep red blood cells (RBC) at 2000 *g* for 5 min and remove supernatant.
3. Wash cells twice with saline, centrifuging as before.
4. To 6.6 ml of CFB add 0.5 ml of packed sheep RBCs and 0.4 ml of neuraminidase.
5. Mix well and incubate in a 37°C water-bath for 40 min, mixing at intervals.
6. Centrifuge at 2000 *g* for 5 min and wash cells twice in saline, centrifuging as before.
7. Resuspend treated sheep RBCs in 18 ml of 0.1% gelatin (Sigma G2625) in CFB. Sensitized sheep RBCs can be used for one week.
8. Centrifuge isolated lymphocytes (from *Protocol 1*) at 400 *g* for 5 min, remove supernatant, and resuspend cells in sufficient volume to give a concentration of 2×10^6 cells/ml in 20% FCS/CFB.
9. To 2×10^6 mixed lymphocytes add 1 ml of neuraminidase treated sheep cells. Add 0.1 ml of Polybrene (Sigma P4515) (5% stock solution diluted 1:14 in CFB).
10. Rotate tubes on rotary mixer for at least 20 min. Centrifuge at 200 *g* for 5 min to induce rosette formation. Gently resuspend cells, and add one drop to two drops of methylene blue to check rosettes in counting chamber.

Those with less than 10% T lymphocytes are HLA typed and if it is decided that they would be a suitable inclusion in the panels, cells are frozen away. We have three CLL panels used monthly by rotation and each composed of eight individuals. Due to the very high B lymphocyte count upwards of 20 or more cryotubes can be frozen resulting in a long term supply of cells. For those laboratories which isolate B lymphocytes from healthy individuals the method of *Protocol 4* can be continued in order to prepare a suspension of B lymphocytes (see *Protocol 5*). Other methods for the isolation of B lymphocytes are detailed in Chapter 3.

One of the major problems in obtaining sera against class II antigens is that it is difficult to decide whether a serum has an antibody to a HLA-DQ specificity, or whether it has a mixture of antibodies reacting against the HLA-DR specificities in strong linkage disequilibrium with the HLA-DQ specificity. For example, has the serum an antibody against HLA-DQ1 or has it antibodies against HLA-DR1, -DR2, and -DR6. If available, cells from individuals with atypical HLA-DR and -DQ associations should be used in the panel to resolve this problem.

Protocol 5. Isolation of B lymphocytes

1. Step **9** of *Protocol 4* is performed using 6×10^6 mixed lymphocytes and 3 ml neuraminidase, and step **10** as in *Protocol 4*.
2. Continue by underlaying the lymphocyte suspension with 5 ml albuminized LSM [70 ml LSM and 5 ml 30% albumin (Ortho 160260)] using syringe and cannula. Centrifuge at 400 *g* for 30 min.
3. Harvest B lymphocytes from interface and wash twice in 20% FCS/CFB. Count cells in trypan blue and prepare concentration of 2.4×10^6 cells/ml for use.

3.2.1 Contamination of class II antibodies with class I antibodies

Since B cells also carry class I antigens a positive B cell reaction does not guarantee the presence of class II antibodies. In addition it has been shown that sera with antibodies to class II are more likely than not to have in addition antibodies to class I (2). As a consequence sera which show positive reactions to a CLL panel on first testing are absorbed with platelets to remove any possible class I antibodies (see *Protocol 6*), and tested against the subsequent two CLL panels. In addition after absorption the sera are tested against the class I lymphocyte panel, to check that absorption is complete, using the extended incubation time of 60 minutes for cells, and 120 minutes for complement, as these are the incubation times at which the sera are screened for class II antibodies.

Protocol 6. Absorption of serum to remove class I antibodies

1. Platelets are obtained as units of platelet concentrate from Blood Transfusion Centres and are processed as follows. Ready to use platelets may also be purchased from Organon Teknika (74555-238). Centrifuge platelets at 200 *g* for 40 min in 50 ml centrifuge tubes. Remove supernatant to a 50 ml centrifuge tube and centrifuge at 2000 *g* for 15 min.
2. Discard supernatant and add 20 ml 0.8% ammonium chloride to the pellet, to lyse red cells, and place on rotary mixer for a minimum of 40 min. (If lysis is incomplete this stage is repeated.)
3. Wash twice with 1% Tris-buffered EDTA/saline at 2000 *g* for 15 min.
4. Check for contamination by leucocytes. If there is more than one leucocyte/μl, centrifuge at 200 *g* for 15 min and keep the platelet rich supernatant.
5. Store at 4°C in 1% Tris-buffered EDTA/saline containing 0.1% sodium azide. Platelet packs received at subsequent times are pooled. Each batch of pooled platelets should contain approximately 100 packs of platelets before being used to absorb antibodies.
6. Each month centrifuge the platelets at 2000 *g* for 20 min, discard the supernatant, and replace with fresh 1% Tris-buffered EDTA/saline containing 0.1% sodium azide. Platelets to be used for absorption should have been stored for more than one month and less than one year.
7. To absorb a serum use a volume of platelets at least four times the volume of serum being absorbed. Sera are usually absorbed in 250 μl amounts, therefore 1–1.5 ml of packed platelets are required for each serum.
8. Wash platelets by transferring sufficient amounts to a 10 ml conical tube and centrifuging at 2000 *g* for 20 min. Remove supernatant, wash once with 1% Tris-buffered EDTA/saline, and twice with CFB, centrifuging as before.
9. Add 50% volume of CFB to packed platelets and mix well.
10. Add 1 ml of above mixture to a pre-labelled microcentrifuge tube (Nycomed 1012-00-0) and centrifuge at 10 000 *g* for 5 min in a microcentrifuge.
11. Remove supernatant, add 250 μl of serum, mix well, incubate at 22°C for 2 h, and centrifuge at 10 000 *g* for 5 min.
12. Repeat this absorption procedure (steps **10** and **11**) adding the same 250 μl of serum to fresh platelets, and increasing the incubation time to overnight at 22°C.
13. Transfer absorbed serum into microcentrifuge tube and centrifuge at 10 000 *g* for 5 min to ensure all platelets are removed.

3.2.2 Use of DNA techniques

Using DNA techniques, both restriction fragment length polymorphism (RFLP) and oligonucleotide typing, (see Chapter 5) we have been able to type the CLL cells for splits of HLA-DR antigens thus ensuring accuracy in the type of the cells we are using for screening. The cells have also been oligonucleotide typed for HLA-DP. However there is less HLA-DP than HLA-DR or -DQ on the surface of cells and to find antibodies against HLA-DP is very rare.

3.3 Screening in small laboratories

Smaller laboratories which do not have the facilities or manpower for large programmes of screening could still collect ante-natal sera and screen against random cells which are being typed on that day in the laboratory. They could send sera with positive reactions to larger laboratories for further characterization and confirmation of specificity of the antibodies.

4. Procurement of large volumes of sera for use as reagents

When a serum is found by screening to contain a suitable antibody, information is obtained from the ante-natal clinic regarding the address of the patient and the address of the patient's own doctor. A check is always made that the pregnancy has not had serious complications. Three methods are used to obtain large volumes of a serum. The method used depends on the specificity of the antibody. The blood is taken after delivery to avoid causing distress to the mother.

(a) From the lady's general practitioner who is requested to take 60 ml clotted bood.

(b) By plasmapheresis which can be performed at a Blood Transfusion Centre, Haematology Clinic, or Renal Unit. Clotting of the plasma is prevented by taking the blood into citrate which binds calcium ions. The plasma is processed to serum by adding calcium to complex with the citrate and leaving sufficient calcium to restore a normal ionized calcium level (see *Protocol 7*). Excess calcium is removed by Duolite which is an ion exchange resin.

(c) Obtaining serum from the placenta (see *Protocol 8*). With this method it may be more convenient to receive as many placentas as possible without prior screening of ante-natal sera and test these *de novo*.

Protocol 7. Conversion of plasma to serum

1. Take blood into a plasmapheresis pack containing acid-citrate-dextrose plus adenine and centrifuge complete pack at 2500 *g* for 30 min. Aspirate plasma into attached transfer pack. At once return red cells to medical

Protocol 7. *Continued*

personnel performing the plasmapheresis. Plasma may also be separated using a cell-separator making the above step obsolete.

2. Transfer plasma from the bag into a clear, labelled dry bottle (Fisons BTF-682-130D).
3. Add 40% $CaCl_2$ (1 ml per 100 ml plasma) to the bottle.
4. Place bottle in 37 °C water-bath for 2 h, after which time the clot should be well formed and retracted.
5. Wash 50 g of Duolite resin (BDH 55046), three times in saline and leave in an additional bottle.
6. Upon complete clot retraction (sample may be left in fridge overnight if this is not complete) filter, using a wire mesh filter into the bottle prepared as in step **5**.
7. Mix serum and Duolite on a shaker for 10 min to aid ion exchange, then re-filter using fresh wire filters, into a clean labelled bottle.
8. Store in appropriate volumes.

Protocol 8. Retrieval of serum from placenta

1. Transfer placenta to a clean yellow biohazard bag set in a basin.
2. Roll the bag from the top outwards to gain easy access to the placenta.
3. With the placenta suspended by the cord, cut through all main blood vessels radiating from the cord.
4. Remove cord and transfer placenta and any blood to a thick plastic bag. Expel excess air from bag and seal using heat sealer.
5. Place in plasma expressor (Travenol BM-1) and leave until all blood has been expressed from the placenta (approximately 5 min).
6. Discard placenta and transfer liquid from plastic bag to 50 ml centrifuge tubes.
7. Centrifuge at 2000 *g* for 10 min.
8. Transfer supernatant to storage vials and freeze at −70°C.
9. Before use centrifuge thawed serum at 2000 *g* for 5 min to remove debris which has precipitated during freezing.

4.1 Storage of serum donation

The serum obtained by any of the methods is stored in decreasing volumes at −70°C in order to prevent excessive freezing and thawing of the antibodies. We store the serum in the following volumes:

- 250 μl in microtest tubes
- 2 ml in Nunc cryotubes
- 50 ml in larger plastic or glass containers

Thus a 50 ml amount is split into a cryotube and a cryotube is split into microtest tubes. The maximum number of freeze and thaw cycles is three.

4.2 Testing of serum donation

The serum is tested against the three lymphocyte panels or the three CLL panels according to what antibody was present in the original screening. If the antibody is found to be suitable after testing against the panels it is *further tested* by adding to the next batch of plates to be used for HLA typing. Until the reagent has been well defined it is not used to determine an HLA type. After testing against 100 individual cell suspensions the serum is ready to be used routinely. The British Society for Histocompatibility and Immunogenetics runs a scheme whereby sera can be submitted from individual laboratories and then screened in other laboratories. As well as acting a quality control exercise this enables the antibodies in a serum to be better assessed. Another method of assessing a serum is to use cells of different ethnic populations provided by the International Cell Exchange run by Professor Terasaki, Tissue Typing Laboratory, University of California, Los Angeles.

4.3 Manipulation of sera to produce useful reagent

4.3.1 Absorption to remove class I antibodies

A serum obtained on the basis of containing class II antibody is absorbed with platelets (*Protocol 6*) and tested against the class I lymphocyte panel, at the extended incubation periods of 60 minutes with cells and 120 minutes with complement, to ensure that any class I antibodies present have been removed. The serum is then tested against the CLL panels. Some HLA antigens (B8, B12, Cw5, Cw7) are expressed only weakly on platelets and are thus more difficult to absorb (3).

During the last three years it has become common practice in this laboratory to use Dynabeads (see Chapters 3 and 4) for the isolation of B lymphocytes for class II typing. Fortunately our experience has been that class II sera obtained with our method of screening are able to be used for tissue typing B lymphocytes isolated by Dynabeads. Although the use of Dynabeads renders the method of typing more sensitive, we use incubation times of 30 minutes with cells and 45 minutes with complement when using Dynabeads. We do not use Dynabeads when screening because of the cost of the beads, bearing in mind that with the use of CLLs we do not need an isolation procedure for B lymphocytes.

4.3.2 Absorption to remove a particular class I antibody

Many sera are found which have 'extra' reactions, i.e. they react with cells not containing the specified antigen. This may be because the serum contains additional antibodies to other class I or II molecules. It can also occur because the antibody is to an epitope which is shared by other antigens (crossreacting). A serum with a mixture of antibodies can be absorbed with specific platelets carrying the antigen to which the unwanted antibody reacts (see *Protocol 6*).

4.3.3 Dilution to remove unwanted antibodies

A serum may also be diluted in the hope that one of the antibodies is present at a higher titre than the other. The serum is diluted in human AB serum, previously tested and found negative for HLA antibodies. Great care is required to ensure consistency of dilution each time the serum is to be used.

4.3.4 Use of sera reacting against more than one antigen

Many sera which react against more than one antigen can be useful. If they recognize an antigen which is difficult to detect due to the lack of monospecific reagents, and in addition an antigen for which there are monospecific reagents, then the serum can be used to identify the difficult antigen in the absence of the other antigen.

4.3.5 Concentrating antibodies

A weak but monospecific antibody found in a donated serum can be concentrated according to *Protocol 9* and further analysed. In mildly acidic conditions, the addition of short-chain fatty acids such as caprylic acid to serum will precipitate most serum proteins with the exception of the IgG molecules. In combination with other purification steps such as ammonium sulphate precipitation, caprylic acid will yield a relatively pure antibody preparation.

Protocol 9. Concentrating antibody using caprylic acid

Day 1

1. Measure the volume of serum to be concentrated (minimum volume 5 ml—maximum volume 50 ml). Transfer to clean beaker, add a 'flea', and place on magnetic stirrer.
2. Add two volumes of 60 mM sodium acetate buffer (pH 4.0). Check the pH and adjust to pH 4.8 if necessary (with 4 M NaOH if pH too acidic, glacial acetic acid if pH too alkali).
3. Add caprylic acid (Sigma C2875) dropwise (0.7 ml caprylic acid/10 ml of original serum volume). Continue stirring for 30 min at room temperature.
4. Centrifuge at 2000 *g* for 30 min. Carefully decant and keep the supernatant.

5. Pass the supernatant through a Sterivex-GS 0.45 μm filter (SLHA 025BS) to remove any remaining particulate material.
6. Transfer the supernatant to dialysis tubing and dialyse against PBS overnight at 4°C on magnetic stirrer. Volume of dialysis fluid (PBS) depends on volume of supernatant being dialysed, e.g. if starting with serum volume of 50 ml dialyse against 5000 ml of PBS at each dialysis step. Before use, dialysis tubing should be boiled in a beaker of double distilled H_2O (ddH_2O) along with two or three spatula measures of EDTA.
7. Prepare a saturated solution of ammonium sulphate $(NH_4)_2SO_4$ by dissolving 200 g of $(NH_4)_2SO_4$ in ddH_2O to a total volume of 325 ml using a boiling water-bath (30 min). Let the solution cool to room temperature and the excess $(NH_4)_2SO_4$ will crystallize. The supernatant is used as the source of saturated $(NH_4)_2SO_4$.

Day 2

8. Remove the dialysed, caprylic acid treated serum from the dialysis tubing using a syringe and needle, withdrawing the supernatant from the top of the tubing by inserting the needle through the tubing wall. Transfer to a clean beaker.
9. Add dropwise, under stirring, an equal volume of saturated $(NH_4)_2SO_4$ to the dialysed, caprylic acid treated serum. For larger volumes use a burette (100 ml capacity) to add the $(NH_4)_2SO_4$. Stir this mixture for 1 h.
10. Centrifuge the mixture at 2000 *g* for 10 min. Carefully discard the supernatant and save the precipitate.
11. Re-dissolve the precipitate in a minimal volume of PBS (20% of original serum volume to give maximal concentration, e.g. 5 ml of original serum—dissolve in 1 ml of PBS).
12. Dialyse the concentrated fraction against PBS overnight.

Day 3

13. Remove the dialysed concentrated serum from the dialysis tubing (as in step **8**).
14. Aliquot and freeze the concentrated serum, and test against panels. If after testing the concentrated serum is now reacting too 'broadly' it can be diluted to a concentration for optimum reactivity.

---

5. Computer analysis

Computer software is available whereby the results of each serum reaction on a tissue typing plate are recorded, the HLA type determined can be added, and a serum's reaction against a batch of cells can be analysed. This applies equally well when a serum is being initially screened, or to when a serum is being used as a reagent to determine the HLA type of an individual. However we normally interpret the screening results by eye which is easy to perform due to the small size of the panels. The 2×2 contingency table has been used to establish and document a serum's pedigree (see also Chapter 4). The four components of the table contain the number of concordant antiserum/panel cell antigen positive (+/+) and negative (−/−) reactions, the number of missed reactions (serum negative, antigen present), and the number of extra reactions (serum positive, antigen absent). A chi-square value is established from the 2×2 tables. The correlation coefficient R can thus be obtained ($R = \text{chi}^2/N$). The quality of a serum increases as the R value approaches unity. For common specificities an R value of 0.95 is sought but for rare specificities the only sera available may have a lower R value.

For sera that are being evaluated as reagents a 'tail' analysis can be performed. The specificity giving the highest chi^2 is assumed to be the major specificity. All cells carrying this specificity in the panel and their corresponding serum reactions are dropped and the remaining cells and serum analysed. The new highest chi^2 determines the second specificity. This approach can be repeated until no further significant chi^2 values are obtained.

6. Records of sera

Records of each serum are kept on a card index. These records contain information on the volume of serum available, where the serum is stored, and 2×2 analysis of the serum when tested against the panels and when used on each batch of routine typing plates. Information on whether the serum needs to be diluted or absorbed are also noted. These records could also be kept on computer.

7. Exchange of sera

7.1 With other laboratories

Once a serum has been assessed to be of use as a reagent it is tested for Hepatitis B and HIV status. This is extremely important as these sera are used in many different laboratories including the originating laboratory and must be known to be safe in their use.

It is not possible in an individual laboratory to obtain the necessary reagents to detect all class I and II antigens. Sera should be exchanged between one laboratory and another with each laboratory circulating a list of sera available for exchange. The exchange is performed by adding sodium azide to the serum to give a final concentration of 0.1% and sending the sera by Airmail post. A laboratory with suitable sera for exchange who cannot, because of financial consideration, continue to screen for new reagents can by exchange obtain good reagents. Sera obtained from other laboratories are initially tested in the recipient laboratory, under the conditions applied in that particular laboratory, before routine use. It does not necessarily follow that a serum which behaves as a good reagent in one laboratory will do so in another laboratory. One reason for this may be the source of rabbit complement.

7.2 With national organizations

National organizations, e.g. United Kingdom Transplant Service Special Authority or Eurotransplant, collect large volumes of sera (>25 ml) sent to them by individual laboratories and then distribute these sera as batches of typing sera in small amounts (250 μl) to requesting laboratories. The American Society for Histocompatibility and Immunogenetics produce annually a catalogue of sera available from individual laboratories. This enables interested laboratories to ascertain the available HLA sera in other laboratories. Sera are also assessed during the International Histocompatibility Workshops that take place every four years. Participation in these workshops is a good method to collaborate with other laboratories and to make contacts for exchange of reagents. Screening sera against panels from different ethnic populations can detect antibodies against antigens rarely found in the originating ethnic population from which the serum was obtained.

7.3 With commercial companies

Sera can also be sold to commercial companies. Usually these companies require sera in minimum volumes of 50 ml. Monies from sera thus sold can help finance further screening. Indeed commercial companies may actually help finance screening in lieu of obtaining part of the reagents procurred. A large volume of serum may also be exchanged with commercial companies for small amounts of different reagents. The serum is initially sent to the companies in 1 ml samples. If the serum is deemed reliable enough to be used as a reagent then bulk amounts of the serum is sent by air freight in dry-ice. It is good idea to insure the serum in this instance as damage can occur to the consignment.

8. Monoclonal antibodies

Monoclonal antibodies (MoAb) are valuable tissue typing reagents which can be used both instead of and to supplement alloantisera on tissue typing plates. In this section we discuss the production, screening, storing, and use of these antibodies.

Before embarking on the rather complex protocols for producing MoAb it may be helpful to discuss the advantages and disadvantages they have, compared with alloantisera as described above.

8.1 Monoclonal antibodies versus alloantisera

8.1.1 Alloantisera

(a) Alloantisera are polyclonal in nature, i.e. they are a heterogeneous mixture of different antibodies. These mixtures may contain antibodies directed at either:

i. the products of different loci. For example an antiserum containing a HLA-DQ1 specific antibody may also contain antibodies to HLA-DR1 or -DR2 which were present in the immunizing donor;

ii. different epitopes present on the same molecule. For example a HLA-DR1 antiserum may contain antibodies to several different epitopes on the HLA-DR1 molecule.

(b) Because alloantisera are polyclonal in nature it is usually impossible to characterize the specific epitopes they recognize.

(c) Alloantisera are always limited in their supply.

(d) Each individual antibody may be at a suboptimal concentration for detection in a cytotoxic assay and consequently only the reactivity common to them all, i.e. HLA-DR1 is commonly observed with odd 'tails' or extra reactions occurring occasionally.

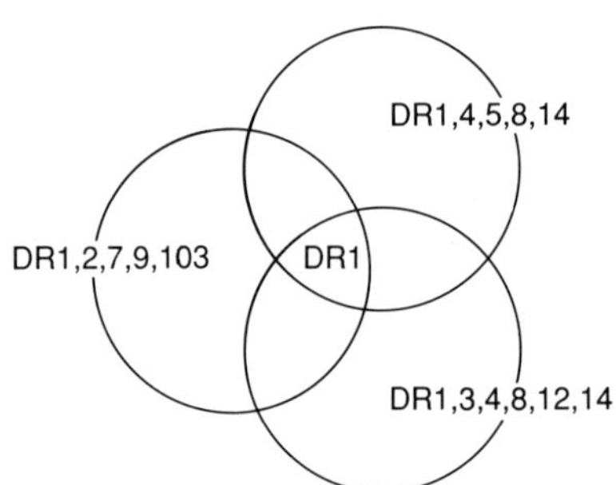

8.1.2 Monoclonal antibodies

(a) Are homogeneous and are directed solely at a single epitope, although this epitope may be found on more than one HLA antigen.

(b) With a MoAb it is possible to define down to the level of a single amino acid the product recognized.

(c) It is possible to grow unlimited supplies of antibody, and to keep stocks of a hybridoma frozen for long-term storage.

(d) The MoAb can be used at a concentration optimal for the detection of the epitope to which it is directed.

Consequently once they have been produced and characterized, MoAb can be widely distributed and used on a long term basis allowing for both standardization between laboratories and an increase in the accuracy of antigen assignment. Some recently recognized variant antigens are currently only definable with MoAb.

8.2 Production of monoclonal antibodies

The production of MoAb was made possible by the work of Kohler and Milstein in 1975 following many experiments in cell hybridization over the previous decade by several laboratories. Although human MoAb have been produced (4) the majority of HLA MoAb have been made in mice or rats.

Two particularly useful practical guides describe the production of MoAb in more detail than is possible here (5, 6).

8.2.1 Technique

i. Fusion

Essentially the technique involves the fusion of a mouse myeloma cell line with spleen or lymph node cells from a mouse previously immunized with the chosen immunogen. The cells are fused *in vitro* to produce hybrids secreting the antibody produced by the immunized spleen cell. The reason for this fusion is that spleen cells alone will not divide and grow in culture and need to be fused with a cell such as a myeloma which does grow in culture, to provide immortalization for the resulting antibody-producing hybrid.

ii. Selective medium

To prevent the dividing myeloma cell from overgrowing the hybrid cells a selective system is used. The myeloma cell line has been selected to lack the enzyme hypoxanthine-guanine-phosphoribosyl transferase (HGPRT) and cannot grow in tissue culture medium supplemented with hypoxanthine, aminopterin, and thymidine (HAT). (Methotrexate can be used instead of aminopterin.) Both these drugs block the main pathway of DNA synthesis. Rescue requires exogenous hypoxanthine and the enzyme HGPRT. Hybrids of the myeloma line and the spleen cells grow in tissue culture medium supplemented with HAT because the myeloma line has the ability to survive in culture and the spleen cells provide the necessary enzyme HGPRT. Polyethylene glycol (PEG) is used as a promoting agent at the time of fusion.

iii. Myeloma fusion partners

There are several variant myeloma lines available, some do not secrete any immunoglobulin (NS1/1-Ag4-1) and some do (P3-X63Ag8).

8.3 Immunization schedules

8.3.1 No 'best' protocol

One point which must be borne in mind in reading the following protocols is that so far no 'best' way has been found for producing MoAb reliably. Different laboratories have found success with different sources of immunogen, different routes and length of immunization schedules, different myeloma fusion partners, different mouse strains, and different fusion protocols. These different routes to success are illustrated in *Table 1* which lists a number of useful HLA MoAb and their production protocols.

8.3.2 General principles for immunization schedules

Adherence to certain general principles may maximize the chance of success but will not necessarily guarantee it.

One overriding general principle which should be noted here is that, as there are so many sources of variability in the production of antibodies, every known variable should be optimized.

(a) Start the immunization programme with inbred, specific pathogen free (SPF) female mice, 6–8 weeks old.

(b) The sera used for growing the hybrids should be of a pre-screened batch shown to promote cell growth optimally, and should not be heat inactivated.

(c) The myeloma line should be growing well at the time of fusion and be mycoplasma free.

(d) Screening and cloning should be done as early as possible to prevent overgrowth of antibody producing cells by other cells.

(e) Aliquots of antibody producing hybridomas should be frozen as early as possible.

8.4 Immunization protocols

8.4.1 Different types of immunogens

As shown in *Table 1*, in most cases the immunizing material used for producing HLA MoAb is of human origin. The essential requirement is that the immunizing material expresses, as strongly or as specifically as possible, the antigen against which the antibody required is to be made.

i. Whole cells

Lymphoid cell lines (LCL) expressing high levels of antigen are very immunogenic and can be produced in large numbers, they are therefore very often used as immunizing cells. The cells should be washed and resuspended in PBS, not adjuvant, for injection. Immunizations with cell lines however may produce a range of different MoAb because the required antigen is present among many others and some of these antigens may be immunodominant. However, as we have shown in *Table 1* there are differences between cell lines which can make some more suitable than others as immunogens.

(a) A LCL expressing HLA-DR13 and -DQ1 was used to produce the HLA-DQ1 antibody TAL4.1.

Table 1. Successful fusion protocols

Antibody name	Specificity	Recipient	Immunizing material	Protocol	Duration of immunization (days)
TAL4.1	DQ1	A-Strong mouse	DR13, DQ1 B cell line	I.P. × 1 I.V. × 1	10
TAL8.1	DR3,11,13,14	BALB/c mouse	DR13 B cell line, coated with mouse antisersa	I.P. × 3 I.V. × 3	63
16.23	DR3,52a(Dw24)	BALB/c mouse	Human melanoma line	I.V. × 1	3
MP1	DQ2	—	Peripheral blood lymphocytes transformed with Epstein–Barr virus	—	—
TAL1B5	DRα monomorphic	BALB/c mouse	Separated α-chains from B cell line	I.P. × 3 I.V. × 3	130
TAL1.2	DQ1	BALB/c mouse	Purified class II from DR2, DQ1 cell line	I.P. × 1 I.V. × 1	11
DP11.1	DPw2,w4	C3H mouse	DPw4 transfected C3H mouse L cell	I.P. × 4 I.V. × 3	116
TAL14.1	DRβ monomorphic	DBA/2 mouse	Hybrid human–mouse clas II transfected into mouse cells	I.P. × 7 I.V. × 1	120

(b) Another LCL was used as an immunogen in the case of the antibody TAL8.1, but this was coated with antiserum prior to injection to block certain sites in order to narrow down the response of the mouse.

(c) LCLs which do not express certain products, for example the Burkitt cell line Daudi which lacks HLA class I, can be used to reduce the possible range of antigens to which the mouse is exposed.

(d) Cell lines known to express high levels of HLA class II have also been used successfully as immunogens as in the case of MoAb 16.23 which was made against a melanoma cell line.

Protocol 10. Immunizations with whole cells

1. Count cells (1:1 with trypan blue), 10^6–10^7 cells are required for injection into each mouse and wash cells 3× in PBS.
2. Resuspend cells in PBS, mix thoroughly to remove any clumps. Volumes to be used:
   Intraperitoneal (I.P.) 100–500 μl/mouse
   Intravenous (I.V.) 100–200 μl/mouse
3. Use a plastic syringe with a 23 gauge needle for injecting.

- wash the cells well as proteins in growth medium are very immunogenic
- frozen cells can be used
- keep cells as sterile as possible
- it is particularly important that there are no clumps in the cell suspension for I.V. injection
- earmark mice so they can be monitored individually

---

ii. Purified antigens

Antigens can be purified to varying degrees. If a highly specific antibody is required then several purification steps may be needed to produce the appropriate antigen. Standard techniques such as column chromatography and protein separations on sodium dodecyl sulphate (SDS)-polyacrylamide gels can be used in the purification steps. Not all purified antigens will elicit a good response from the mouse but the only way to find out is to try it and see!

(a) Purified HLA class II antigen from a HLA-DR2, -DQ1 cell line produced HLA-DQ antibody TAL1.2. The disrupted cells were initially passed down a lentil lectin column followed by passage down a column containing the HLA class I specific antibody W6/32 to deplete the material of HLA class I protein.

(b) Separated chains of an HLA class II product were used as an immunogen in the production of TAL1B5. Since this antigen was denatured the resulting MoAb has been particularly useful in techniques where recognition of denatured material is required, for example the staining of fixed tissue sections (see Chapter 8). However, antibodies made in this way may not bind to the native antigen.

Protocol 11. Immunization with purified antigen

1. Add Freund's adjuvant to microtube.
2. Vortex while adding an equal volume of purified protein soution. Volumes to be used:
   subcutaneous (S.C.) 10–50 μg purified protein in 50 μl PBS mixed with 50 μl adjuvant, i.e. 100 μl total/injection site. Maximum two injection sites/mouse.
3. Vortex until a thick emulsion is formed.
4. Use a glass Luer lock syringe and a 19 gauge needle to complete mixing.
5. Change needle to a 23 gauge before injection.

- Freund's adjuvant can be replaced by other adjuvants
- Freund's adjuvant is potentially harmful, care needs to be taken when it is handled
- neither adjuvants nor toxic chemicals should be used for I.V. injections
- complete Freund's adjuvant should only be used for the primary injection, incomplete Freund's adjuvant should be used for subsequent boosts

---

iii. Transfectants

Mouse cells into which human genes have been transfected have proved to be useful immunogens in the production of MoAb since if the immunized mouse is of an H-2 strain syngeneic to that of the transfectant, the only immunizing stimulus, apart from a viral component, will be the human gene product against which an antibody is required.

The required antigen can be presented in isolation on a transfectant but is not in a denatured form.

The lines can be maintained in tissue culture, frozen and thawed, and prepared for injection in essentially the same way as for LCL.

(a) DP11.1 was made by injecting a DPw4 transfected C3H mouse L cell into a C3H mouse.

(b) TAL14.1 was also made using a mouse transfectant but the genes were transfected to engineer the expression of a hybrid human/mouse class II molecule on the surface, thus presenting only a portion of the molecule as 'foreign' to the recipient mouse.

(c) Cell membranes from transfectants may also be used as immunogens. In this way large amounts of antigen can be injected whilst not increasing the cell bulk.

iv. Synthetic peptides

Since the DNA sequences of various proteins are now known, synthetic peptides have been used to produce antibodies against these proteins. They can be coupled to carrier proteins for injection.

Synthetic peptides may not produce a good immune response from the mouse and it may be more difficult to raise antibodies which recognize the native protein. This may also be the case when highly purified proteins are used as immunogens.

8.4.2 Amount of immunizing material

In general a range of 10^6–10^7 cell equivalents is required. The number of cells available for injection may be limited, depending on their source, as these same cells are required for both immunizing and subsequent screening.

8.4.3 Routes of immunization

A variety of routes of immunization can be used over varying lengths of time. These include, intraperitoneal (I.P), intravenous (I.V.), subcutaneous (S.C.), intrasplenic, footpad injection, and skin graft.

8.4.4 Length of immunization period

The length of time over which the immunization is carried out varies from three days (one I.V. or one footpad injection) as in the case of antibody 16.23, to 130 days as in the case of TAL1B5. As a general rule shorter schedules tend to produce IgM antibodies and longer schedules IgG antibodies, but even this generalization has exceptions.

8.4.5 Choice of mouse recipient

The strain of mouse used for immunization varies. BALB/c mice are often used as this is the strain from which a number of the most commonly used mouse myeloma lines were derived, e.g. NS1/1-AG4-1, NSO/1, and Sp2/O-Ag14.

C3H mice are often used for immunization with L cell transfectants. The reason for this is that the L cell line is a continuously growing cell line derived from a C3H mouse. As mentioned above, by using either a C3H mouse or another mouse syngeneic with C3H, i.e. carrying the same mouse H-2 histocompatibility genes, the range of immune response differences is limited.

8.5 Example of a fusion schedule

For the sake of illustration let us suppose we wish to make an antibody to HLA-DQ1. We select a LCL which expresses HLA-DQ1, in addition to other HLA products. To make the task easier we select a cell line which is homozygote for HLA, thus reducing by half the range of possible stimuli the mouse will see. Such homozygote lines, often from the offspring of first cousin marriages, are readily available (*European Collection of Animal Cell Cultures*) having been selected for use as typing reagents for the mixed lymphocyte culture (MLC) (see Chapter 6).

We decide to immunize BALB/c mice and use as a fusion partner the non-secreting myeloma line NS1/1-Ag4-1 so that all immunoglobulin produced by the hybrids will be from the immunized spleen cell.

We use an immunization protocol which involves giving four I.P. injections at two-weekly intervals, and a series of I.V. injections to ensure a rapid and strong response just before fusion. We plan to give multiple injections because we want an IgG antibody, which can be used for immunochemistry and other techniques as well as tissue typing. A single dose can produce MoAb, but this regime tends to select for IgM producing hybridomas.

We start with more than one mouse (three to six) as we want to be able to select the mouse whose serum before sacrifice has a high antibody titre, and if the first fusion fails have other mice already partly primed for a further attempt.

The following protocols tell us the materials that will be necessary and the steps to be taken from the initial immunization to the fusion, plating, and screening of the resulting hybrids.

Note. The following protocols involving the use of animals are all acceptable under the current UK Guidelines on the Use of Animals [Animals (Scientific Procedures) Act 1986]. However guidelines change from time to time and care should be taken to consult the latest guidelines before starting a series of immunizations. Protocols which apply in one country may not be relevant in another.

Protocol 12. Immunization schedule

Day 1 I.P.—5×10^6 whole cells into each of four BALB/c mice.
Day 15 I.P.—5×10^6 whole cells into each of four BALB/c mice.
Day 29 I.P.—5×10^6 whole cells into each of four BALB/c mice.
Day 43 I.P.—5×10^6 whole cells into each of four BALB/c mice.
Day 47 Tail bleeds.
Day 48 Check serum titre (see *Protocol 13*).
Day 50 I.V.—5×10^6 whole cells into best responder.
Day 51 I.V.—5×10^6 whole cells into best responder.
Day 52 I.V.—5×10^6 whole cells into best responder.
Day 54 Fusion (see *Protocol 14*).

It is very important that the serum titres of the immunized mice are tested in order to select the mouse with the most positive serum titre to use in the fusion. Mice not used at this stage may be further boosted and used for a fusion at a later date.

Protocol 13. Checking the serum titre of immunized mice

1. Tail bleed mice and remove serum.
2. Dilute test serum in PBS ($1/10–1/10^8$).
3. Dilute control serum from non-immunized mouse ($1/10–1/10^8$).
4. Test by conventional screening method.

- the serum from the mouse will be polyclonal
- monitor the serum titre from each individual mouse and compare with the titre of the control mouse
- use a screening method consistent with that used for screening antibodies following fusion as it is also a check of the assay system itself
- a low serum titre can be either due to an inadequate response from the mouse or to a screening assay which is too insensitive
- if serum titre is positive at a dilution of 1/1000 proceed with fusion, however, if the serum titre is only positive at a dilution of around 1/200 further immunizations should be planned
- tail bleed the mice three to four days following last injection

---

When the mouse or mice to be used for the fusion have been selected preparations should be made for the fusion itself. It is important to check that all reagents are available before starting the procedure.

Protocol 14. Fusion of spleen cells with melanoma cells

Prior to fusion the following are required:

- polyethylene glycol (PEG) 4000 (Merck 9727). Weigh out and autoclave in a glass Universal. Add RPMI while PEG is still hot and liquid in the ratio 1 g PEG: 1 ml RPMI. Leave in CO_2 incubator with lid loose to adjust pH to around 7.5.
- NS1/1-Ag4-1 mouse myeloma fusion partner (ICN/FLOW 05–530). Recover from frozen at least a week before fusion, feed every day with RPMI 1640 medium supplemented with 10% heat-inactivated FCS and 20 μM 6-thioguanine (6TG) (Sigma A4882).
- immunized mice
- mice for feeder layers

1. Bleed out the mouse to remove serum.
   (a) Anaesthetize the mouse using CO_2.
   (b) Open the mouse over the heart, and cut the diaphragm.
   (c) Use a 1 ml syringe to remove blood from the heart.
   (d) Leave blood for 30 min at room temperature to clot, then at 4°C for the clot to retract.
   (e) Spin the clot and remove serum.
2. Remove spleen into a 20 ml Universal containing 5 ml of serum-free RPMI.
3. Remove spleen from non-immunized mouse for feeder cells, into another Universal containing 5 ml serum-free RPMI.
4. Disrupt spleen for fusion using metal grid and syringe barrel, into RPMI (10 ml) in 90 mm Petri dish. Transfer cell suspension to a Universal leaving large clumps behind.
5. Count spleen cells (1:9 with 4% acetic acid).
6. Count NS1 cells: (1:1 with trypan blue) 10^7 NS1 cells are required for every 10^8 spleen cells. Anticipate $\sim 2 \times 10^8$ spleen cells per well-immunized mouse.
7. Spin spleen cells and NS1 separately for 5 min at 175 *g*. Gently resuspend spleen cells and NS1 in 25 ml fresh serum-free RPMI, still separately, spin for 5 min at 175 *g*.
8. Add spleen cells to NS1 leaving lumps behind, in a ratio of 10^8 spleen cells: 10^7 NS1 and make up volume to 25 ml with serum-free RPMI. Spin for 5 min at 175 *g*.
9. Remove supernatant, disrupt pellet by flicking tube well, warm to 37°C. Add 800 μl warm PEG over 1 min, shake tube during addition, and keep at 37°C for 1 min.
10. Add 1 ml RPMI slowly over 1 min, shake tube during addition. Add 20 ml RPMI over 3–5 min, shake tube during addition. Spin gently for 15 min at 120 *g*.
11. Resuspend in RPMI 1640, 20% FCS, hypoxanthine (15 μg/ml) (Sigma H9377), methotrexate (0.3 μg/ml) (Lederle 4587-60), and thymidine (5 μg/ml) (Sigma T9250), (HMT) plus feeders (spleen cells or macrophages).
12. Plate in 96 well flat-bottom tissue culture plates (Dynatech M129A), at different dilutions—maximum 4×10^7 immunized mouse spleen cells/plate, 200 μl/well.
13. Feed on day four by aspirating approximately 100 μl medium from each well and adding 100 μl of fresh RPMI 1640, 20% FCS, and HMT.

- viability of the myeloma cells should be greater than 95%
- spleen cells and fusion mixture should always be treated gently

8.6 Screening of hybridoma supernatants following fusion

A series of screens will be needed to characterize fully the antibody produced. Limit your first screen to a small number of informative cells as you will have a lot of wells to screen and a limited amount of material and time! For example, to screen for a HLA class II MoAb, in your first screen include the immunizing cell, a cell of a different HLA class II type, and a T cell line which does not express HLA class II.

Subsequent screens after hybrids have been frozen and supernatants stored can be carried out on a larger cell panel (see Section 9.2 for storage conditions).

8.6.1 Cloning of hybridomas

There are a number of different cloning methods.

(a) Limiting dilution—a suspension of single hybridoma cells are counted accurately and diluted so that a known number of cells can be deposited per well, for example, one, five, or ten cells per well.

(b) Cloning in agar—hybridomas are counted accurately and plated in soft agar at a dilution so that the resulting single cell colonies can be easily transferred to tissue culture plates.

(c) Fluorescent activated cell sorter (FACS)—antigen is coupled to a fluorescent label and mixed with the hybridoma cells. Hybridomas secreting the appropriate MoAb will bind to this fluorescently labelled antigen and can be separated from non-secreting cells; single cell clones can then be deposited in tissue culture plates.

(d) Single cell cloning—single cells are isolated and pipetted into wells as described in *Protocol 15*.

Protocol 15. Single cell cloning on to mouse spleen feeders

1. Remove spleen from mouse, disrupt through metal gauze, wash once in serum-free RPMI, and resuspend in 100 ml of medium.
2. Add 200 μl of spleen cell mixture to each well of a 96 well plate.
3. Pipette cells to be cloned into a single cell suspension, add one drop from a Pasteur pipette into 10 ml of RPMI serum-free medium already in a 90 mm Petri dish.
4. Looking down the microscope transfer single cells to each well by mouth-pipetting using a drawn out 100 μl capillary pipette (Dade 709044) connected to a mouth-pipetting device with a 0.2 μm filter fitted in the line.
5. Leave for four days untouched and check there are single colonies per well.

8.6.2 Screening using the ELISA

As indicated above, many methods can be used for screening the antibodies, depending to some extent on the purpose for which the antibodies are required. A particularly suitable method is an enzyme-linked immunosorbent assay (ELISA) since it can identify antibodies of all classes, is very sensitive and does not use complement or radioactivity. Cells are used as test antigens which can be fixed on to plates in advance, reducing the work required at screening time, which is always a very busy period in the production schedule. The plates are prepared as described in *Protocol 16* and a β-galactosidase/anti-β-galactosidase (GAG) ELISA assay is described in *Protocol 17*. A micro ELISA has been developed involving the use of 72 well typing trays. The advantages of this micro assay are that only 5 μl/well of test supernatant is required and fewer target cells. For an example of this type of assay see Sadler *et al.* (7).

Protocol 16. Attachment of cells to 96 well plates for ELISA

1. Cover base of wells with poly-L-lysine (Sigma P1274) 0.1 mg/ml in PBS. Leave for 60 min at room temperature.
2. Flick out and wash 2× in PBS then blot dry.
3. Add target cells in 50 μl PBS after washing 3× in PBS (between 5×10^4 to 5×10^5 cells/well). Spin for 5 min at 400 *g* and stand at room temperature for 15 min.
4. Do not remove supernatant.
5. Add glutaraldehyde 0.025% in PBS carefully to almost fill wells (200 μl/well) and stand at room temperature for 15–30 min.
6. Flick out and wash 3× in PBS then blot dry.
7. Fill wells with gelatin 0.1% in PBS and 0.02% sodium azide (200 μl/well). Cover with plate sealers (ICN Flow 77-400-05) and store at 4°C until required (for up to three months).

Protocol 17. Screening method using the GAG ELISA

The following reagents are required.

- 0.2% Casein (Oxoid L41) in PBS
- 0.2% Casein and 0.2% Tween 20 in PBS
- β-galactosidase/anti-β-galactosidase (GAG) complex:
  take 100 μl β-galactosidase *E. coli* (5000 U in 1 ml 200 mM Tris pH 7.4) (Sigma G5635)
  add 150 μl (purified IgG) of an anti-β-galactosidase MoAb (4C7) (made in laboratory)

Protocol 17. *Continued*

add 750 μl 200 mM Tris pH 7.4
incubate at 4°C at least overnight (this keeps for several months at 4°C when stored in 0.02% sodium azide)
For use, dilute 1:1000 in 200 mM Tris pH 7.6 and 10% normal human serum.

- rabbit anti-mouse (RAM) (Dako Z259)
  dilute 1:100 in 200 mM Tris pH 7.6
- substrate (make fresh before use):
  dissolve 80 mg $MgCl_2$ in 400 ml PBS
  add 3.1 ml β-mercaptoethanol (Sigma M6250)
  add 4-methylumbelliferyl-β-D-galactoside (Sigma M1633) to the required volume of the above at a concentration of 1 mg/ml to make a saturated solution
  stir for 10 min
  filter

1. Flick out gelatin.
2. Wash (by pouring carefully over plate, making sure wells are filled, and flicking out) 2× with PBS Casein.
3. Add 50 μl of hybridoma supernatant (1st antibody)—leave for 60 min at room temperature.
4. Flick out and wash 3× with PBS/Casein/Tween.
5. Add 50 μl of RAM leave for 60 min at room temperature.
6. Flick out and wash 3× with PBS/Casein/Tween.
7. Add 50 μl GAG complex leave for 60 min at room temperature.
8. Flick out and wash 3× with PBS/Casein/Tween.
9. Add 100 μl substrate leave for 30 min at room temperature.
10. Read on Microfluor automated microplate fluorometer (Dynatech Am140).

- 25 μl volumes instead of 50 μl, can be used
- 15 min incubation with substrate may be enough
- can be adapted for use with 72 well typing trays

8.7 The cloned, screened, monoclonal antibody

By this stage you would hope to have identified at least one, and possibly several, MoAb with the specificity required and possibly other antibodies which look 'interesting'. At this point a decision has to be made as to whether to pursue 'interesting' antibodies or to stick single-mindedly to the original aim of the production. Whichever decision is made it is worth freezing down hybrids producing 'interesting' antibodies as well as those with the required specificity as they can then be further analysed at a later date.

For those which are to be followed up immediately it is important to have several aliquots frozen down at early stages so that 'lost' antibodies can be retrieved if necessary. Further and more extensive screening should then be carried out, if possible by different methods, to refine the specificity of the antibody and to explore the range of use to which the antibody can be put, for example, cytotoxicity assays, staining of histological slides, etc. In some cases reactivity of the antibody can be enhanced by producing a different form of antibody, for example by concentrating the culture supernatant or by producing purified immunoglobulin from the supernatant. If you are very fortunate you may have produced a strong stable antibody with a well defined specificity which can go on to become a standard antibody for the definition of a particular epitope. Such antibodies will then go on to be used in many assays, as for example the cytotoxicity assay described below.

9. The use of monoclonal antibodies in cytotoxicity assays

There are several differences in the way MoAb need to be handled compared with alloantisera in cytotoxic assays (see Chapter 3). Adherence to the technical precautions discussed here allows MoAb to be used with alloantisera on routine tissue typing plates.

9.1 Form of monoclonal antibody

Although the more usual form of a MoAb as a typing reagent will be that of a hybridoma culture supernatant, it is also possible to obtain a MoAb in either mouse ascitic fluid or as purified immunoglobulin.

9.2 Bulk storage

All MoAb preparations can be stored for long periods of time. It is necessary to add 0.1% sodium azide to prevent deterioration due to bacterial or fungal contamination if they are to be stored refrigerated. Culture supernatants are best stored in the dark at 4°C in either glass or plastic containers. These can be stored in large volumes (up to one litre) providing care is taken to dispense them in a sterile environment such as a tissue culture hood. Ascites and purified immunoglobulin are best stored frozen at −80°C, in small aliquots, to prevent the need for unnecessary thawing and freezing.

It is particularly important not to store culture supernatants frozen, as freeze-thawing their relatively lower protein concentration (10%) results in a significant loss of antibody activity.

9.3 Absorption

As MoAb are '*monoclonal*' they only contain one antibody, thus it is never necessary to platelet absorb them as they never contain contaminating antibodies.

9.4 Dilution of antibody

The optimum concentration for a MoAb must be determined by first screening the antibody at a variety of dilutions, as different antibodies and preparation can vary considerably in their strength. A suitable range for culture supernatants is 1×, $^1/_5$, $^1/_{10}$, $^1/_{20}$, $^1/_{50}$, and $^1/_{100}$. When screening ascites or purified immunoglobulin dilutons from $^1/_{100}$ to $^1/_{5000}$ are more appropriate.

A balance must be achieved between the need to have the highest antibody titre possible to counteract the effect of any deterioration of the antibody while stored long-term on the plate, and a dilution which minimizes the problem of antibody carryover.

9.5 Carryover

The extremely high titre of many MoAb requires extreme caution to be used to prevent the contamination of one antibody with another. It is necessary to use disposable plastic pipette tips for the transfer of antibody, glass 'Hamilton' syringes should not be used. When using an automatic serum dispensing machine for plate making it is important to perform extensive washing steps betwen the dispensing on to different batches of plates to ensure that all traces of a MoAb have been removed.

When using a conventional glass syringe (Hamilton 1705RN) to dispense cells or complement on to a tissue typing plate the tip of the syringe needle is placed under the layer of oil and into the antibody. If a well contains a MoAb a problem can occur with the antibody from one well being 'carried over' to subsequent wells on the tip of the needle. This cross contamination is not usually a problem when using alloantisera only, which have a much lower antibody titre. There are several ways of combating this problem. A procedure common to all of these solutions is always to add the cells to every well on the plate in the same pre-defined order, so that the direction of possible carryover can be clearly identified.

9.5.1 Use of carryover wells

When designing tissue typing plates, MoAb are diluted to a titre where the antibody will only carryover into one or two subsequent wells. The two to three wells immediately following the MoAb are then filled with AB serum or RPMI medium. Then if there is any carryover it is only into these wells and there is no contamination of any of the other antibodies on the plate. This procedure is effective and allows the cells and complement to be added in the normal manner, however it does tend to be wasteful of both reagents and plate space.

9.5.2 Rinse and wipe technique

In the rinse and wipe method the reagents are added in the normal way, but between each well the syringe needle is dipped into PBS and wiped on a tissue to remove any antibody. This procedure is, however, extremely slow and tedious.

9.5.3 No-touch technique

In this procedure the tip of the Hamilton needle is allowed to touch the surface of the oil in the well, but does not penetrate into the antibody, thus preventing carryover.

9.5.4 Shooting technique

Hamilton syringes are available which have especially fine-bore needles. Using these it is possible to shoot the cells or complement through the oil into the antibody, while at no time coming in contact with the contents of the well. This is the most successful of all the manual reagent addition techniques, being both fast and effective in alleviating the problem of carryover.

9.5.5 Automated cell dispenser

The easiest and fastest method of adding cells or complement without incurring carryover is the use of an automated cell dispenser (Biotest, One Lambda). These will add reagents to each well by shooting them through the layer of oil directly into the antibody. Highly concentrated forms of the MoAb may be used on the plates to extend their shelf life. By using such a machine, carryover can be eliminated, and the relatively slow process of adding reagents to plates manually is speeded up dramatically.

9.6 Rabbit complement

If rabbit complement is to be used with both alloantisera and MoAb, a MoAb should be included in the initial screening procedure for complement activity. This ensures that the rabbit serum contains complement which is able to bind to a mouse antibody in addition to human antibodies.

10. Use of monoclonal antibodies for definition of sequence epitopes

As a MoAb consists of one antibody directed against a single epitope rather than containing a mixture of different antibodies as in an alloantisera, it is possible to identify exactly the epitope which is recognized by a MoAb. The epitope may be defined to the level of a single amino acid residue or to a cluster of residues, known as a sequence motif. The procedure for identifying both HLA class I and class II epitopes is identical, with the exception of the cells used in the assay.

10.1 Initial monoclonal antibody screening

As with alloantisera, MoAb must be screened on a panel of well characterized cells. A convenient source which is particularly useful in the characterization of HLA class II antibodies is a panel of homozygous LCL lines (see Section 8.5). As each cell is homozygous it only expresses one set of HLA-A, -B, -C, -DR, and -DQ antigens. In cases where the cells are also consanguineous this may extend as far as HLA-DP. Cell lines that have been extensively serologically typed in successive workshops and more recently oligotyped for the HLA class II alleles, DRB, DQA1, DQB1, DPA1, and DPB1, are recommended.

Screening on LCL has the same requirements as screening on other cell types (see Section 3) except that the rabbit complement used must have been selected to give low background toxicity on this cell type. However, the incubation times used are 30 minutes for cell lines plus antibody, followed by 60 minute incubation with rabbit complement, rather than the more normal 60 minutes with antibody followed by 120 minutes with complement used when testing peripheral B lymphocytes.

The panel should contain as wide a range as possible of different antigens and if possible several cells expressing each of the antigens. A panel of some 120 cell lines allows for all identified HLA-class II antigens to be tested with at least four cells (where available) of each HLA DR-DQ haplotype.

Screening can also be performed on panels of *heterozygous* CLL cells or peripheral B lymphocytes, although the initial assignment of antibody reaction pattern may not be as easy. Once this initial screen has been performed it may be possible to define the major antibody specificity. However, due to the co-expression of different alleles on the same cell, and because alleles are held together in haplotypes by linkage disequilibrium, it may not be possible to identify exactly which chains are being detected.

10.2 Further screening

To understand more fully which of the different HLA class II chains are being recognized it may be necessary to screen the antibodies on further panels.

10.2.1 Mutant cell panel

A number of mutant cells expressing differing levels of HLA-DR, -DQ, and -DP are available and these can be useful in determining to which locus products the antibodies are directed. In addition screening with cells such as Daudi which lacks HLA class I expression can exclude the possibility that an antibody is directed at HLA class I.

10.2.2 Transfectant cell panel

Both mouse, and mutant human HLA class II negative cells, have been transfected to express single class II molecules. These are especially useful in understanding which HLA-DRβ chain is recognized in cells where a haplotype expresses more than one chain, i.e. DR2 haplotypes which express both DRβ1 and DRβ5 chains. They are also useful when screening for HLA-DQ, in determining whether the antibody detects either the DQα or DQβ chain, both of which are polymorphic. These transfectants may be used in cytotoxic, ELISA or FACS assays.

10.2.3 Peripheral B lymphocyte cell panel

Screening of MoAb on a panel of peripheral B lymphocytes is recommended as these will be the cells on which the antibodies will usually be tested against in a routine tissue typing laboratory. Some MoAb react with the same specificity on all cell types, other MoAb with perhaps lower avidity fail to react as clearly on cells expressing less antigen.

10.3 Correlation of antibody reaction pattern with sequence

Having first determined the reactivity of the MoAb it is then possible to speculate on the amino acid sequence responsible for its binding pattern (8). This is done by comparing an alignment of relevant sequences with the antibody reaction pattern, then looking for either single residues or sequence motifs that are conserved in all the alleles with which the antibody detects, but are absent in all other alleles.

10.3.1 Epitope mapping of HLA-DR reactive monoclonal antibody

The MoAb TAL8.1 reacts with all cells in a panel expressing HLA-DR3, -DR11, -DR13, and -DR14, i.e. cells expressing the alleles DRB1*0301, *0302, *1101, *1102, *1103, *1104, *1301, *1302, *1303, *1401, *1402, and *1403. Cells expressing the DRB1*1404 allele are not recognized by TAL8.1.

Listed in *Table 2* are the HLA-DRB first domain protein sequences, allele names given in bold type indicate alleles which are detectable by TAL8.1 binding.

There is an amino acid motif 'EYSTS' residues 9–13 which is unique to all of these alleles. Moreover the haplotype expressing the DRB1*1404 allele is identical to the haplotype expressing DRB1*1401 allele except for the residues 9 and 16 in the DRB1 allele sequence where DRB1*1404 is identical to the DRB1*12 and DRB1*08 sequences. This indicates conclusively that these residues are responsible for TAL8.1 binding.

Table 2. HLA-DRB sequences

```
          1       10        20        30        40        50        60        70        80
          .        .         .         .         .         .         .         .         .
DRB1*0101 GDTRPRFLWQLKFECHFFNGTERVRLLERCIYNQEESVRFDSDVGEYRAVTELGRPDAEYWNSQKDLLEQRRAAVDTYCRHNYGVGESF
DRB1*0102 ------------------------------------------------------------------------------------AV---
DRB1*0103 ------------------------------------------------------------------I--DE------------------
DRB1*1501 ----------P-R------------F-D-YF---------------F-------------------I---A--------------V---
DRB1*1502 ----------P-R------------F-D-YF---------------F-------------------I---A------------------
DRB1*1503 *****-----P-R------------F-D-HF---------------F-------------------I---A--------------V---
DRB1*1601 ----------P-R------------F-D-YF-----------------------------------F--D-------------------
DRB1*1602 ----------P-R------------F-D-YF--------------------------------------D-------------------
DRB1*0301 --------EYSTS------------Y-D-YFH----N---------F-----------------------K-GR--N--------V---
DRB1*0302 --------EYSTS------------F---YFH----N---------------------------------K-GR--N------------
DRB1*0401 --------E-V-H------------F-D-YF-H---Y---------------------------------K------------------
DRB1*0402 --------E-V-H------------F-D-YF-H---Y-----------------------------I--DE--------------V---
DRB1*0403 --------E-V-H------------F-D-YF-H---Y------------------------------------E-----------V---
DRB1*0404 --------E-V-H------------F-D-YF-H---Y------------------------------------------------V---
DRB1*0405 --------E-V-H------------F-D-YF-H---Y-------------------S--------------------------------
DRB1*0406 --------E-V-H------------F-D-YF-H----------------------------------------E-----------V---
DRB1*0407 --------E-V-H------------F-D-YF-H---Y------------------------------------E---------------
DRB1*0408 *************------------F-D-YF-H---Y--------------------------------------------------**
DRB1*1101 --------EYSTS------------F-D-YF-----Y---------F----------E--------F--D-------------------
DRB1*1102 --------EYSTS------------F-D-YF-----Y---------F----------E--------I--DE--------------V---
DRB1*1103 --------EYSTS------------F-D-YF-----Y---------F----------E--------F--DE--------------V---
```

```
DRB1*1104 --------EYSTS------------F-D-YF-----Y---------F-----------E--------F--D---------------V---
DRB1*1201 --------EYSTG--Y-------------HFH----LL--------F----------V--S------I--D--------------AV---
DRB1*1301 --------EYSTS------------F-D-YFH----N---------F--------------------I--DE--------------V---
DRB1*1302 --------EYSTS------------F-D-YFH----N---------F--------------------I--DE------------------
DRB1*1303 --------EYSTS------------F-D-YF-----Y--------------------S---------I--DK------------------
DRB1*1401 --------EYSTS------------F-D-YFH----F--------------------A--H---------R---E-----------V---
DRB1*1402 --------EYSTS------------F---YFH----N-----------------------------------------------------
DRB1*1403 --------EYSTS------------F---YFH----N---------------------------------D---L---------------
DRB1*1404 *****---EYSTG--Y---------F-D-YFH----F--------------------A--H---------R---E-----------V---
DRB1*0701 ---Q------G-YK----------QF---LF-----F--------------------V--S------I--D--GQ---V-----------
DRB1*0702 ---Q------G-YK----------QF---LF-----F--------------------V--S------I--D--GQ---V-----------
DRB1*0801 --------EYSTG--Y---------F-D-YF-----Y--------------------S---------F--D---L------------***
DRB1*0802 --------EYSTG--Y---------F-D-YF-----Y------------------------------F--D---L---------------
DRB1*0803 *****---EYSTG--Y---------F-D-YF-----Y--------------------S---------I--D---L---------------
DRB1*0901 ---Q----K-D--------------Y-H-G------N--------------------V--S------F--R---E---V--------***
DRB1*1001 --------EEV------------------RVH----YA-Y------------------------------R-------------------
DRB3*0101 --------ELR-S------------Y-D-YFH----FL-------------------V--S----------K-GR--N------------
DRB3*0201 --------EL--S------------F---HFH----YA-------------R-------------------K-GQ--N--------V---
DRB3*0202 --------EL--S------------F---HFH----YA-------------R-------------------K-GQ--N------------
DRB3*0301 --------EL--S------------F---YFH----F--------------------V--S----------K-GQ--N--------V---
DRB4*0101 ---Q----E-A-C----L------WN-I-Y------YA-YN--L---Q----------------------R---E------Y----V---
DRB5*0101 --------Q-D-Y------------F-H-D------DL-----------------------------F--D-------------------
DRB5*0102 --------Q-D-Y------------F-H-G------N------------------------------F--D-------------------
DRB5*0201 -----C--Q-D-Y------------F-H-G------N------------------------------I---A-------------AV---
DRB5*0202 -----C--Q-D-Y------------F-H-G------N------------------------------I---A-------------AV---
```

10.3.2 Epitope mapping of HLA-DQ reactive monoclonal antibodies

When MoAb react to the HLA-DQ products, epitopes on both the DQA and DQB chains have to be considered, as both are polymorphic. However, as DQA and DQB chains pair in different combinations on different haplotypes it is usually possible to identify which chain is being recognized. *Table 3* lists the reaction patterns for three HLA-DQ monoclonal antibodies against a panel of cell types each expressing a different combination of DQA1 and DQB1 alleles.

Table 3. HLA-DQ monoclonal antibody reaction patterns

HLA Type			**Monoclonal Antibody**		
Serological type	**DQA1 allele**	**DQB1 allele**	**TAL4.1**	**SFR20-DQα5**	**HU-23**
DR1 Dw1 DQ5(1)	*0101	*0501	+	−	−
DR15(2) Dw2 DQ6(1)	*0102	*0602	+	−	−
DR15(2) Dw12 DQ6(1)	*0103	*0601	+	−	−
DR16(2) Dw21 DQ5(1)	*0102	*0502	+	−	−
DR17(3) Dw3 DQ2	*0501	*0201	−	+	−
DR4 Dw4 DQ7(3)	*0301	*0301	−	−	+
DR11(5) Dw5 DQ7(3)	*0501	*0301	−	+	+
DR13(6) Dw18 DQ6(1)	*0103	*0603	+	−	−
DR14(6) Dw9 DQ5(1)	*0101	*0503	+	−	−
DR7 Dw17 DQ2	*0201	*0201	−	−	−
DR7 Dw11 DQ9(3)	*0201	*0303	−	−	−
DR8 Dw8.1 DQ4	*0401	*0402	−	+	−
DR8 Dw8.3 DQ7(3)	*0601	*0301	−	+	+
DR9 Dw23 DQ9(3)	*0301	*0303	−	−	−
DR10 DQ5(1)	*0101	*0501	+	−	−

The antibody TAL4.1 is seen to be reactive with all the cells typing as HLA-DQ1. This includes those expressing DQA1*0101, *0102, *0103 or DQB1*0501, *0502, *0503, *0601, *0602, *0603 alleles. It would be impossible to distinguish whether this reactivity was due to the DQA or DQB chain, if it was not possible to test the antibody on a transfectant cell expressing an unusual DQA/DQB combination. For this antibody it has been shown that the reactivity is due to the DQB chain. It is possible then by sequence comparison to map the sites on the DQB1 alleles which are conserved in all DQB1*05 and *06 alleles and different in all others. There are three areas which fit these criteria: residues 53–56 'QGRP', 84–85 'EV', and 89–90 'GI'. It has not been possible to further refine this epitope at the present (*Table 4*).

The reaction pattern of SFR20-DQα5 (9) quite clearly indicates that this antibody recognizes an epitope on the DQA chain, specifically with all cell

Table 4. HLA-DQA1 and -DQB1 sequences

```
          1        10        20        30        40        50        60        70        80        90
          .        .         .         .         .         .         .         .         .         .
DQB1*0501 RDSPEDFVYQFKGLCYFTNGTERVRGVTRHIYNREEYVRFDSDVGVYRAVTPQGRPVAEYWNSQKEVLEGARASVDRVCRHNYEVAYRGILQR
DQB1*0502 --------------------------------------------------------S------------------------------------
DQB1*0503 --------------------------------------------------------D------------------------------------
DQB1*0601 --P-----L---AM-----------Y---Y------D-------------------D--------DI--RT--EL-T---------F------
DQB1*0602 --------F----M-----------L---Y-------A------------------D-------------T--EL-T---------F------
DQB1*0603 -------------M-----------L-----------A------------------D-------------T--EL-T---------F------
DQB1*0604 -------------M-----------L-----------A-------------------------------RT--EL-T--------G-------
DQB1*0201 -------------M-----------L-S-S------I--------EF----LL-L-A--------DI--RK--A---------QLEL-TT---
DQB1*0301 *****-------AM-----------Y---Y-------A------E-------L-P-D------------RT--EL-T------QLEL-TT---
DQB1*0302 -------------M-----------L---Y-------A--------------L-P-A------------RT--EL-T------QLEL-TT---
DQB1*0303 -------------M-----------L---Y-------A--------------L-P-D------------RT--EL-T------QLEL-TT---
DQB1*0401 --------F----M--------L------Y-------A--------------L--LD--------DI--ED-----T------QLEL-TT---
DQB1*0402 --------F----M---------------Y-------A--------------L--LD--------DI--ED-----T------QLEL-TT---

          1        10        20        30        40        50        60        70        80        90
          .        .         .         .         .         .         .         .         .         .
DQA1*0101 EDIVADHVASCGVNLYQFYGPSGQYTHEFDGDEEFYVDLERKETAWRWPEFSKFGGFDPQGALRNMAVAKHNLNIMIKRYNSTAATNEVPEVT
DQA1*0102 ---------------------------------Q-----------------------------------------------------------
DQA1*0103 ------------------------F--------Q------K---------------------------------------------*******
DQA1*0201 ----------Y------S------F-------------------V-KL-L-HRL R----F--T-I--L------L---S-------------
DQA1*0301 ----------Y------S-------S------------------V-QL-L-RR-RR----F--T-I--L------V---S-------------
DQA1*0302 ****------Y------S-------S------------------V-QL-L-RR-RR----F--T-I--L------V---S-------------
DQA1*0401 ----------Y------S---------------Q-----G----V-CL-VLRQ- R----F--T-I--T------L---S-------------
DQA1*0501 ----------Y------S---------------Q-----G----V-CL-VLRQ- R----F--T-I--L-----SL---S-------------
DQA1*0601 *****************S------F--------Q-----G----V-CL-VLRQ- R----F--T-I--T------L---S*************
```

types expressing the DQA1*0401, *0501, and *0601 alleles. This correlates with the presence of two amino acid motifs which are conserved in these alleles: 'G' at 40 and 'CLPVLRQ' at positions 47–53 (*Table 4*).

The antibody HU-23 (10) has a very simple reaction pattern, recognizing only the DQB1*0301 chain. A comparison of the DQB1 allele sequences indicates that only one residue, 'E' at position 45, is unique to this allele and must be responsible for antibody binding (*Table 4*).

This brief description of epitope mapping indicates the way in which MoAb can help us to identify specifically epitopes which can be seen serologically and which are important functionally.

10.4 Further reading

In this chapter we have described the steps, with accompanying protocols, necessary for the acquisition, production, screening, characterization, and storage of alloantisera and MoAb for HLA typing. Further and more detailed information can be found in the references listed below.

Acknowledgement

We would like to thank Amanda Sadler for her contribution to this chapter.

References

1. Middleton, D., Nelson, S. D., and Martin, J. (1978). *U. Med. J.*, **47,** 162.
2. Vassalli, P., Jeannet, M., de Moerloose, P., and Chardonnens, X. (1979). *Tissue Antigens,* **13,** 77.
3. Aster, R. H., Szatkowski, N., Leibert, M., and Duquesnoy, R. J. (1977). *Transplant Proc.,* **9,** 1695.
4. Kolstad, A., Bratlie, A., and Hannestad, K. (1989). *Tissue Antigens,* **33,** 542.
5. Harlow, E. and Lane, D. (ed.) (1988). *Antibodies, A Laboratory Manual.* Cold Spring Harbor Press, Cold Spring Harbor, NY.
6. Burden, R. H. and van Knippenberg, P. H. (ed.) (1984). *Monoclonal Antibody Technology.* Elsevier, Amsterdam.
7. Sadler, A. M., Krausa, P., Marsh, S. G. E., Heyes, J. M., and Bodmer, J. G. (1992). *J. Immunol. Methods,* **149,** 11.
8. Marsh, S. G. E. and Bodmer, J. G. (1989). *Immunol. Today,* **10,** 305.
9. Amar, A., Radka, S. F., Holbeck, S. L., Kim, S. J., Nepom, B. S., Nelson, K., and Nepom, G. T. (1987). *J. Immunol.,* **138,** 3986.
10. Kasahara, M., Takenochi, T., Ogasawara, K., Ikeda, H., Okuyama, T., Moriuchi, J., Wakisaka, A., Kikuchi, Y., Aizawa, M., Kaneko, T., Kashiwagi, N., Nishimura, Y., and Sasasuki, T. (1983). *Immunogenetics,* **17,** 485.

3

Clinical HLA typing by cytotoxicity

CHRISTOPHER DARKE and PHILIP DYER

1. Introduction

The microlymphocytotoxicity test is universally employed for serological HLA-A, -B, -C, -DR, and -DQ typing. Peripheral blood lymphocytes (PBL) (consisting of approximately 80% T and 10–15% B cells in healthy subjects) or T cells are generally used as targets for HLA-A, -B, and -C typing. B cell enriched suspensions are used for typing HLA-DR and -DQ antigens which are not expressed on resting T cells.

Lymphocytes are tested with a set of HLA specific alloantibodies which are polyclonal antisera or monoclonal antibodies (see Chapter 2) each selected to react with one (monospecific), or sometimes two or more (polyspecific/multispecific) HLA antigens. For full HLA typing the antibody set should cover as many of the various HLA antigens as possible. Reagents for HLA typing should ideally be selected to give strong reliable positive and negative results. The interpretation of a weak or doubtful positive requires an intimate knowledge of the reactivity and specificity of the antisera. It is not possible to produce sets of HLA-A, -B, -C, and -DR, -DQ typing antisera that will detect every single combination of the known HLA antigens. Antisera are obtainable commercially, by local 'screening' programmes, and by exchanges with other tissue typing laboratories (see Chapter 2). There are differences in antigen frequencies between races so the ethnic backgrounds of the subjects must be taken into account when selecting antibody specificities. Individuals might be:

- HLA-A, -B, -C, -DR, -DQ typed, e.g. patients awaiting transplantation
- HLA-A, -B typed only, e.g. donors of platelets for transfusion
- typed for a single antigen only, e.g. patient typing for HLA-B27 (an antigen strongly associated with ankylosing spondylitis)

Lymphocytotoxicity is a three-stage, complement-dependent reaction.

- viable lymphocytes are first incubated with HLA specific antibody
- rabbit serum is added as a source of complement
- live and complement damaged (dead) cells are differentiated

Antibody bound to HLA antigen on the cell membrane activates the complement which damages the lymphocyte membrane making the cells permeable to vital dyes, e.g. eosin-Y, trypan blue, or fluorochromes, such as ethidium bromide and propidium iodide. A positive test indicates the presence of the antigen which the reagent antibody is directed towards. The absence of the specific antigen is indicated when the cells remain viable and unstained. The degree of damage to lymphocytes can also be assessed by the loss of fluorescein diacetate or acridine orange from pre-labelled viable lymphocytes (1).

2. Target cells, storage, and transport

The final major requirements for lymphocyte suspensions used in cytotoxicity assays are:

- viability—should normally exceed 90% (determination by dye exclusion, see *Protocol 1*)
- purity—preparations should be essentially free from red cells, platelets, granulocytes, and monocytes; B cell enriched preparations should be at least 85% pure
- lack of contamination by bacteria or yeasts

Protocol 1. Determination of cell concentration and viability

Lymphocyte preparations should also be assessed for contamination by platelets, granulocytes, and red blood cells during this procedure by observing cell morphology.

- 3% aqueous acetic acid containing 0.025% w/v crystal violet dye (Sigma C-3886)
- 5% aqueous eosin-Y (BDH 26086) or 0.4% trypan blue dye (Sigma T8154)
- improved Neubauer cell counting chamber and coverslip

1. Dilute cell suspension with RPMI 1640 (Gibco, 041-2400) 'by eye' so that it appears just cloudy.
2. Dilute 10 μl of suspension with an equal volume of the acetic acid/dye solution.
3. Fill counting chamber, place in wet chamber, and allow cells to settle for 5 min.
4. Count number of cells covering 1 mm^2 (1/10 mm^3) and divide by 50. The answer is equal to the number of cells in 1 ml of suspension $\times 10^6$, e.g. cells counted = 150; cell concentration = 3×10^6/ml or 3000/μl.

5. Adjust to required lymphocyte concentration in RPMI. For cytotoxicity testing cells are used at a concentration of $1\text{–}2 \times 10^6$/ml.

$$\frac{\text{concentration of suspension}}{\text{concentration required}} = \text{factor necessary to adjust volume of suspension}$$

   e.g. cell concentration = 3×10^6/ml
   required concentration = 2×10^6/ml
   dilute 1 ml of suspension to 1.5 ml to give a working concentration of 2×10^6 cells/ml or 2000/μl.

6. Repeat steps **1–3** but use eosin or trypan blue in place of acetic acid dye solution.
7. Separately count the stained (dead) and non-stained (viable) cells covering 1 mm^2.
8. Calculate percentage cell viability of suspension:

$$100 - \frac{\text{number of dead cells} \times 100}{\text{number of viable + number dead cells}}$$

2.1 Preparation of lymphocyte suspensions for HLA-A, -B, -C typing

Lymphocytes are usually obtained from peripheral blood by discontinuous gradient centrifugation using Ficoll/sodium metrizoate solution at a density of 1.077 at 18–22°C (*Protocol 2*). This lymphocyte separation medium (LSM) is more dense than either platelets, lymphocytes, or monocytes, but less dense than red cells or granulocytes. Diluted whole blood is layered on to the medium and centrifuged, the red cells and granulocytes pass through the separation medium whereas platelets, lymphocytes, and monocytes float on top of the separation medium at the plasma/separation medium interface. Platelets are removed from mononuclear cells by differential washing. A variation in technique for isolation of lymphocytes is set out in Chapter 2.

Protocol 2. Preparation of lymphocytes from whole blood

- anti-coagulated (preservative free heparin, acid citrate dextrose, EDTA) or defibrinated blood
- Ficoll/sodium metrizoate lymphocyte separation medium (LSM), density 1.077 at 18–22°C, (e.g. Nycomed, Lymphoprep); other lymphocyte separation media are available, (e.g. Biotest, Fresenius, One Lambda)

1. Dilute blood with an equal volume of Hanks' balanced salt solution (HBSS) (Gibco, 041–4020).
2. Carefully layer the diluted blood over half its volume of LSM. Avoid mixing.

Protocol 2. *Continued*

3. Centrifuge at 800 *g* for 20 min at 18–22°C. Avoid mixing during centrifugation by ensuring that the speed is increased slowly and that no braking occurs.
4. Remove cells from the plasma/LSM interface.
5. Thoroughly resuspend cells in HBSS and wash by centrifugation for 5 min at 500 *g*. Decant supernatant containing platelets.
6. Repeat wash step **5** twice.
7. Resuspend cells in RPMI and determine purity, viability, and standardize for testing (*Protocol 1*), storage, or freezing (see Chapter 2).

Typing peripheral blood lymphocytes from cadaveric organ donors may prove difficult due to poor reactivity caused by the drug treatment of the patient, low cell viability, or a paucity of cells, especially B lymphocytes. Many of these problems are avoided if immunomagnetic beads are used to isolate lymphocytes from the peripheral blood of organ donors. Spleen tissue in particular is an excellent source of lymphocytes (see *Protocol 3*) and often contains at last 50% B cells, so HLA class II typing can be performed on splenic lymphocytes without further purification provided the proportion of B lymphocytes is monitored.

Protocol 3. Lymphocyte preparation from lymph node and spleen tissue

- non-toxic heat inactivated (56°C for 30 min) fetal calf serum (FCS) (Gibco, 013-6290)

A. *Preparation from lymph node*

1. Trim off any excess fatty tissue from the lymph node and agitate in distilled water in a Petri dish to remove tissue debris and lyse any contaminating red cells.
2. Place lymph node in a Petri dish containing 5 ml of RPMI and puncture several times with syringe needle. Slowly inject medium into the node so that lymphocyte suspension leaks out. If the yield of cells appears small the node may be sliced up with scissors to ensure maximum cell yield.
3. Transfer cell suspension to a plastic tube and allow any tissue debris/cell aggregates to settle out for 2–3 min.
4. Transfer the supernatant cell suspension to a fresh tube, add RPMI containing 5% FCS, thoroughly resuspend, and centrifuge for 5 min at 500 *g*. Decant supernatant.
5. Check lymphocyte purity, viability, and standardize for use (*Protocol 1*). If the cell viability is unacceptable it may be improved (see *Protocol 5*).

B. *Preparation from spleen tissue*

1. Place a small piece of spleen tissue (approximately 0.5 cm^3) into a Petri dish containing 5 ml of RPMI and cut into small fragments. Firmly press each small piece of tissue in the end of a pair of forceps to expel cells.
2. Transfer the cells in RPMI to a plastic tube and allow any tissue debris/cell aggregates to settle out for 2–3 min. Transfer supernatant to a fresh tube.
3. Separate lymphocytes (see *Protocol 2*) then adjust to typing concentration (see *Protocol 1*). For spleen lymphocytes the number of wash steps can be reduced since platelet contamination is not a problem.

A concentrate of white cells and platelets ('buffy coat') obtained from whole blood, may be used as starting material for lymphocyte preparations (see *Protocol 4*).

To help remove phagocytic cells whole blood can be incubated, at 37 °C for 30 min, with carbonyl iron (Sigma C3886) (250 mg/10 ml blood) which is ingested by monocytes and granulocytes. This increases their density so they easily pass through the separation medium.

Protocol 4. Preparation of 'buffy coat' for use as starting material for lymphocyte preparation

1. Centrifuge whole anti-coagulated blood sample at 500 *g* for 10 min.
2. Remove concentrate of white cells and platelets ('buffy coat') from the plasma/red cell interface.
3. Dilute 'buffy coat' to 8 ml with HBSS solution. Layer diluted buffy coat on LSM (see *Protocol 2*).

2.1.1 Common problems occurring with density gradient separation

Impaired separation, low cell yield, and inadequate lymphocyte purity can be caused by the following.

(a) Whole blood or cells mixing with the LSM during the separation procedure. Care should be taken, especially during the layering and centrifugation steps, to avoid cell/LSM mixing.

(b) Addition of excessive whole blood or cells, e.g. obtained from spleen, to the LSM.

Excessive frothing of whole blood or cells can be a cause of low cell viability. If lymphocyte viability falls below approximately 85% it can often be improved, with some cell loss, by removing damaged and non-viable cells by re-centrifugation on LSM (see *Protocol 5*). A low viability cell preparation can also be treated with DNase prior to re-separation, this improves its viability by lysing damaged cells (see Chapter 2).

Protocol 5. Improving the viability of a lymphocyte preparation

1. Centrifuge cell suspension at 500 *g* for 5 min, decant supernatant, and resuspend in 2 ml of RPMI containing 20% FCS.
2. Gently layer suspension over 3 ml of LSM in 75 mm × 12 mm plastic tube and centrifuge at 1500 *g* for 5 min.
3. Harvest the cells from the interface and wash by centrifugation (see *Protocol 2*).
4. Resuspend cells in RPMI and determine cell viabiity (see *Protocol 1*), if still unsatisfactory perform DNase treatment (see Chapter 2).

2.2 Preparation of B lymphocytes for HLA-DR/DQ typing

HLA-DR and -DQ antigens have a restricted tissue distribution but can be readily detected on B lymphocytes. There are several methods in use to B cell enrich a lymphocyte preparation.

(a) Differential adherence of B cells to nylon wool columns (see *Protocol 6*).
(b) Positive selection of cells by magnetizable microspheres coated with a specific monoclonal antibody (immunomagnetic beads) (see *Protocol 7* and Chapter 4). This method can also be used to separate T cells.
(c) Lysis of non-B cells by monoclonal antibodies and complement (Lymphokwik-B cell, One Lambda).
(d) Depletion of T cells by binding (rosetting) to sheep red blood cells (2) (see Chapter 2).
(e) Using a high concentration of peripheral blood lymphocytes containing fluorescein conjugated antibody labelled B cells in the two colour fluorescence test (3).

Protocol 6. B lymphocyte preparation using nylon wool column

- polythene drinking straw (about 15–20 cm long and 6 mm in diameter) (Sweetheart Company)
- combed, scrubbed, nylon wool fibre (Robbins)

A. *Preparation of nylon wool column*

1. Gently tease apart about 10 mg of nylon wool to form a flat loose mat about 8 cm long and 3 cm wide.

2. Pack the wool into the straw to form a loose even column with no tangles about 8 cm in length and 5 cm from one end of the straw.
3. Gently wash the column through with 5 ml of RPMI containing 5% FCS taking care not to dislodge the wool. Ensure that the wool is well wetted and remove any air bubbles by gently squeezing the column.
4. Store the prepared columns at 4°C if they are to be used the same day, or heat seal ends and store frozen at −20°C.

B. *T and B lymphocyte separation*

1. Pre-warm RPMI/FCS to 37°C.
2. Immediately prior to use drain column and wash through with 5 ml of RPMI/FCS. Clip off at the bottom of the wool with artery forceps and add sufficient RPMI/FCS to just cover the wool. Warm column to 37°C.
3. Thoroughly resuspend the lymphocytes in 0.5 ml of RPMI/FCS.
4. Unclip forceps and allow column to drain. Clip column off in same place as before.
5. Pipette cell suspension on to column and allow to soak into the wool. Add sufficient RPMI/FCS to just cover the wool and prevent if from drying out. Incubate column at 37°C for 30 min.
6. Recover non-adherent cells (mainly T lymphocytes). Hold column vertically over a plastic tube, unclip forceps, and allow to drain. Wash non-adherent cells out by allowing about 20 ml of RPMI/FCS to drip through the column. If the last 2–3 ml of the effluent is cloudy continue washing until it becomes clear.
7. Centrifuge diluted cells for 5 min at 500 *g*. Decant supernatant, resuspend lymphocytes in medium, and determine purity, viability, and standardize for testing, storage, or freezing.
8. Recover adherent cells (mainly B lymphocytes) by adding 1–2 ml of RPMI/FCS to column and repeatedly squeeze the straw and the wool vigorously as the effluent drips through. Repeat until 5 ml of effluent has been collected. Follow step **7**.

Protocol 7. Preparation of T or B lymphocytes using immunomagnetic beads

- phosphate-buffered saline (PBS) containing 0.2% w/v EDTA (PBS/EDTA)
- Dynabeads M-450 HLA-class I coated with CD8 antibody (Dynal, 210.03); immunomagnetic beads are also available from Biotest (Lymphobeads), and One Lambda (Fluorobeads)

Protocol 7. *Continued*

- Dynabeads M-450 HLA-class II coated with antibody specific for HLA class II beta chain monomorphic epitope (Dynal, 210.04)
- 6-carboxyfluorescein diacetate (CFDA) (Calbiochem, 216276)—1 ml of 0.01% w/v CFDA in acetone added to 31.25 ml of PBS; store in small aliquots at −40°C, use on the day thawed
- magnetic particle concentrator (MPC) (Dynal)

1. Cool all reagents to 2–4°C.
2. Resuspend the lymphocytes in 1 ml of PBS/EDTA containing 1% FCS at a concentration of 5×10^6/ml. Mix thoroughly and cool to 2–4°C in 13 mm × 100 mm tube (Gibco, 1-45470).
3. Add 10 µl of appropriate immunomagnetic beads for each 60 well tray of antisera to be tested (up to a maximum of six trays of HLA-A, -B, -C sera and four of HLA-DR/DQ sera).
4. Mix gently by inversion and immediately place on a roller mixer at 4°C for 4 min.
5. Remove from mixer, add 5 ml PBS/EDTA, and mix gently by inversion. Place tube in the MPC for 2 min. The cell/bead complex will migrate towards one side of the tube.
6. With the tube still in the MPC remove the bead-free supernatant which contains bead-free lymphocytes.
7. Remove the tube from the MPC and carefully pipette 6 ml of PBS/EDTA down the opposite side of the tube to the cell/bead complex. Mix gently by inversion and replace tube in the MPC for 2 min. Discard the supernatant.
8. Repeat step **7** twice using PBS in place of EDTA/PBS.
9. Draw the tube against the magnet of the MPC to collect the cell/bead complex to the bottom of the tube. Remove the supernatant and re-suspend the cell/bead complex in 0.5 ml of CFDA solution. Incubate in a 37°C water-bath for 5 min.
10. Using the MPC concentrate the cell/bead complex to the bottom of the tube and remove the supernatant CFDA solution.
11. Add 70 µl of RPMI for every 10 µl of immunomagnetic beads added in step **3** and gently resuspend cell/bead complexes.
12. Use the cells in the fluorescence cytotoxicity test.

2.3 Storage

The success of the cytotoxic assay is dependent on the use of viable target cells. Lymphocytes obtained from fresh or up to 12 hour old blood are generally >90% viable. When blood or cell preparations cannot be processed or tested immediately carry out the following steps.

(a) Store whole blood at room temperature. If storage for more than 12 hours is planned dilute with an equal volume of tissue culture medium, e.g. RPMI 1640, to help curtail declining cell viability.
(b) Lymph node and spleen tissue samples are kept at 4°C in tissue culture medium. Viable cells can be obtained several days after removal of such tissues from a cadaver.
(c) Freshly separated sterile cells in tissue culture medium supplemented with 5% non-toxic fetal calf serum, (e.g. Park-Terasaki medium—ICN 16-921-49) will generally remain >90% viable for at least one week at room temperature (4). Overnight stored samples should preferably be as separated lymphocytes in tissue culture medium.
(d) Lymphocytes kept for long periods should be frozen at a controlled rate and stored in the vapour phase of liquid nitrogen (see Chapter 2). After thawing cells may be more sensitive than fresh cells to complement mediated damage.

2.4 Transport

Material for HLA typing is best transported as separated lymphocytes although blood or tissue may be transported over long distances (see *Protocol 8*).

The final viability of lymphocytes will vary considerably depending on the patient's health, the transport conditions, and the time elapsed between taking the blood or tissue and cell separation.

(a) Packaging must conform to local regulations for the transport of pathological specimens.
(b) Packaging must contain sufficient absorbent material to prevent leakage if the immediate sample container is damaged.
(c) Transportation of solid carbon dioxide (dry-ice) is often covered by special regulations.
(d) Pack sample containers in polystyrene blocks which in addition to protecting against impact damage will help to even out large fluctuations in temperature.

Protocol 8. Transportation

1. Separated lymphocytes can be transported as the following.
   (a) Sterile cells suspended in medium supplemented with 5% non-toxic heat inactivated fetal calf serum, e.g. Park-Terasaki medium (ICN Inc, 16-921-49).
   (b) Previously frozen cells packed in solid carbon dioxide (dry-ice).

Protocol 8. *Continued*

2. For transporting blood or tissue over long distances.
   (a) Take blood into acid citrate dextrose, EDTA, preservative free heparin, or defibrinate.
   (b) Add tissue culture medium, e.g. RPMI 1640, containing antibiotics (2 ml of penicillin/streptomycin (ICN 16-700-49) to 100 ml of medium), to lymph node and spleen tissue. It may also be added to whole blood to help prevent infection and maintain cell viability.
3. Journey time for blood and tissues should ideally not exceed 24–48 h. The ideal temperature for whole blood and cell suspensions is about 20°C. Lymph node and spleen tissue should be transported on melting ice.

3. Cytotoxicity assays

3.1 Technique

Although the micro-dye exclusion lymphocytotoxicity technique has many variations the National Institutes of Health (NIH) method is generally accepted as the standard procedure for cytotoxicity testing (*Protocol 9*). One useful modification is the use of fluorochromes in place of dye to differentiate live and dead cells (*Protocol 10*). Incubation times are usually longer for HLA-DR, -DQ typing, using B lymphocytes, than for HLA-A, -B, -C testing. Incubation is often shorter when T or B cells have been prepared using immunomagnetic beads (*Table 1*).

Table 1. Incubation times (in minutes) for cytotoxicity testing

Typing for HLA-	A, B, C		DR, DQ	
Lymphocytes:	PBL or T cells	IMB-T cells	B cells	IMB-B cells
1st stage antibody + lymphocytes	30	30	60	30
2nd stage + rabbit complement	60	45	120	45

PBL peripheral blood lymphocytes
IMB immunomagnetic beads

Protocol 9. Standard 'NIH' micro-dye exclusion cytotoxicity test

You will need polystyrene Terasaki trays (Robbins, 1003-00-0) primed with 1 μl volumes of antisera under non-toxic oil (Biotest 824040). Sets of trays primed with antisera are prepared in bulk and stored at −30°C. Trays stored for longer than three months should be sealed in airtight bags.

1. Gently mix lymphocyte suspension and standardize to contain $1\text{–}2 \times 10^6$ cells/ml RPMI (see *Protocol 1*).
2. Thaw typing trays at 22°C immediately prior to use and check for the presence of serum in each well against a black background. Alternatively add 1 µl phenol red to 100 µl bulk typing serum when preparing typing trays to visualize serum in wells.
3. Dispense 1 µl of cell suspension into each well of the typing tray using a manual Hamilton microdispenser or automated equipment (see Section 4.1). Avoid carryover and check that the antibody and cell suspension have mixed. Ensure that mixing has occurred after the addition of each subsequent reagent.
4. Incubate first stage of test at 22°C.
5. Add 5 µl of freshly prepared complement to each well.
6. Incubate second stage of test at 22°C.
7. Add 2 µl of 5% aqueous eosin-Y to each well.
8. After 2–5 min add 5 µl of fixative (formaldehyde adjusted to pH 7.2 with 0.5 M KOH) to each well.
9. Stand tray at room temperature for a minimum of 30 min to allow cells to completely settle in wells, or centrifuge plates for 1 min at 200 *g*.
10. Carefully lower a coverslip onto wells. The tests are now ready for reading.

Protocol 10. Fluorescence cytotoxicity test

- stock solution of 0.1% acridine orange (Sigma A6014) in PBS, store in small aliquots at −30°C
- acridine orange/ethidium bromide 'cocktail'—1 volume of acridine orange stock solution, 4 volumes ethidium bromide stock solution, and 1.5 volumes of PBS/EDTA
- ethidium bromide (Sigma E8751) stock solution 0.1% in PBS, store in small aliquots at −30°C: working solution dilute to 0.025% in 5% w/v EDTA in PBS (PBS/EDTA)
- quenching agent (10% black calligraphy ink, Staedtler, 745, in PBS)

A. *Using PBLs, T, or B lymphocytes prepared by* Protocols 2, 3, *or* 6

1. Follow steps **1** to **6** in *Protocol 9*.
2. Add 1 µl of acridine orange/ethidium bromide 'cocktail'.
3. Add 1 µl of quenching agent.
4. Read tests within 4 h.

Protocol 10. *Continued*

B. *Using cells separated by immunomagnetic beads (Protocol 7)*

1. Follow steps **1** to **6** *Protocol 9*.
2. Add 1 μl of ethidium bromide working solution.
3. Follow steps **3** to **4** above.

3.1.1 Technical considerations in cytotoxicity testing

Causes of false positive reactions include:

- poor viability of the lymphocyte preparation
- contamination of a negative test with positive antiserum (carryover)
- strong anti-species antibody in rabbit complement
- over incubation
- incorrect dilution of antiserum
- incubation temperature too high
- reagent preparation problems, e.g. pH, isotonicity, or bacterial or yeast contamination
- high granulocyte contamination (granulocytes are killed by rabbit complement but can be distinguished from lymphocytes by eye)
- increased sensitivity of test due to damaged lymphocytes (can occur using frozen cells)
- increased sensitivity of cells caused by homozygosity for an HLA antigen
- missing fixative
- incomplete absorption of HLA-A, -B, -C antibodies from DR/DQ antisera (see Chapter 2)

Causes of false negative reactions include:

- excessively high platelet or granulocyte contamination (cells compete with lymphocytes for antibody)
- incubation temperature too low
- incubation time too short
- missing antiserum or complement in test
- inadequate mixing of cells with antiserum and/or complement (too much oil in well can contribute to this)
- antiserum deterioration (inactive antibody, and/or anti-complementarity, and/or bacterial contamination)
- incorrect dilution of antiserum
- low pH of antisera caused by CO_2 from dry-ice during transport
- excessively high target cell concentration (reduces test sensitivity)

- failure to allow damaged cells to settle after completion of test
- effects of patient's drug treatment on cells
- high T cell contamination of B cell preparations for HLA-DR, -DQ typing

3.1.2 Controls

Although HLA typing antisera themselves serve as controls, positive and negative control wells should be included on every set of typing trays. Controls must be treated identically to tests and at least one negative control should be present on each typing tray to indicate cell quality and viability. Negative control sera consist of a pool of sera from non-transfused male donors lacking lymphocytotoxic activity. Positive controls indicate if complement has been added to the tray and in HLA-DR/DQ typing can be a guide to the purity of the B cell preparation. Sources of antibodies for positive controls include the following.

(a) To indicate complement addition:
  i. mixture of Bw4 and Bw6 specific antisera
  ii. pool of antisera that react with high frequency against a random lymphocyte donor panel
  iii. HLA class I monomorphic monoclonal cytotoxic antibody, e.g. W6/32 (Dako)

(b) To indicate the purity of a B lymphocyte suspension:
  i. pool of platelet absorbed HLA-DR antisera (including anti-DRw52 and anti-DRw53)
  ii. HLA class II monomorphic monoclonal cytotoxic antibody (L243, ATCCHB55)
  iii. CD19 monoclonal cytotoxic antibody, e.g. T26-1-4 (C-Six)

3.2 Complement

Rabbit serum is the most effective source of complement for the lymphocytotoxic assay. It supplies two factors, complement and naturally occurring anti-species antibody that reacts with human lymphocytes. These two components act together to determine the strength of the complement source and therefore the sensitivity of the asssay. If the natural antibody is too strong then the rabbit serum alone may injure the target cells and cause a high background cell death.

3.2.1 Standardization

Since rabbit serum represents one of the most important variables in determining the sensitivity of the cytotoxicity assay it must be carefully standardized for use.

(a) Rabbit serum should not damage test cells in the absence of HLA specific antibody. Different mononuclear cell types vary in their susceptibility to injury by rabbit complement alone, i.e. susceptibility of T cells, <B cells, <B cells from patients with chronic lymphocytic leukaemia, <cultured B lymphoid cell lines. Cells that have been stored, especially frozen, tend to be more susceptible to rabbit serum damage than freshly prepared cells of the same type. To test for natural cytotoxicity in rabbit serum see *Protocol 11*.

(b) Non-cytotoxic rabbit serum should be examined for its specific complement activity (see *Protocol 11*). Different complement batches may need to be selected for HLA-A, -B, -C, and HLA-DR, -DQ typing and their reactivity with monoclonal antibodies will also need to be assessed.

Protocol 11. Evaluation of rabbit complement

A. *To test for natural cytotoxicity in rabbit serum*

1. Use the exact assay the complement reagent will finally be used in.

2. Assay at least five different examples of the cell type(s) to be routinely tested, and include stored or frozen examples if appropriate.

3. Test against at least five different samples of serum from non-transfused male donors known to be cytotoxin free (in place of antiserum).

B. *Studies on new batches of non-toxic complement*

These must include existing complement batch(es) for comparison.

1. Dilute HLA antisera against appropriate positive cells, e.g. an anti-HLA-A2 with a titre of about 1:16 against A2 positive lymphocytes. Make serial dilutions to 1:32 in RPMI containing 20% fetal calf serum.

2. Select lymphocytes from subjects who are heterozygous for the appropriate allele.

3. Repeat step **1** with dilutions of the complement batches (chequer-board titration). Make serial dilutions to 1:32 in complement-fixation test buffer (CFB) pH 7.2 (CFB diluent tablets, Unipath BR16).

4. Test HLA antisera known to weakly cross react with a particular antigen (see *Table 3*) with cells possessing the cross reactive antigen, e.g. an HLA-DR1, weak DR10 antiserum and B lymphocytes from a heterozygous DR10 subject (who is DR1 negative).

5. Evaluate activity on routine HLA typing trays.

3.2.2 Sources, storage, and use of rabbit complement

This reagent may be obtained, usually freeze-dried, from commercial suppliers, (e.g. C-six), usually in 1 ml, 2 ml, and 5 ml aliquots. Suppliers will provide a sample from several different batches for evaluation of natural cytotoxicity and complement 'strength' prior to purchase. Alternatively serum can often be obtained locally from either donor rabbits or animals destined for the meat market. The natural antibody and complement levels varies between different rabbits so pool material from a large number of animals (>50) to even out these components. Ideally, serum rather than whole blood should be pooled. The following should be considered when storing and using rabbit complement.

(a) Store freeze-dried material at −20°C and freshly frozen serum at or below −80°C.
(b) Reconstitute or thaw immediately prior to use, store on melting ice, and do not re-freeze for future use.
(c) Take care to reconstitute freeze-dried complement thoroughly and avoid frothing which can promote degradation.
(d) After being thawed or reconstituted it should not be turbid, or contain any sediment, gel, or particulate matter visible on ×200 microscopy.

4. Automation, reading, and interpretation

4.1 Automation

A range of hand held dispensers is available, (e.g. Hamilton, Robbins) that deliver microlitre quantities of reagent. These usually consist of a dispenser, which delivers 1/50th of the volume of the attached syringe (other sizes are available). Multi-dispensers hold six syringes which allows reagent addition to a whole row of a microtest trays in a single action. Important considerations in their use include the following:

(a) Keep dispensers in perfect working order.
(b) Keep syringes scrupulously clean.
(c) Store syringes clean and filled with distilled water.
(d) The needle of the syringe must not become contaminated with antiserum which could 'carryover' into the next well and cause a false positive reaction —avoid this by using a 'shooting' or 'soft drop' technique (*Protocol 12*).
(e) Do not allow bubbles of air to become trapped between the end of the syringe plunger and the column of reagent, this will have a detrimental affect on the accuracy of dispensing.
(f) Avoid vigorous flushing of cell suspensions in the syringe as this may injure the cells.

Protocol 12. Use of syringe dispensers

A. *'Shooting' method*

This is particularly suitable for the addition of lymphocytes to tray wells containing antibodies.

Protocol 12. *Continued*

1. Ensure repeating syringe dispenser or multiple syringe dispenser is filled with lymphocyte suspension (or other reagent) and that there are no air bubbles trapped in the syringe(s) or needle(s).
2. Hold the syringe dispenser vertically over the test well(s).
3. Do not allow the needle(s) to come into contact with the surface of the oil.
4. Operate the dispensing button with a firm and regular pressure.
5. Check, against a black background, that the contents in the wells have mixed. If not:
   (a) Carefully mix with the tip of a needle attached to a syringe. The needle should be clean and filled with saline (to prevent reagents entering the needle).
   (b) Alternatively, centrifuge trays at 120 *g* for 30 sec.
   (c) Alternatively, gently hold the tray against the head of a vortex mixer for 2–5 sec.

B. *'Soft drop' method*

This is particularly suitable for the addition of complement, dyes, and fixative.

1. Operate the dispensing button with a gentle pressure so that the reagent 'falls' on to the surface of the oil and slowly mixes with the contents of the well.

Stream splitters, which attach to adjustable volume dispensers, (e.g. Robbins), allow the addition of oil, dye, or fixative to a row of wells at a time. In addition to hand held dispensers equipment is available to automate or semi-automate:

(a) dispensing oil, antisera, test cells, and other reagents into the wells of a microtest tray (e.g. Biotest Autosera Dot, One Lambda Lambda Dot, Robbins serum dispenser)

(b) control inverted microscope stage movement and allow keypad addition of manually read cytotoxic test results to computer file (e.g. Biotest Orak, One Lambda Lambda Scan, Robbins Autoscope)

(c) fluorescence cytotoxicity test reading (e.g. Astroscan, One Lambda Lambda Auto Scan, Leica Patimed)

Serological tissue typing involves the testing of several hundred antisera to obtain a single HLA type. This creates a considerable number of results that require interpretation to identify the subject's HLA type, and analysis to monitor ongoing reactivity and identify any additional specificities in the antisera.

Raw cytotoxicity scores, antiserum details (identification, major specificities) and, for further analysis, subjects' HLA phenotype, are conveniently kept in computer files. A variety of software is available, (e.g. Astroscan, Biotest, One Lambda, Robbins) to manage data of this type including programmes to:

(a) Prepare and print test results for subsequent manual interpretation of HLA phenotypes.
(b) Interpret HLA phenotypes from the reactivity of lymphocytes with antisera of known specificity.
(c) Monitor the reactivity of HLA reagents using a variety of statistical methods.

4.2 Reading of tissue typing trays

Tests employing dye exclusion to differentiate live and complement damaged cells are read on an inverted phase contrast microscope using a ×10 magnification objective and a ×12.5 eyepiece, e.g. Wilovert (Leica). Tests using fluorochromes are evaluated on an inverted fluorescence microscpe fitted with filters appropriate for the fluorochromes used (acridine orange/ethidium bromide—excitation filter 495/515 nm), e.g. Fluovert (Leica).

Under phase contrast microscopy viable lymphocytes appear small, bright, and refractile. Damaged cells stained with vital dye flatten so appear larger than live cells and are dark and non-refractile. Test reactions are scored by determining the percentage of damaged cells in negative control tests and establishing whether an increased proportion of stained cells is present in test wells. Results should be recorded using the accepted International Histocompatibility Workshop (IHW) scoring system (*Table 2*).

When cell death in the negative control tests exceeds 10% a judgement can often be made as to whether the test well is negative, i.e. containing a similar number of dead cells as the negative control (score 1–2), a doubtful positive reaction (score 4), or a positive reactive (score 6–8). Considerable skill and experience is involved in accurately assessing percentage cell death, especially against high background cytotoxicity and/or with, for example, leukaemic patients' preparations containing some non-mononuclear cells.

Occasionally it is not possible to obtain a lymphocyte preparation of high viability or good purity. These 'difficult' cell preparations may come from

Table 2. IHW cytotoxicity scoring system

Percent cell death	Score	Interpretation
0%–10%	1	Negative
11%–20%	2	Probably negative
21%–50%	4	Weak positive
51%–80%	6	Positive
81%–100%	8	Strong positive
	0	Non-readable or invalid

uraemic, leukaemic patients or cadaveric organ donors. Some drugs, e.g. steroids, anti-inflammatory agents, cause patients' cells to become poorly reactive in the cytotoxicity test.

Good practices and points that should be considered when reading cytotoxicity tests:

(a) Reading must be performed by experienced laboratory staff; interpretations must be checked by a supervisor.

(b) Regularly check the adjustment of microscopes.

(c) When reading a 'difficult' cell preparation continually refer to the negative controls to aid interpretation of a weak test well score. Some computer controlled microscope stages allow the operator to move to the negative control from any test well on the tray with a single key stroke.

(d) Learn to recognize the different effects produced by:
   i. a missing reagent, e.g. complement.
   ii. platelet, red blood cell, non-mononuclear cell, immature leucocyte, yeast/bacterial contamination.
   iii. a poorly adjusted microscope, e.g. phase contrast.

(e) Scoring must be standardized between different laboratory staff; check inter-laboratory scoring regularly.

(f) Tests on 'difficult' preparation or 'unusual' phenotypes should be read by two different individuals.

4.3 Interpretation

Accurate serological HLA typing requires several factors:

- lymphocytotoxicity testing performed with the utmost care and attention to technical detail
- well selected and characterized antisera able to discriminate as many HLA antigens as possible
- a thorough knowledge of the reactivity and specificity of the antisera

Further important considerations are difficulties with, or errors in, phenotype interpretation.

4.3.1 Cross reactivity and 'extra' antibodies

HLA antisera often react with two or more different HLA antigens (see Chapter 2). This may be due to either the presence in the antiserum of separate antibodies that react with different HLA antigens, e.g. anti-B8 and anti-DR3 or anti-B13 and anti-B14, or a single antibody that reacts with a shared epitope (see Chapter 2). When an HLA typing antiserum reacts with an epitope which is shared then the antigens are termed cross reactive. Studies of the reactivity of numerous HLA antisera have identified cross reactive groups (CREGs) consisting of two or more HLA antigens (*Table 3*).

Table 3. Examples of cross reactive groups of HLA antigens

HLA-A	HLA-A/B
A1,A36	A2,B17
A1,A3,A11	A9,A25,A32,BW4
A1,A10,A11	
A2,A28	**HLA-B/C**
A2,A9,A28	B46,CW3
A10,A11	
A10,A32	**HLA-C**
A10,A28,A33	CW4,CW6
A29,A30,A31,A33	CW5,CW8
A30,A31	
HLA-B	**HLA-DR**
B5,B15,B17,B18,B21,B35,B53,B70	DR1,DR10
B7,B22,B27,B40,B42,B48	DR3,DR5,DR6
B8,B14,B18,B39	DR4,DR7,DR9
B12,B13,B21,B47	DR5,DR8
B13,B21,B40,B41,B45,B47,B48	
B15,B17,B21,B35,B70	
B27,B13,B47	

'Shorter' patterns of cross reactivity are often encountered within HLA-B groups

Many CREGs have been confirmed by the reactivity of monoclonal antibodies to HLA antigens (5) and sequence studies (6) (see Chapter 7). To avoid errors in phenotype interpretation it is of paramount importance to be fully aware of the HLA CREGs. HLA alloantisera contain several populations of antibodies that react together to produce the final specificity of the antiserum under standard cytotoxicity test conditions (operational specificity) (see Chapter 2).

Antisera of defined specificity (monospecific or polyspecific) may react with further HLA antigens under certain circumstances.

(a) An increase in the overall sensitivity of the cytotoxicity test due to:
  i. poor viability of the lymphocyte preparation
  ii. lymphocytes damaged by frozen storage
  iii. rabbit complement with higher activity than previous batches
  iv. over incubation

This may detect additional specificities in the antisera that follow the cross reactive group of the main specificities, e.g. a 'monospecific' anti-B18 may cross react with B14 or a B7, B27 antibody may cross react with B22. Alternatively, an increase in the sensitivity of the assay may detect other, non-cross reactive, antibodies in the reagent. In antisera stimulated by pregnancy these will correspond to antigens encoded by the paternal HLA haplotype inherited by the fetus.

(b) Increased 'sensitivity' of cells caused by homozygosity for an HLA antigen. A subject who is homozygous for a particular HLA allele will have a 'double dose' of the corresponding antigen on the cell surface. These cells may detect hitherto unknown cross reactive or additional antibodies directed towards this antigen. This may cause particular interpretation problems when it involves an infrequent antigen or an antigen that is difficult to define, e.g.
   i. A28 antisera cross reacting with cells from a homozygous A2 subject.
   ii. antisera containing anti-DR11 cross reacting with homozygous DR8.
   iii. cells from a B44 homozygous subject reacting with A29 antisera (A29 antisera may contain weak B44 antibodies when they are stimulated by antigens encoded by the HLA-A29, B44 haplotype).

(c) Antisera may give 'false' positive reactions with lymphocytes possessing two antigens belonging to the same cross reactive group. This effect is similar to the increased sensitivity caused by a 'double dose' of HLA antigen since the two antigens probably possess the same shared epitope, e.g. A1, A3 positive cells reacting with A11 antisera.

4.3.2 Definition by exclusion

Some HLA antigens fail to stimulate (or rarely stimulate) operationally monospecific antisera and are only definable by exclusion, e.g.

(a) A32—often defined by reactivity of cells with A10, A32 antisera in the absence of A25, A26, and A34 antigen subdivisions of A10.

(b) DR11, DR12, DR13, DR14—often defined by a variety of crossreactive antisera (7).

4.3.3 Use of HLA-Bw4 and -Bw6 specific antisera

Antisera recognizing the Bw4 and Bw6 epitopes, that divide the HLA-B antigens, are a useful aid to accurate HLA-B antigen assignment (see Chapter 11).

(a) Several antigens are subdivided on the basis of their possession of either a Bw4 or Bw6 epitope, e.g. B12, B21, B15, B16.

(b) The presence of Bw4 or Bw6 in the apparent absence of a Bw4 or Bw6 associated antigen may indicate:
   i. a technical or interpretative error.
   ii. that the panel of HLA-B antisera is not able to resolve a particular HLA-B specificity.

4.3.4 Broad and split specificities

Clinical HLA typing necessitates defining HLA phenotypes as precisely as possible. It is usually not sufficient to type for the 'broad' HLA antigens only, e.g. A9 rather than A23 and A24. However, antibodies reacting with 'broad' rather than 'split' HLA antigens provide important typing information by:

(a) Identifying a low frequency antigen subdivision in the absence of specific antisera, e.g. A34 or A66 positive cells can be identified by their reactivity with A10 antisera and their lack of reactivity with A25 or A26 antisera.

(b) Confirming an antigen assignment, e.g. cells assigned as HLA-B45 positive would be expected to react with all HLA-B12 antisera.

4.3.5 Linkage disequilibria

A knowledge of positive linkage disequilibria (see Section 5.1) between HLA-A/B, -B/C, -B/DR, and -DR/DQ alleles in the typed subject's ethnic group often proves helpful in phenotype interpretation by:

(a) Providing additional *confirmation* of a 'difficult' phenotype assignment. For example, the subdivisions of HLA-A19 (A29, A30, A31, A32, A33, A74) are sometimes difficult to assign. A subject possessing a HLA-B allele known to be in positive linkage disequilibrium with a particular A19 allele will provide additional support for the phenotype assignment. Examples of positive linkage disequilibria between HLA-A19 and HLA-B alleles in European Caucasoids are:
   i. A29 and B44, B45
   ii. A30 and B13, B18
   iii. A31 and B51, B60
   iv. A32 and B64
   v. A33 and B58, B65

(b) Prompting additional testing of apparent homozygous individuals or unusual phenotypes. For example, HLA-B37 is in strong linkage disequilibrium with DR10 in Caucasoids. A phenotype possessing B37 in the absence of DR10 (especially if only a single DR antigen had been assigned) would merit further testing.

(c) Indicating possible reasons for 'false' positive reactions of an antiserum. For example, the reactivity of lymphocytes from an HLA-B44 homozygous subject with a B35 antiserum: this could be caused by a weak anti-Cw4 antibody in the B35 antiserum. The explanation for this being that B35 and Cw4 alleles are in strong linkage disequilibrium. Cells stimulating the B35 antibody would probably possess the Cw4 antigen so could stimulate an additional weak Cw4 antibody. The weak Cw4 component might go undetected until it was tested against B35 negative cells expressing a double dose of the Cw4 antigen (HLA-B44 is in linkage disequilibrium with Cw4). The test cells could be from an individual who is homozygous for both B44 and Cw4.

5. Family studies, antigen and allele frequencies, allelic association, ethnic groups

The term HLA *antigen* is used to indicate the protein product of a HLA gene which has been detected using serological techniques, and the term *allele* is used to indicate a gene sequence which is one possible polymorphism coded for by a HLA gene and which is usually detected using molecular biological techniques.

5.1 Family studies

Studies of families using both serological and molecular techniques to define major histocompatibility complex (Mhc) polymorphisms provide significantly more information than population studies. Even if information relating to only one individual is needed it is often necessary to perform a full family study. From family studies it is possible to derive information on:

- establishment of HLA haplotypes
- associations and linkages
- exclusion of parentage
- recombination events
- homozygosity
- chimerism and triplets

A suitable approach to performing a family study is shown in *Protocol 13*.

Protocol 13. Design of a family study

1. Obtain ethical approval from relevant committee.
2. Carefully plan the number of family members to be studied. This could be limited by availability and some revision may be needed during the study.
3. Contact families to obtain informed consent for blood samples to be taken and arrange dates for venepuncture.
4. Ensure laboratory is informed of dates and times of specimen delivery.
5. Minimize blood volumes needed by arranging correct specimen storage and types of specimens taken. For example, heparinized blood for lymphocyte extraction can also be used to provide DNA from residual granulocytes.
6. Obtain blood specimens and transfer to laboratory within 24 h. Specimen containers must be clearly and carefully labelled with donor's details.
7. Obtain a clear statement of how individuals within a family are related. Draw up a family tree using recognized symbols.

8. Collate HLA typing results with family tree and interpret data with senior laboratory staff. Careful interpretation is essential to clearly establish haplotype data. Experienced laboratory staff will advise on technical problems and interpretation of serology and molecular typing results.

9. Assign haplotypes to each family member. It may become clear that further family members must be tested to complete the study.

10. Examine completed family trees for recombination events and exclusions of paternity. Confirm by repeat testing to exclude the possibility of mislabelling or unintentional mixing of specimens and by widening the family study. Consider testing for polymorphic markers on other chromosomes to establish claimed relationships and to exclude non-paternity.

Analysis of family study data reveals the pattern of inheritance of HLA alleles within the relationship. Those alleles which are inherited together are termed a *haplotype*; for HLA class II alleles in particular, the *linkage* is tight because of the small genetic distance between genes. Haplotypes extend throughout the Mhc but with increasing genetic distance, recombination events become more likely. The approximate frequency is 1% of meioses for HLA-A/-B and HLA-B/-DR recombinations.

Some haplotypes are highly conserved but their frequency can be common or rare. The haplotypes established in *Figure 1* illustrate some notable features which might be encountered. Examples are:

- HLA-A30, B18, Bw6, Cw5, DR3, DR52, DQ2 (individual I.1, haplotype a) most frequent in Mediterranean populations, otherwise rare
- HLA-A1, B8, Bw6, Cw7, DR3, DR52, DQ2 (I.2, c) commonest haplotype in European Caucasoids
- HLA-A25, B18, Bw6, Cw7, DR2, DQ1 (II.1, j) carries the deficiency allele (C2Q0) of complement C2
- HLA-B37, DR10 (I.1, b)
  both rare antigens usually found on the same haplotype

In population studies, it is possible to calculate the *allelic association* (sometimes termed *linkage disequilibrium*) of two antigens or alleles which occur together more often than expected by chance. This phenomenon arises because of conserved HLA haplotypes, which only family studies can confirm. Sometimes the association of the antigens or alleles is strong enough to indicate linkage at the level of the genes. A good example is the linkage of HLA-DRB1 alleles to HLA-DRB3 and -B4. In European Caucasoids haplotypes with the antigens HLA-DR3 or -DR11 or -DR6 invariably carry the HLA-DR52 antigen, while HLA-DR4 and -DR7 carry HLA-DR53. It should always be understood that exceptions to these linkages are possible and all are worthy of detailed study.

I	1 ab	2 cd	3 ef	4 gh			
II	1 ij	2 ac	3 fg	4 fh	5 kl		
III	1 ai	2 ci	3 aj	4 cj	5 af/g	6 fm	7 fk

Individual	HLA- A	B	Bw	Cw	DR		DQ
I.1	30,31	18,37	4,6	5	3,10	52	2
I.2	1,2	8,44	4,6	5,7	3,7	52,53	2
I.3	24	35	6	4	1,8	52	1
I.4	23,33	27,14	4,6	1,3	1,6	52	1,3
II.1	3,25	7,18	6	7	2	–	1
II.5	11,28	51,60	4,6	3	4	53	7
III.5	24,30	18,27	4,6	3,5	1,3	52	1,2
III.6	24,29	35,44	4,6	4,5	4,8	52,53	7

Haplotypes	HLA- A	B	Bw	Cw	DR		DQ
(a)	30	18	6	5	3	52	2
(b)	31	37	4	?	10	–	–
(c)	1	8	6	7	3	52	2
(d)	2	44	4	5	7	53	–
(e)	?	?	?	?	1	?	?
(f)	24	35	6	4	8	52	–
(g)	23	27	4	3	1	–	1
(h)	33	14	6	1	6	52	3
(i)	3	7	6	7	2	–	1
(j)	25	18	6	7	2	–	1
(f/g)	24	27	4	3	1	–	1
(k)	11	51	4	?	4	53	7
(l)	28	60	6	3	?	?	?
(m)	29	44	4	5	4	53	7

? antigen cannot definitely be assigned to haplotype, possible homozygote
– no antigen identified

Figure 1. A family study.

When analysing family data it is essential not to assign antigens or alleles to haplotypes unless there is *unequivocable* evidence to do so. In individual II.1 (*Figure 1*) both haplotypes are passed to children (i to III.1 and III.2, and j to III.3 and III.4) so it can be established that the HLA-DQ1 antigen occurs on both i and j haplotypes; II.1 is proven homozygous for HLA-DQ1. In contrast individual I.1 expresses only HLA-DQ2 but since haplotype b is not passed on it cannot be proven that I.1 is homozygous for DQ2; he may express a DQ specificity which has not been detected. In practice, previously unrecognized antigens are rarely detected in this way and if high quality typing antisera are used or if molecular typing is carried out this possibility can be excluded. The frequency of undefined alleles is highest for the HLA-C locus.

A recombination event is seen in individual III.5 since the HLA-A24 antigen from haplotype e has been exchanged for A23 on haplotype g. Alternatively but less likely an explanation is exclusion of maternity.

Exclusion of paternity is shown in individual III.6 since haplotype m is absent in parents II.4 and II.5. Although studies of HLA typing are a rigorous test of paternity it is important to follow accepted practice in cases of disputed parentage (see Chapter 9).

Given random segregation, one in four siblings will share two or zero haplotypes, and one in two siblings one haplotype. Complete transmission of haplotypes has occurred in children III.1–4 from parents II.1 and II.2. Non-random segregation of haplotypes (*segregation distortion*) should always be looked for and if established would have implications for the study findings.

There are very few reports of individuals who express more than two HLA haplotypes. In the case of healthy adults they have been shown to be chimeric for their lymphocyte population but triploid neonates, who have survived for only a few days, have been described.

5.2 HLA antigen and allele frequencies

In all populations studied the incidence of different HLA antigens and alleles varies within a given population and between different populations. The method for establishing antigen frequencies is set out in *Protocol 14*.

Protocol 14. Calculation of antigen and gene frequencies

The *antigen frequency, f,* of a specific antigen is the number of times that antigen is found to occur in a population; it is expressed as a percentage or decimal.

1. List each different antigen able to be detected in the study population.
2. Count the number of individuals positive for each antigen detected.
3. Count the number of individuals positive for only one antigen; i.e. the apparent homozygotes.
4. Sum the counts in **2** and **3**. Check the sum equals twice the number of individuals tested.
5. Calculate values of f by dividing each antigen count by the total number of individuals tested. Express values of f as a percentage or decimal.

Protocol 14. *Continued*

6. Calculate the *gene frequency, p* (allele frequency) for each antigen detected using each value of f in the formula:

$$p = 1 - \sqrt{(1 - f)}$$

When molecular techniques have been used to establish phenotypes then counting of alleles will give gene frequencies directly and true homozygotes will be evident.

The efficiency of serological typing methods can be measured by the single antigen frequency; this will increase when there is a reduced antigen detection rate due to poor methodology. The acceptable figure for individuals typing for only one antigen in European Caucasoid populations is less than 15%, and this is confirmed in studies using molecular techniques when homozygotes can be clearly detected.

Differences in antigen and allele frequencies between populations can be detected by comparing values of f and p for each antigen or allele. This is usually done to test for differences between a normal control population and a group of disease study patients.

5.3 HLA antigens and alleles in different ethnic populations

Antigen frequencies and the derived gene frequencies for several populations are shown in Chapter 11. Some HLA antigens and alleles are exclusive to one ethnic population; for example the antigen HLA-B54 has a frequency of about 15% in Japanese but is absent from other populations. It is more usual to find population specific variations in frequencies (see Chapter 11). Each IHW report contains frequency data on all workshop populations for all antigens detected. In particular, the Eleventh IHW (8) concentrated on studies of many different ethnic groups and IHW data tables should be reviewed when performing analysis of data from specific populations.

Since there is racial variation in the occurrence of HLA antigens then alloantibodies to some antigens will only be found in specific populations. The exchange of typing antisera between laboratories serving different ethnic populations should therefore be encouraged to facilitate serum typing sets covering all known specificities (see Chapter 2).

6. Statistical tests

The statistical significance of a difference between frequencies is tested for in the *chi-square test* derived from a *2 × 2 contingency* table as shown in *Table 4*. The chi-square is next converted to a p value using reference tables, and a significant difference is indicated when the value of p is less than 0.05. This

Table 4. The 2 × 2 contingency table, chi-square, and Relative Risk statistics

Disease	**Antigen or gene**		**row totals**
	present	**absent**	
present	a	b	s
absent	c	d	t
column totals	u	v	n

$$\text{chi-square} = \frac{\{(ad - bc)^2\}}{(s \times t \times u \times v)} \times n$$

$$\text{Relative Risk} = \frac{a \times d}{c \times b}$$

means that the odds of the observed difference occurring by chance alone is less than 1 in 20. When the contingency table contains small numbers (<10) it is advisable to use modifications of the chi-square test such as *Yates' correction* or *Fisher's exact test*. All of these statistical tests are commonly available as personal computer software, (e.g. Minitab, Microstats, SPSS PC+). If a significant difference is found it must be corrected (p-corr.) by multiplication of the p value by the number of comparisons made. This is usually the number of antigens or alleles tested for. This avoids a statistical error which is introduced by multiple testing. If 20 comparisons are made then by chance one of them could give a significant difference which has no biological relevance. A better approach is to perform a second study on a different group of patients to confirm preliminary findings and correction of p values will not be required. Further pitfalls arise when dealing with small numbers of individuals and it is usual to study a minimum of 50 patients.

7. Studies of HLA and disease

This large topic is not easily covered in detail; the interested reader should refer to the comprehensive text of Tiwari and Terasaki (9) which lists data from studies of over 500 diagnoses. An outline approach to performing a study of HLA and disease is shown in *Protocol 15*.

Protocol 15. A design for the study of HLA and disease

1. Assess the need for the study by a review of previous published studies from reference texts and by computerized literature searches.

Protocol 15. *Continued*

2. Proceed if no previous or recent study found, if disease definition has changed, or if antigen/allele definition is new.
3. Define criteria for inclusion of patients by clearly defined clinical diagnostic criteria.
4. Obtain ethical approval.
5. Set total patient numbers for completion of study.
6. Define patient data to be collected and samples required.
7. Set study start and end dates.
8. Define control data.
9. Decide on methods of data analysis.
10. Decide authorship of study publication and preferred journal.
11. Define laboratory tests to be completed.
12. Collect patient and control samples and record clinical data.
13. Carefully interpret laboratory results and enter data to computer.
14. Analyse data.

The strength of an association is measured by the *Relative Risk* also known as the *Odds Ratio*. The Relative Risk indicates the number of times more (or less) frequently the disease occurs in individuals possessing a given antigen than in those without that antigen. A positive Relative Risk indicates increased susceptibility while a negative Relative Risk indicates protection.

It is particularly important to carefully select control data for the study. This is usually achieved by testing a non-patient population who should be healthy and from the same ethnic background as the patient group. In addition the same techniques must be used to test both patients and controls.

HLA and disease studies have established associations between an antigen and a disease in different ethnic groups, between different antigens and the same disease in different ethnic groups, associations involving more than one antigen which are often the result of allelic association, and strong and weak associations.

8. Reporting results

Testing for HLA antigens or alleles is an expensive and time consuming exercise. It is therefore of great importance that all results are carefully documented and efficiently reported. This applies equally to clinical and research studies. The current WHO Nomenclature (see Chapter 11) must always be used in a clear and informative format.

8.1 Clinical testing

The ASHI has documented standards for reporting results to clinicians, and in the UK laboratory accreditation via CPA (UK) Ltd. enforces adherence to documented criteria for reporting. It is essential that a clear and informative report is made to those directing patient care and this can only come about through close liaison between the laboratory and the clinic.

Due to the complex nature of the HLA system it is advisable to furnish a full interpretative report setting out, for example, cross reactions and implications for patient care; e.g. when matching for organ transplantation, mismatched but cross reacting antigens may be a preferable sitution to more widely disparate antigen mismatches. In family studies recombination events and exclusions of paternity are examples of interpretative information vital to clinical applications.

8.2 Journal publication

All studies of HLA in disease populations should be published for the benefit of colleagues. It is most frustrating to embark on a large study to discover that the same study has been completed but left unpublished. Thus even studies with negative findings should be documented, if only in a brief format.

In any publication where HLA data is included it is essential that the current WHO Nomenclature is used; journals should not accept any other. It is also important that authors ensure that the HLA data in a manuscript is carefully checked by an expert in the HLA field to avoid pitfalls in interpretation. Some studies are invalidated by incorrect assessment of HLA data such as the use of inappropriate control data.

Journals which publish papers in the field of histocompatibility and immunogenetics include *Disease Markers, Human Immunology, Immunogenetics, European Journal of Immunogenetics,* and *Tissue Antigens.*

References

1. Bodmer, W., Tripp, M., and Bodmer, J. (1967). In *Histocompatibility Testing 1967* (ed. E. S. Curtoni, P. L. Mattiuz, and R. M. Tosi), pp. 341–50. Munksgaard, Copenhagen.
2. Fauchet, R., Bouhallier, O., and Genetet, B. (1980). In *Histocompatibility Testing 1980* (ed. P. I. Terasaki), p. 289. UCLA Tissue Typing Laboratory, Los Angeles.
3. van Leeuwen, A. and van Rood, J. J. (1980). In *Histocompatibility Testing 1980* (ed. P. I. Terasaki), pp. 278–9. UCLA Tissue Typing Laboratory, Los Angeles.
4. Lau, L., Terasaki, P. I., Park, M. S., and Barbetti, A. (1989). In *Clinical Transplants 1989* (ed. P. I. Terasaki), pp. 447–56. UCLA Tissue Typing Laboratory, Los Angeles.
5. Colombani, J., Lepage, V., Raffoux, C., and Colombani, M. (1989). *Tissue Antigens,* **34,** 97.

6. Parham, P., Lawlor, D. A., Salter, R. D., Lomen, C. E., Bjorkman, P. J., and Ennis, P. D. (1989). In *Immunobiology of HLA* (ed. B. Dupont), Vol. 2, pp. 10–33. Springer-Verlag, New York.
7. Schreuder, G. M. Th. (1989). *Serological Definition of HLA-DR and -DQ Polymorphisms.* Thesis, University of Leiden, (available from C-Six).
8. Tsuji, K., Aizawa, M., and Sasazuki, T. (ed.) (1992). *HLA 1991*. Oxford University Press.
9. Tiwari, J. L. and Terasaki, P. I. (1985). *HLA and Disease Association.* Springer-Verlag, New York.

4

Antibodies and crossmatching for transplantation

SUSAN MARTIN and FRANS CLAAS

1. Screening for HLA alloantibodies

1.1 Introduction

Alloantibodies directed against HLA antigens can be produced in response to three different types of stimuli:

- pregnancy
- blood transfusion
- transplantation

The method of antigen presentation, the dose of antigen, and hence the nature of the response differs for each of these stimuli which can all have occurred in a potential transplant recipient.

1.1.1 Alloimmunization of pregnancy

HLA antibodies can be produced during pregnancy, most commonly during second or subsequent pregnancies, as a result of stimulation by paternal HLA antigens expressed by the fetus. Primary stimulation occurs when high levels of antigen enter the maternal circulation during the birth of the first child. During subsequent pregnancies HLA alloantibodies can be produced as part of a secondary response to low levels of stimulating antigen which cross the placenta. This response, occurring in a usually healthy woman, gives rise to high titre, high affinity IgG antibodies of restricted HLA specificity which are the main source of HLA typing reagents (see Chapter 2).

1.1.2 Blood transfusion

The end stage renal failure patient receiving a blood transfusion is unhealthy and has an immune system that may be somewhat depressed as a result of uraemia. The stimulating HLA antigens are introduced intravenously in small and possibly overlapping doses. The source of the antigen(s) is the lymphocyte content of the transfused blood. Although lymphocytes can remain viable, even after storage of the blood (1), multiple transfusions are usually required to elicit a response in non-parous, non-transplanted patients. Patients requiring multiple transfusions will usually receive blood from different donors with different HLA phenotypes. Antibodies are therefore most often produced against public specificities expressed by several donors, or against those antigens most common in the population. This is illustrated in the situation of an individual from one ethnic group receiving transfusions from donors of a different ethnic group for which the most frequently encountered HLA antigens may differ. There is then a greater chance that the patient will be exposed to non-self HLA antigens and hence produce an antibody response.

1.1.3 Transplantation

An allograft recipient is a chronically ill individual who is also receiving immunosuppressive therapy. Their immune response is therefore immature and may be variously affected by different immunosuppressive drugs or combinations thereof. The antigenic stimulus is presented as a single bolus but the number of stimulating antigens will depend on the degree of HLA mismatching between donor and recipient.

The antibody response may be of limited specificity, directed against one or two donor HLA antigens, or polyclonal with antibodies directed against cross reactive groups, or even antigens encountered during a previous blood transfusion, pregnancy, or transplant.

The time course of the response may vary considerably, with antibodies appearing weeks, months, or years post-transplant in association with late onset acute or chronic rejection (2, 3). Although antibody titres may rise following a transplant nephrectomy, they are often detectable prior to removal of the rejected organ (3). Whilst antibodies are usually IgG, transient IgM antibodies can also be detected.

Occasionally, lymphocytotoxic activity is detected in the serum of non-transfused, non-transplanted males. If it has been definitely established that no transfusions have occurred, then this activity may be attributed to auto-reactive lymphocytotoxic antibodies (see Section 3.1).

1.2 Screening for alloantibodies in relation to organ transplantation

The target cell used to screen for HLA alloantibodies is the lymphocyte. In order to design a protocol for the detection and identification of HLA allo-antibodies one must consider their properties.

(a) Antibodies can be directed against HLA class I or class II antigens. Whilst class I antigens are expressed by all types of lymphocytes, class II antigen expression is restricted to B lymphocytes and activated T cells.

(b) Class I antibodies can be directed against a private specificity, e.g. HLA-A2 or an epitope shared by a group of antigens, e.g. HLA-Bw4, and a cross reacting group of antigens (CREG), e.g. HLA-B5, 35, 53, 18 (see Chapters 2 and 3).

(c) Class II antibodies may appear to be directed against a public determinant but actually be against an antigen in strong linkage disequilibrium with certain HLA-DR specificities, e.g. HLA-DQw2 with HLA-DR3 and HLA-DR7.

(d) IgM lymphocytotoxic antibodies may occur which are also reactive with autologous lymphocytes (see Section 3.1).

The principles of screening in relation to transplantation are the same as for reagent procurement (see Chapter 2) but the approach differs because the aim is not to acquire reliable, monospecific antibodies, but rather to completely define the lymphocytotoxic antibody content of a patient's serum. This should include information on antibodies to HLA class I private and public specificities, HLA class II specificities, and autoreactivity. This leads to the identification of those antigens to be avoided in transplantation and allows intelligent crossmatching, avoiding both wasting time in the acute situation and wastage of limited volumes of patient sera. Conversely, for highly sensitized patients, antigens to which there is no sensitization can also be identified. These antigens provide 'windows' in the patient's sensitization profile and donors can be selected for a given recipient on the basis of having an HLA type which falls into the 'windows' for that patient (see Section 5.3). Monitoring of HLA alloantibodies in patients awaiting transplantation is of major importance as hyperacute rejection is inevitable for patients with circulating donor specific antibodies.

In relation to post-transplant screening, the identification of specific anti-donor reactivity can be indicative of rejection (3).

1.2.1 Selection of cell panels

A cell panel should be carefully selected to represent all frequently encountered HLA specificities and rarer ones when possible. Antigen combinations should avoid common haplotypes to facilitate antibody definition. Panel cells should be well characterized for both HLA class I and class II antigens, including private and public specificities and antigen splits. The size of the panel will influence the quality of the analysis. A 30–40 cell panel will be adequate for initial screening and identification of broad specificities, but a larger panel is needed for more detailed antibody assignment. The objective of the screening is to rule out unsuitable donors prior to the crossmatch but not to substitute for it. If a patient has an antibody of rare specificity which cannot easily be included in the cell panel, it follows that it will be encountered only infrequently in a donor, but if it does occur then it can be detected by the crossmatch.

i. HLA class I cell panel

HLA class I antigens are expressed by all peripheral blood lymphocytes (PBL) which can therefore be used after separation from anti-coagulated blood (see Chapters 2 and 3) without further separation into T and B cells. However, the use of T cells (see *Protocol 1*) is sometimes preferred in order to avoid weak positive reactions which could be due to class II antibodies reacting with the small percentage of B cells present.

ii. HLA class II cell panels

B cell panels must be used for the assignment of HLA-DR, DR52/53 and -DQ specific antibodies. These can be separated from a population of lymphocytes using immunomagnetic beads (see *Protocol 1*, and for a variation of the technique see Chapter 3). Chronic lymphocytic leukaemia (CLL) is usually a B cell leukaemia, and PBLs from patients with CLL can be used as a source of B cells without further separation being necessary. When using PBLs from patients with CLL it is essential to check the percentage of B cells since this varies with the stage of the disease (see Chapter 2).

1.2.2 Random cell panels

If sera are screened against a random cell panel then the percentage of the panel with which a serum reacts (panel reactive antibodies, %PRA) will be predictive of how that serum might react when crossmatched against donors from the random population. This is a guide as to how long a patient might have to wait for a negative crossmatch and hence a transplant. However, the use of random cell panels results in over-representation of more common antigens and haplotypes which can make it more difficult to define antibody specificities. In general, the use of carefully selected panels is preferred (see Chapter 2).

Protocol 1. Isolation of T and B cells using immunomagnetic beads

See also Chapter 3 for an alternative method.

1. Isolate peripheral blood lymphocytes as in Chapters 2 and 3. All subsequent procedures must be carried out on ice.
2. Aliquot 4×10^6 cells in suspension into a 4 ml cryotube (Northumbria Biologicals, C2051) and centrifuge for 5 min at 200 *g*.
3. Carefully aspirate the supernatant and resuspend the cell pellet in 100 μl PBS.
4. Cool the tube on ice, and then add 15 μl Dynabeads class I (Dynal, 210.03) for T cells, or class II (Dynal, 210.34) for B cells.
5. Leave on ice for 5 min with frequent swirling of the tube to ensure mixing of cells and beads.

6. Add 3 ml cold PBS, mix gently, and place tube on a rare earth magnet (Dynal, 120.02) for 3 min.
7. Pipette off the supernatant.
8. Remove the tube from the magnet and repeat steps **6** and **7**.
9. Resuspend the rosettes in 75 μl Hanks' balanced salt solution (HBSS) for use.

1.2.3 Serum samples

The specimen required for testing is obtained from 5 ml clotted blood, taken into a tube without anti-coagulant.

Serum samples are collected monthly from patients on the renal transplant waiting list when the risk of sensitization following a blood transfusion is relatively high. Samples should also be collected two weeks following a blood transfusion. For other transplants, such as hearts where blood transfusions are unlikely, only one sample for screening prior to transplantation may be necessary. Whilst an individual laboratory may wish to establish its own protocol for collecting specimens, it is of key importance that liaison is established between the histocompatibility laboratory and the dialysis centres, the medical staff, and the clinics to ensure procurement of appropriate samples.

Serum samples are also collected monthly from recipients during their first year post-transplant, when the risk of failure is greatest and less frequently thereafter, to coincide with their follow-up appointments. In this way, any potentially graft damaging antibodies can be detected and an antibody profile of the patient obtained as an aid to finding the most suitable re-transplant, should that be necessary.

1.2.4 Screening methods

The chosen method for screening and crossmatching may vary between laboratories but the methods used for screening should always *reflect and be at least* as sensitive as those used for crossmatching. Traditionally, sera are batched and then dispensed in one microlitre aliquots on to Terasaki trays and screened in a microlymphocytotoxicity assay (see Chapter 3) using fresh or frozen cell panels. More recently, the advent of cell panels frozen on to Terasaki trays has allowed greater flexibility and speed, with a result obtainable in less than four hours of specimen receipt. This is particularly applicable to heart transplantation when an unscreened patient may be in urgent need of a transplant. A serum sample can then be rapidly screened for deleterious HLA specific alloantibodies. Tray frozen cell panels are available commercially or can be prepared in the tissue typing laboratory (4) (see *Protocol 2*). Serum screening methods are essentially the same whether using tray frozen cell panels or fresh/frozen cell panels (see *Protocol 4* and Chapter 2).

Protocol 2. Preparation of PBL cell panels frozen on Terasaki trays

A cell panel is selected from previously HLA typed volunteers. Either three 20 cell panels or two 30 cell panels will fit on one 60 well Terasaki tray. Design a protocol so that the position of each panel cell on the tray is clearly identified.

It is crucial that the cell suspensions and the trays remain cold throughout this procedure. If the cells do not remain cold once DMSO has been added there will be a reduction in viability because DMSO is toxic at room temperature.

1. Use fresh PBL isolated as in Chapters 2 and 3 and adjust cell concentration to 10×10^6/ml in Park and Terasaki medium (Flow Laboratories, 16-921-49) plus 20% fetal calf serum (FCS) (Flow Laboratories, 29-101-49). Store 30 min at 4°C to cool. To prepare 500 trays with a 20 cell panel a minimum of 1.1 ml of each cell suspension will be required.
2. Add paraffin oil (Hillcross Pharmaceuticals) to each well of the trays.
3. Pre-cool the Terasaki trays at 4°C on aluminium trays.
4. Pipette 35 μl of each cell suspension into the appropriate position of a serum reservoir (Dynatec, 860173).
5. Using a serum dispenser, add 35 μl HBSS plus 14% DMSO (Sigma, D58791) pre-cooled to 4°C, mix and immediately dispense 1 μl cell suspension into each oiled well of the Terasaki trays. Discard excess cell suspension. This gives 5×10^3 cells/well. To avoid warming, there should be no more than 50 trays/batch.
6. Cool the plates inside the top of a −80°C freezer for 5 min, replace the lids, and stack on a polystyrene tile inside the freezer overnight. Next day pack in plastic boxes for storage for up to 2 months at −80°C.

Protocol 3. Preparation of CLL panels frozen on Terasaki trays

A panel of ten CLL cells selected from HLA typed cells stored in liquid nitrogen (see Chapter 2) is required. Since CLL cells are more robust than PBLs it is possible to re-freeze them as screening panels.

1. Remove selected CLL cells from liquid nitrogen, thaw rapidly in a 37°C water-bath, and dilute in 10 ml HBSS plus 20% FCS.
2. Centrifuge at 400 *g* for 10 min, and wash twice in HBSS to remove all traces of cryopreservative.
3. Resuspend to 10×10^6/ml in Park and Terasaki medium plus 20% FCS.
4. Cool. Proceed as in *Protocol 2*, step **2**.

Protocol 4. Serum screening using frozen cell panels

The procedure is the same for both frozen PBL and CLL trays.

- frozen PBL and CLL trays: if there are three PBL panels and six CLL panels on the respective trays, to screen six sera you will need two PBL and one CLL tray
- 50 μl and 250 μl Hamilton syringes (Hamilton)
- staining cocktail: india ink, 0.2 mg/ml ethidium bromide (Sigma, E8751), 0.1 mg/ml acridine orange (Sigma, A6014)

All batches of screening should include positive and negative controls. Fetal calf serum or serum from a non-transfused blood group AB male donor are suitable negatives which give a base line for cell viability. The positive control should be an alloantiserum of known specificity, or a cytotoxic monoclonal antibody reacting with an HLA class I shared determinant, e.g. W6/32 (Serotec, MCA818), or with HLA-DR antigens, e.g. L243 (ATCC HB55).

1. Organize sera in a rack in the desired order for dispensing on to the trays. Carefully note the order on a protocol.
2. Remove the required number of trays from the −80 °C freezer and thaw at 37 °C for 3 min. Add test serum within 30 min since cell viability will be lost. It is recommended that trays are handled in batches of no more than 15–20.
3. Add 1 μl of serum/well to each panel cell, ensuring there is mixing by centrifuging the trays at 200 *g* for 1 min.
4. Incubate the trays at 22 °C for 1 h.
5. Add 4 μl complement/well and incubate for a further 1 h at 22 °C.
6. Add 1 μl staining cocktail/well.
7. Read reactions using inverted fluorescent microscope and record the percentage cell death/well. Dead cells stain orange and viable cells stain green.

1.2.5 Interpretation

Patient sera will contain a mixture of antibodies because the response is not monoclonal. Even when apparently monospecific, there will be a mixture of antibodies directed against the different epitopes that comprise the broad specificity. An individual's immune response is unique, giving rise to a heterogeneous population of antibody molecules with different affinities for antigen. The result is that it is unlikely that two sera will give exactly the same reaction pattern and discrepancies will arise and a certain degree of uncertainty be experienced. Furthermore, HLA antigens may be expressed to differing levels on cells of the same phenotype. Serological reaction patterns can therefore be analysed using statistical methods which allow for a measure of this uncertainty. The most frequently used statistic is the two by two contingency table and the chi-square statistic (see Chapter 3).

Protocol 5. Determination of HLA specificities in a test serum

See Chapter 2 for an alternative technique.

1. The results with the negative control give a baseline for the viability of each panel cell.
2. Note the antigens present on the panel cell where a positive reaction is seen.
3. If a serum reacts with more than half of the cells expressing a particular antigen then the corresponding antibody may be present.
4. For the assigned specificity, count the number of true positives (+/+), false positives (+/−), false negatives (−/+), and true negatives (−/−).
5. 'Eyeballing' the results will identify monospecific sera and antibodies to CREGs. If statistical analysis is required, calculate a chi-square value using a 2×2 contingency table for which various computer packages are available.

Problems.

(a) If a serum is multispecific it may be necessary to screen it against further cell panels and/or test it in dilution before specificities can be assigned.

(b) If HLA class II alloantibodies are suspected in the presence of class I alloantibodies then pooled platelets can be used to absorb out the latter before re-testing (see Chapter 2).

(c) Reactivity with the PBL but not the CLL panel may be due to either autoantibodies (see Section 3.1), or antithymocyte globulin (ATG) or OKT3 in the serum of post-transplant patients being treated with, or having recently finished, a course of ATG, or OKT3 immunosuppressive therapy.

(d) Poor cell viability could be due to:
- warming of cells during freezing of cell panels on trays
- delays in addition of sera to thawed trays
- poor PBL isolation technique

2. Antibody classes

HLA alloantibodies are usually IgG although transient IgM antibodies can be detected following blood transfusions and organ transplantations. IgM lymphocytotoxic antibodies will often also react with autologous lymphocytes in the cytotoxicity assay when they are referred to as autoantibodies (see Section 3.1). As these antibodies give rise to false positive crossmatch results it is vital that they are distinguished from IgG and IgM HLA alloantibodies. A modification of the lymphocytotoxicity assay using dithiothreitol (DTT) or dithioerythritol (DTE) provides a simple and quick method for distinguishing reactions due to IgM from those due to IgG. Further studies will then confirm whether IgM antibodies are autoreactive (see Section 3.1).

DTT and DTE are sulphydryl compounds which can reduce sulphides quantitatively according to the type of bonds. The intersubunit bonds of the IgM pentamer are more susceptible to cleavage than the interchain bonds of IgG or the IgM subunits; DTT at the appropriate concentration will therefore inactivate IgM without affecting IgG activity.

Protocol 6. DTT modification of the lymphocytotoxicity assay

DTT is added to the test serum prior to screening against frozen cell panels

1. Prepare a stock solution of 0.1 M DTT (Sigma, D632) in HBSS, store in 100 μl aliquots at −40°C, and thaw when required.
2. In a Beckman tube, mix 36 μl serum with 4 μl 0.1 M DTT to give serum containing 0.01 M DTT with only minimal dilution of the serum: the actual volume of serum and 0.1 M DTT can be adjusted as required provided their relative proportions remain constant.
3. Screen DTT treated sera against frozen PBL and CLL panels as in *Protocol 4*. The final concentration of DTT in the assay (1 μl treated serum plus 1 μl cell suspension) is 0.005 M.
4. Include DTT treated negative and IgG positive controls to respectively monitor cell viability and confirm that no IgG is reduced.
5. Interpretation of results.
   (a) Abrogation of cytotoxicity: reaction due to IgM antibodies.
   (b) Partial reduction of cytotoxicity: reaction due to IgG and IgM antibodies.
   (c) No reduction of cytotoxicity: reaction due solely to IgG antibodies.
   (d) Enhancement of cytotoxicity: possible IgM blocking antibodies in the serum.
6. Problems. Partial inhibition of cytotoxicity could be due to incomplete reduction of high titre IgM antibodies: pre-incubate serum plus DTT for 30 min at 37°C. Negativity of the IgG positive control could be duc to:
   (a) Too high a concentration of DTT: repeat with lower DTT concentrations.
   (b) Inactivation of the rabbit complement by the DTT: repeat the assay, adding 0.002 M cysteine (Sigma, C8755) to the rabbit complement before use to inhibit further action by the DTT.

---

3. Screening for non-HLA antibodies

3.1 Autoantibodies

Autoreactive lymphocytotoxic antibodies react with lymphocytes from random donors and autologous lymphocytes. They are almost always IgM antibodies and are usually considered not to be detrimental to transplant function (5).

As these antibodies result in a false positive crossmatch, it is important that they are identified prior to the crossmatch situation.

Protocol 7. Identification and absorption of autoreactive lymphocytotoxic antibodies

A. *Screening*

1. Screen sera using PBL and CLL cells (see *Protocol 4*). A positive result with PBL and negative or weak with CLL cells suggests the presence of autoantibodies.
2. Repeat the screening with and without the addition of DTT (see *Protocol 6*). Abrogation of the reaction by DTT indicates the presence of an IgM antibody. Proceed to auto-crossmatching.

B. *Auto-crossmatching*

1. Isolate autologous PBLs from 20 ml heparinized blood (see Chapters 2 and 3).
2. Separate lymphocytes into T and B cells using immunomagnetic beads (see *Protocol 1* and Chapter 3).
3. Aliquot 1 μl of each test serum into Terasaki trays, including positive and negative controls: prepare six identical trays.
4. Dispense 1 μl of cells into each well, using T cells for three trays and B cells for the remaining three trays.
5. Incubate one T cell tray and one B cell tray at each of 4°C, 22°C, and 37°C for 45 min. The use of cells separated with immunomagnetic beads allows a reduction in incubation time.
6. Add 4 μl rabbit complement to each well and incubate all the trays at 22°C for a further 45 min.
7. Add 1 μl staining cocktail (see *Protocol 4*) and read using an inverted fluorescent microscope.
8. Autoantibodies react most strongly at 4°C but weakly at 37°C.

C. *Auto-absorption*

The presence of autoantibodies can be further confirmed by absorbing them with autologous lymphocytes. Approximately 5×10^6 cells are required for each absorption. The large number of cells needed to adequately absorb several sera from one patient are best obtained by establishing a lymphoblastoid cell line (LCL) from that patient (see Chapter 6).

1. Count the lymphocytes obtained from an LCL culture or separated from heparinized blood (see Chapter 3).
2. Transfer 5×10^6 lymphocytes to an Eppendorf tube (Sarstead, 72.6901) and microcentrifuge for 5 min. Store remaining cells at 4°C.
3. Mix 60 μl serum with the cell pellet and incubate overnight at 4°C.
4. Microcentrifuge as in step **2** and remove serum supernatant.
5. Repeat steps **2** to **4**.
6. Screen absorbed serum against autologous lymphocytes, PBL, and CLL cells to check for removal of autoantibodies and identify any residual alloreactivity.

Once it has been clearly shown that only IgM autoantibodies have been identified in a potential transplant patient's serum, that serum can be subsequently crossmatched using the auto-absorbed sample, or in the presence of DTT to avoid a false positive result. When a serum contains a mixture of IgM allo and IgM autoantibodies, the auto-absorbed serum should be used for crossmatching as the use of DTT will also remove IgM alloantibodies.

3.2 Monocyte antibodies

Antibodies directed against monocytes that are clinically significant in relation to transplantation react with a characteristic epitope shared by peripheral blood monocytes, macrophages, and endothelial cells. The failure of HLA identical transplants has been linked to the presence of these antibodies (6). Screening for monocyte antibodies follows the same principles as screening for lymphocytotoxic antibodies but employs a panel of well characterized peripheral blood monocytes or cultured endothelial cells. These should be selected so that any HLA reactivity can be distinguished from reactivity specifically with monocytes. Alternatively, HLA antibodies can be absorbed out of the sera prior to testing. As monocytes express HLA class II antigens in addition to class I, LCL or CLL cells should be used for the absorptions as they express both classes of antigens. Methods for isolating monocytes from peripheral blood and for isolating endothelial cells are clearly described by Stastny in the ASHI Laboratory Manual (7, 8).

3.3 Epithelial cell antibodies

IgM epithelial cell antibodies have been detected in association with paediatric transplant failure (9, 10). They can be detected by screening in a standard microcytotoxicity assay (see *Protocol 9*) or using a flow cytometric technique (10) with the epithelial cell line A549 as the target cell.

Protocol 8. Culture of epithelial cell line, A549

All procedures should be carried out in a sterile cabinet, using sterile equipment and reagents.

- Dulbecco's modification of Eagle's medium (DMEM) (Northumbria Biologicals M392)
- L-glutamine, 200 mM (glu) (Northumbria Biologicals M902)
- penicillin and streptomycin solution, 10 000 IU/ml (pen/strep) (Northumbria Biologicals M974)
- 0.5% (w/v) trypsin plus 0.2% Na_2EDTA in PBS (Northumbria Biologicals M916)

1. Add 5 ml glu and 5 ml pen/strep to 500 ml DMEM.
2. Prepare DMEM plus 10% FCS for cell culture as required.
3. Maintain epithelial cell line A549 (ATCC, CLL185) in culture in 10 ml DMEM plus 10% FCS in 50 ml culture flask (Costar, C30551) at 37°C, 5% CO_2. The cells adhere to the base of the flask giving a 'cobblestone' appearance. When confluent (after approximately three days), they can be removed from the flask and transferred into fresh medium. Cells will survive approximately seven days without a change in medium.
4. Remove medium from cells with a pastette and wash cells once with DMEM to remove traces of serum.
5. Add 2.5 ml trypsin/EDTA for 2 to 3 min at 37°C. The cells lift off the flask: gently resuspend them with a pastette and transfer to a conical tube. Add 1 ml FCS to inactivate the trypsin.
6. Make up to 10 ml with DMEM and centrifuge at 200 *g* for 5 min to pellet the cells, remove supernatant.
7. Repeat step **6** and resuspend the pellet in 2.5 ml DMEM plus FCS.
8. One flask yields 5 to 10 $\times$ 10^6 cells. Return 0.5 ml cell suspension with 9 ml DMEM plus FCS to the flask for culture.

Protocol 9. Epithelial cell microcytotoxicity assay

- 50 μl repeating Hamilton syringe with Luer attachment and tips (V. A. Howe, 31331)
- rabbit complement (pre-screened in the cytotoxicity assay to ensure it is not toxic to A549)

1. Use sterile Terasaki trays. Trays can be sterilized by immersing in 70% methanol for at least 30 min. Drain and air-dry in the sterile cabinet.
2. Take 0.5 ml cell suspension prepared from a three to four day confluent culture (see *Protocol 8*) and make up to 5 ml with DMEM plus 10% FCS. It is not necessary to count the cells every time: if the given volumes and dilutions are adhered to, the number of cells/well remains sufficiently constant. If the cells have been allowed to overgrow, they should be counted and standardized to 10^5/ml.
3. Using the repeating Hamilton syringe and tips, plate out 1 μl cells/well into the required number of Terasaki trays. Allow sufficient for each serum to be screened in triplicate.
4. Incubate in a humid box for 24 h at 37°C, 5% CO_2.
5. Immediately prior to plating out the sera wrap tray in a paper towel and vigorously flick off all the medium: the cells will adhere to the bottom of the well.
6. Take an aliquot of each test serum and treat with DTT (see *Protocol 6*).
7. Test serum and DTT treated serum in parallel. Add 1 μl serum or serum plus DTT/well, testing each in triplicate.
8. Incubate 60 min, 4°C in a humid box.
9. Add 3 μl complement/well and repeat 60 min incubation at 22°C.
10. Add 1 μl stain/well, read, and record percentage cell killing.
11. A reproducible cell killing of >20% which is abrogated by DTT treatment indicates the presence of an IgM antibody reactive with epithelial cells.

Problems.

(a) IgM autoreactive lymphocytotoxic antibodies (see Section 3.1) also react in this assay which can therefore only be correctly interpreted with a full knowledge of the lymphocytotoxic antibody profile of the serum.

(b) HLA antibodies also react with A549, but these are usually IgG and not DTT sensitive. If IgM HLA class I alloantibodies are suspected they can be removed by platelet absorption (see Chapter 2). A549 has not been found to express HLA class II antigens.

3.4 Anti-idiotype antibodies

The immunglobulin molecule is made up of two pairs of chains, each pair having one light and one heavy chain. Each of the four chains have a variable (V) and a constant region. The V regions of the four chains together form the antibody binding site which permits recognition of a particular epitope. As the antigen binding site of a particular antibody molecule exhibits a unique protein structure, it is itself antigenic and those antigenic sites which distinguish one V domain from another are termed *idiotypes*. Each antibody molecule therefore expresses a unique idiotype which can act as an antigen to stimulate the production of a second antibody, referred to as an anti-idiotype antibody. This process cascades to create a network which plays an important part in the regulation of immune responses.

Sophisticated assays are available to determine the presence of anti-idiotype antibodies (11, 12). However, these assays are technically difficult which is why many investigators remain sceptical about the existence and functional significance of anti-idiotype antibodies relating to HLA. The principle of these assays involves assessing whether a serum that is negative for a particular specificity has the ability to *specifically* block the activity of other sera with that specificity either from the same or other individuals. The method is described by Reed and Suciu-Foca in the ASHI Laboratory Manual (12).

If a patient has previously had antibodies against a particular HLA specificity, e.g. HLA-A24 but they are no longer detectable, it may be that they have subsequently produced anti-idiotype antibodies directed against the HLA-A24 antibodies which specifically block their reactivity. If these can be demonstrated, it may be possible to transplant that patient across an A24 mismatch with a peak positive/current negative crossmatch, on the assumption that the anti-idiotype antibodies will block hyperacute or accelerated graft rejection.

4. Crossmatch procedures

4.1 Introduction

Many patients who are waiting for a transplant have been in contact with allogeneic cells via previous blood transfusions, pregnancies, or failed transplants.

These patients may have formed antibodies against the foreign HLA antigens present on these allogeneic cells. In the early days of transplantation, it was recognized that the presence of pre-formed antibodies to the HLA antigens of a kidney donor, in the serum of the recipient could lead to hyperacute rejection of the graft (13).

This knowledge has led to the introduction of the crossmatch in clinical transplantation, which is designed to prevent such hyperacute rejection. In the crossmatch the serum of the recipient is tested with lymphocytes of the potential organ donor. When anti-donor antibodies are present, as detected by a positive crossmatch, the transplant is not performed. Despite the long history of crossmatching, the clinical relevance of the crossmatch is ambiguous since not all antibodies reacting with a target molecule on a lymphocyte are detrimental to the transplanted organ (14).

It is essential to determine the exact specificity and immunoglobulin class (see Sections 2 and 3) of the antibodies causing the positive crossmatch. This is the main reason why regular antibody screening is performed with the sera of patients on the transplant waiting list (see Section 1). A positive crossmatch due to autoantibodies, which are mainly of the IgM class, may not lead to graft failure. By performing a crossmatch both in the presence and absence of DTT, these antibodies can be distinguished from HLA alloantibodies, which are usually IgG but may be IgM. Only autoreactive IgM antibodies may be clinically insignificant. It is generally accepted that IgG antibodies directed against the HLA class I antigens of the donor are detrimental and are likely to induce hyperacute rejection. Antibodies directed against HLA class II antigens are sometimes associated with hyperacute rejection (15), but this is not always so. Possibly, the titre of the antibody and the expression of the class II antigen on the target organ play an important role. Separate crossmatches with donor T and B lymphocytes as targets (see *Protocol 1*) may differentiate recipient antibodies with HLA class I and class II specificity.

The relevance of antibodies in non-current sera is also a point of much debate. Some centres disregard antibodies in historical sera and claim that graft survival is not significantly worse when the historical crossmatch is positive (16). However, again, the specificity and antibody class seem to be important. Historical sera containing IgG antibodies against HLA class I antigens of the donor are strongly associated with a very poor graft survival (17). It is likely that these antibodies themselves are not responsible for the early graft failure but that the switch from IgM to IgG antibody production is associated with activation of T cells, which later will mediate acute graft failure (18).

Not only the interpretation of the crossmatch results, but also the technical aspects of crossmatching are open to debate. For years, the complement-dependent lymphocytotoxicity assay (see Chapter 3) was considered to be the 'gold standard'. However, not all antibody classes are able to fix complement and therefore modifications of this assay, i.e. the antiglobulin assay (see *Protocol 10*), or assays not based on complement fixation like flow cytometry (see *Protocol 13*) have been introduced, although the relevance of these more sensitive assays is not well established. For every assay, case reports are available which show a negative lymphocytotoxicity assay and a positive crossmatch in the more sensitive assay involved in a patient with early graft failure. However, many cases can be found in which such crossmatch results do lead to an excellent graft survival.

In addition to assays developed to increase the sensitivity of serological crossmatching, modifications of the complement-dependent cytotoxicity assay have been applied in order to determine the actual specificity of the antibody, e.g. specific for donor HLA class I causing the positive crossmatch. This is particularly applicable to mixtures of antibodies or in the absence of proper screening results. Nevertheless, it should be stressed that optimal screening of patient sera against well selected panels is of considerable help for the interpretation of a crossmatch.

4.2 Technical aspects

The principles of crossmatch assays are very similar to the ones described for the screening of antibodies, including the use of positive and negative control sera (see Section 1). In complement-dependent assays patient sera are tested both undiluted and diluted (1:4) in order to detect a possible prozone effect. Target cells can be either spleen cells, lymph node cells, or peripheral blood cells. Separation into T and B lymphocyte populations (see *Protocol 1*) may be necessary to differentiate recipient HLA class I and class II specific antibodies. Lymphocytes are isolated by lymphocyte separation medium (LSM) density gradient centrifugation as described in Chapters 2 and 3. Several studies have shown that splenocytes are more sensitive to the effect of cytotoxic antibodies compared to peripheral blood lymphocytes. One version of the complement-dependent cytotoxicity assay is described in Chapter 3. It is essential that crossmatch procedures are at least as sensitive as recipient antibody screening (see Section 1.2.4), and particular attention should be paid when deciding on suitable incubation times and methods for staining assays as use of fluorochromes is more sensitive than eosin. Here modifications of this assay and other crossmatch assays not based on complement activation are described.

The inability of the complement-dependent cytotoxicity assay to detect non-complement fixing antibodies can lead to false negative crossmatches. By adding a heterologous complement-fixing antibody directed against the light chain of the human antibody (or an anti-Fab), one can reveal the reactivity of non-complement fixing or weak complement-fixing antibodies in the complement-dependent cytotoxicity assay (see *Protocol 10*).

Protocol 10. Antiglobulin assay

- heterologous complement-fixing antibodies directed against human kappa-light chain (Kent, 117002) or against human Fab in order to detect also the lambda-light chains

1. Add to each well of the oiled Terasaki tray 1 μl test serum and 1 μl of target cells (4 × 10^6/ml).
2. Spin the trays for 5 min at 150 *g*.
3. Incubate serum and cells for 30 min at 22°C.
4. Add 10 μl HBSS, centrifuge the tray (5 min, 150 *g*), and remove the fluid by flicking the tray. Repeat this wash step three times.
5. Add 1 μl of diluted (1:100) anti-human kappa-light chain to each well.
6. Incubate for 2 min at 22°C.

7. Follow *Protocol 4* (step **5** onwards).

Interpretation of the results.
When standard complement-dependent cytotoxicity is negative and anti-globulin assay is positive, the serum contains either non-complement fixing antibodies or a non-cytotoxic amount of complement fixing antibodies.

Several kinds of antibodies may lead to a positive crossmatch. In order to establish whether the antibodies are directed against HLA class I or HLA class II alloantigens of the donor, spleen, lymph node, or PBL, target cells can be pre-incubated with non-complement fixing monoclonal antibodies (MoAb) to either HLA class I or HLA class II (see *Protocol 11*). When pre-incubation with, for instance a MoAb to HLA class I is able to block the reactivity of the patient serum, one can reasonably conclude that the positive crossmatch was caused by alloantibodies to HLA class I. Using this method it is possible to determine the target molecule of the antibody causing the positive crossmatch in the absence of complete screening results.

Protocol 11. Blocking complement-dependent cytotoxicity with monoclonal antibodies

- non-complement fixing MoAb directed against HLA class I, e.g. PA 2.6 (17)
- if non-complement fixing antibodies are not available, the combination of a complement-fixing IgG MoAb (Immunohaematology Leiden B11 G6) and a $F(ab)_2$ of a heterologous anti-mouse IgG (Fc) antibody (Sigma M-1522, F 8850) can be used
- wash fluid: HBSS or PBS plus 1% bovine serum albumin (BSA)

1. (a) Incubate 2×10^6 lymphocytes with 0.5 ml titrated non-complement fixing antibody for 1 h at 22°C in a centrifuge tube.

or

(b) Incubate the lymphocytes, as in (a), with the complement-fixing antibody for 1 h at 22°C. Add 10 ml wash fluid to the cells and centrifuge 5 min at 800 *g*. Repeat wash step once. Incubate lymphocytes for 1 h with the $F(ab)_2$ of a heterologous anti-mouse IgG, diluted 1:100 in PBS, at 4°C.

2. Wash the lymphocytes twice and adjust cell count to 4×10^6/ml.

3. Add 1 μl test serum, diluted in series from neat to 1:16, and 1 μl pre-treated target cells to each well of a Terasaki tray.

4. Incubate for 30 min at 22°C.

5. Follow *Protocol 4* from step **5** onwards.

Protocol 11. *Continued*

Interpretation.
When pre-incubation with the monoclonal antibody leads to significant reduction in titre of the cytotoxicity, the alloantibodies in the serum are directed against HLA class I antigens.

Problem.
When the sera contain a mixture of antibodies to HLA class I and class II, the specificity can be difficult to assess.

In standard lymphocytotoxicity crossmatching the effector mechanism used to detect antibody reactivity is complement-dependent cytotoxicity. It is likely that *in vivo* other antibody-dependent effector mechanisms like antibody dependent cellular cytotoxicity (ADCC) play a role in graft destruction. In ADCC, killer cells will be activated via their Fc receptors by antibodies bound to cell surface antigens and cause destruction of the target cells.

Protocol 12. Antibody dependent cellular cytotoxicity (ADCC)

- culture medium: RPMI 1640 plus 10% FCS
- carboxyfluorescein diacetate (CFDA) (Sigma, C5041): stock solution 10 mg/ml in acetone AR, stored at −20°C in glass tubes with rubber stoppers
- glass tubes: protein coated by rinsing with culture medium
- Terasaki trays with cylindrical wells, volume 40 μl (Greiner, 007726180)
- ink diluted 1:300 in 4.9% Na_2EDTA in distilled H_2O, pH 7.0, stored at room temperature

1. Add 2×10^6 donor cells to 1 ml freshly prepared CFDA working solution (25 μl CFDA stock in 5 ml PBS).
2. Incubate for 15 min in the dark at room temperature.
3. Wash target cells once and adjust to 2×10^5/ml culture medium.
4. Dispense 10 μl target cell suspension to 18 wells of a Terasaki tray.
5. Add 10 μl patient serum neat, and serially diluted 1:2 to 1:16 in culture medium, and 10 μl negative control serum each in triplicate.
6. Add 10 μl patient effector cell suspension, using concentrations appropriate for the required effector:target (E:T) ratio (100:1, 50:1, 25:1, 12.5:1).
7. Centrifuge the trays for 5 min at 150 *g*.
8. Measure the pre-incubation values (*b*) of the fluorescence in the target cells in an automated microscope, e.g. Leica, Patimed.
9. Incubate for 3 h at 37°C in a humidified CO_2 incubator.

10. Centrifuge the trays for 5 min at 400 *g*.

11. Add 10 μl ink suspension to each well to quench the fluorescence released into the medium. Do this with care.

12. Assay 10 min later by measuring the remaining fluorescence in target cells in the automated microscope (*a*).

13. Correct for variable target cell numbers by dividing the assay values (*a*) by the pre-incubation values (*b*).

14. Calculate percentage lysis using the formula:

$$(1 - x/y) \times 100\%$$

x = median fluorescence value of (target cells + antibodies + effector cells)
y = median fluorescence value of (target cells + medium + effector cells)

15. A crossmatch is positive when percentage lysis of the target cells in the presence of test serum exceeds 10%.

In some centres it is standard practice to use a crossmatch technique, which is more sensitive than the standard lymphocytotoxicity. One of these assays is the antiglobulin assay (see *Protocol 10*) and the other, which is more widely applied, is flow cytometry (see *Protocol 13*). This assay may be clinically relevant in patients waiting for a re-transplant although in first transplants, the assay may be too sensitive (false positives). In flow cytometry, binding of (allo)antibodies to membrane antigens is demonstrated by the reaction of fluoresceinated heterologous antibodies directed against the human immunoglobulin. This assay detects the binding of non-complement fixing antibodies as well as complement-fixing IgG antibodies in concentrations lower than mandatory for complement-mediated cytotoxicity. Only in the case of IgM antibodies is the complement-dependent cytotoxicity assay more sensitive.

Protocol 13. Flow cytometry

- FITC conjugated $F(ab)_2$ rabbit anti-human IgG (Dakopatts, F 315)
- phycoerythrin (PEY) conjugated monoclonal anti-human CD3 to label T cells (Becton Dickinson, anti-Leu 4-PE, 7349)
- 0.5% paraformaldehyde in PBS
- flow cytometer, e.g. FACScan (Becton Dickinson), Epics Profile (Coulter)

1. Adjust donor lymphocytes (preferentially spleen cells) to 10^7 cells/ml in PBS.

2. Add 100 μl cell suspension to a sample tube.

3. Centrifuge for 5 min at 800 *g*.

Protocol 13. *Continued*

4. Discard supernatant.
5. Add 100 μl recipient serum to the cell pellet and resuspend.
6. Incubate serum and cells for 30 min at 22°C.
7. Wash cells three times with 10 ml cold PBS/BSA by centrifugation.
8. Add 20 μl FITC conjugated $F(ab)_2$ rabbit anti-human IgG (1:10 in PBS), followed directly by 5 μl PEY anti-human CD3 (1:10 diluted in PBS) to perform the T cell crossmatch.
9. Incubate 30 min at 4°C.
10. Wash three times as in step **7**.
11. Resuspend cells in 200 μl cold PBS/azide, and add 300 μl paraformaldehyde in PBS.
12. Store cells at 4°C prior to FACS analysis. The cells can be stored for 2–3 days.
13. By gating the CD3 positive T cells, one can determine the fluorescence intensity on these cells alone. Normal limits of this fluorescence intensity should be established by testing 20–25 sera from unsensitized individuals. T cell peak displacement with significant channel shifts can be considered to be a positive crossmatch. The degree of channel shift accepted as giving a positive result should be established using local controls.

Note. Similarly B cell crossmatches can be performed using PEY conjugated anti-CD19 (Dakopatts, R 808) instead of anti-CD3 in step **8**.

The advantage of leucoagglutination, which was the standard assay in the early days of HLA, is the ability to detect non-complement fixing antibodies with a low affinity for the target antigen. In contrast to the antiglobulin assay and flow cytometry the agglutination assay does not require any washing steps after incubation of target cells and antisera.

Protocol 14. Leucoagglutination

- 10 ml fresh EDTA, citrate, or defibrinated blood
- 5% dextran solution
- siliconized 10 ml glass tubes

1. Add 2 ml of 5% Dextran to the blood.
2. Incubate for 20 min in a 37°C water-bath.
3. Aspirate the upper plasma layer containing lymphocytes, granulocytes, and monocytes.

4. Add 2 μl of this cell rich plasma to 2 μl heat inactivated patient or control serum in a Terasaki tray under paraffin oil.
5. Incubate for 135 min at 37°C in the incubator.
6. Read the agglutination using an inverted microscope.
7. Agglutination can be scored according to the number of leucocytes included in the clumps from 0 (negative) to 5. A cell suspension, in which leucocytes are evenly distributed, is called negative. The presence of clumps with almost no free cells means a strong positive reaction (score 5).

---

5. HLA matching and recipient selection

5.1 Renal transplantation

The majority of studies (see Chapter 1) have shown matching for HLA-B and -DR to be more beneficial than matching for HLA-A, although the Collaborative Transplant Study (CTS) shows a significant effect of HLA-A, especially when the splits rather than the broad antigens are considered (21). The differential effect of HLA-A versus HLA-B mismatches is also reflected in the cytotoxic T cell precursor frequency (see Chapter 6), which is significantly lower for HLA-A. These cytotoxic T cells are considered to play an important role as effectors of HLA class I directed graft rejection.

Significant effects of matching for HLA-DR have been shown in many multi-centre studies. Some centres claim the effect of HLA-DR matching is the most pronounced in the first months after transplantation, whereas the effect of matching for HLA-B is also visible in the long-term (22, 23). This might be explained by the fact that the HLA class I alloantigens will remain expressed on the donor organ, whereas the HLA class II antigens, are constitually expressed on antigen presenting cells from bone marrow origin which with time, will be replaced by antigen presenting cells from the recipient. In Eurotransplant there is a hierarchy of matching, HLA-DR > -B > -A, (see *Protocol 15*) but other centres may have a different emphasis.

Protocol 15. Matching for renal transplantation

This is an outline scheme that can be adapted for donor/recipient matching. Information on all patients on the waiting list with regard to HLA type, ABO blood group, urgency, and other clinical data are included in a computerized database.

1. Input donor HLA type and ABO blood group to computer, and run matching programme.
2. Initial selection of recipients is on the basis of ABO identity or compatibility. Blood group O donors are preferentially transplanted in blood group O recipients.

Protocol 15. *Continued*

3. Matching based on HLA-A, -B, -DR identity or no donor/recipient mismatch.
4. Hierarchy for other matches is HLA-DR > -B > -A.
5. Antibody positive patients have a higher urgency than antibody negative patients.
6. Other factors which are considered during selection are clinical urgency and waiting time.
7. Special options: matching exclusively for paediatric patients and priority for highly sensitized patients (see *Protocol 16*).
8. Following selection of potential recipients on the basis of HLA matching, they should next be crossmatched with the donor target cells using the chosen assay, and recipient sera carefully selected on the basis of lymphocytotoxicity antibody screening results.

5.2 The need for quality controls in HLA typing and matching

Due to the complexity of serological HLA typing it can be difficult to show significant effects of HLA matching in large multi-centre studies. Recent DNA typing techniques (RFLP as described in Chapter 5) showed that 25% of the transplants that were reported as HLA-A, -B, -DR compatible according to serological typing were found to be HLA-DR mismatched (25). A comparison of HLA types performed in different centres and the reference centre, as happens within Eurotransplant, is a good measure of individual laboratory performance. A follow-up of multiple typings is essential. In addition, typing using molecular methods should be compared to results using serological methods.

5.3 Matching in highly sensitized patients

Patients who have formed multispecific antibodies against foreign HLA antigens following previous blood transfusions, pregnancies, or failed mismatched transplants, are very difficult to transplant. The crossmatch will be positive with many potential donors and, once a crossmatch is found to be negative, these sensitized patients (especially when they are waiting for a re-graft) have often been found to have a poorer prognosis than non-sensitized patients. It is usual to look for an HLA compatible donor for such patients so that the crossmatches will be negative and graft survival will be high. The problem, however, is the extensive polymorphism of the HLA system, which makes the chance of finding such compatible donors very small.

Within Eurotransplant when such patients are transplanted across a negative crossmatch performed with both historical and current sera, an HLA matching effect is seen. A very strong effect of matching for HLA-DR is found but hardly any effect of matching for HLA-A and -B. The patients may have formed antibodies against the HLA-A and -B antigens, towards which they easily make an immune response, while the remaining HLA-A and -B antigens are less immunogenic.

This observation was the basis of the acceptable mismatch programme (26), which aims to increase the number of suitable donors for highly sensitized patients by determining the HLA-A and -B antigens towards which the patients have never formed antibodies. The patients are transplanted with an HLA-DR compatible donor. There may be several HLA-A and -B mismatches as long as the patient has never formed antibodies against these mismatches. Graft survival is excellent and also in this patient group no effect of matching for HLA-A and -B is seen.

Protocol 16. Acceptable mismatches

1. Determine the acceptable HLA-A and -B mismatches by:
   (a) considering the HLA mismatches of the negative panel donors in standard lymphocytotoxic antibody screening.
   (b) testing the sera from each patient in complement-dependent cytotoxicity against a panel of lymphocyte donors selected so that each donor has only one HLA-A or -B mismatch with the patient. In case of negative reactions with all patient sera the mismatched HLA-A or -B antigen is considered to be an acceptable mismatch.
2. Include the acceptable mismatches in the computer database together with the patient's own HLA-A, -B, and -DR antigens.
3. Given the HLA-A, -B, and -DR antigens of every potential organ donor, the computer matching programme will select potentially crossmatch negative recipients on the basis of compatibility of donor antigens with the recipient's own HLA-A, -B, and -DR antigens in combination with the acceptable HLA-A and -B mismatches.
   Thus, the donor is always optimally matched for HLA-DR but may have several HLA-A and -B mismatches.

Two European schemes (HIT-trial by Opelz and SOS-scheme by UKTS) include the circulation of patient sera to many tissue typing centres in order to extend the pool of potential donors. Crossmatches are performed with all ABO blood group compatible donors.

Patients are transplanted based on negative crossmatches with these potential donors. The difference with the acceptable mismatch programme is that not all centres include historical sera, and thus not all alloantibodies ever produced by the patient are tested in the crossmatch.

5.4 Matching for other solid organs and tissues

Prospective matching for HLA, as routinely performed in renal transplantation, is not yet fully introduced for other solid organs. Retrospective studies in hearts (27) and crossmatching in livers (28) show, however, an effect of HLA matching on graft survival and/or rejection incidence in most cases (see Chapter 1). It is important to provide HLA typing results from potential organ donors before organ retrieval to facilitate prospective matching when ischaemia times are limited as in the case of hearts and livers. This is possible with serological assays using the most up-to-date techniques such as isolation of donor target cells using immunomagnetic beads.

5.4.1 Bone marrow transplantation

The need for HLA matching in bone marrow transplantation is well recognized (see Chapter 1) and studies on 'the windows' of T cell reactivity as analysed via precursor frequencies (29) may become of pivotal importance in the future.

Acknowledgements

We would like to thank Phil Evans, Marian Witvliet, Marijke Spruyt-Gerritse, Ingrid Curiël, and all tissue typing laboratory staff who have assisted in the development of the techniques described.

References

1. Martin, S., Dyer, P. A., Harris, R. H., Manos, J., Mallick, N. P., Gokal, R., and Johnson, R. W. G. (1985). *Transplantation,* **39,** 256.
2. Solomon, L. R., Martin, S., Short, C. D., Lawler, W., Gokal, R., Johnson, R. W. G., and Mallick, N. P. (1986). *Transplantation,* **41,** 262.
3. Martin, S., Dyer, P. A., Mallick, N. P., Gokal, R., Harris, R., and Johnson, R. W. G. (1987). *Transplantation,* **44,** 50.
4. Sinnott, P. J., Kippax, R. L., Sheldon, S., and Dyer, P. A. (1985). *Tissue Antigens,* **26,** 318.
5. Chapman, J. R., Tayor, C. J., Ting, A., and Morris, P. J. (1986). *Transplantation,* **42,** 608.
6. Cerilli, J., Bay, W., and Brasile, L. (1983). *Hum. Immunol.,* **7,** 45.
7. Stastny, P. (1990). In *American Society for Histocompatibility and Immunogenetics Laboratory Manual* 2nd Edition (ed. A. A. Zachary and G. A. Teresi), p. 107. ASHI, Lenexa, KS.
8. Stastny, P. (1990). In *American Society for Histocompatibility and Immunogenetics Laboratory Manual* 2nd Edition (ed. A. A. Zachary and G. A. Teresi), p. 110. ASHI, Lenexa, KS.
9. Martin, S., Brenchley, P. E., Postlethwaite, R. J., and Johnson, R. W. G. (1991). *Tissue Antigens,* **37,** 152.

10. Harmer, A. W., Haskard, D., Koffman, G. C., and Welsh, K. I. (1990). *Transplant. Int.,* **3,** 66.
11. Sucia-Foca, N., Reed, E., D'Agati, V. D., Ho, E., Cohen, D., Benevisty, A. I., McCabe, R., Brensilvcr, J. M., King, D. N., and Hardy, M. A. (1991). *Transplantation,* **51,** 593.
12. Reed, E. and Suciu-Foca, N. (1990). In *American Society for Histocompatibility and Immunogenetics Laboratory Manual* 2nd Edition (ed. A. A. Zachary and G. A. Teresi), p. 272. ASHI, Lenexa, KS.
13. Kissmeyer-Nielsen, F., Olsen, S., Petersen, V. P., and Fjeldborg, O. (1966). *Lancet,* **2,** 662.
14. Ting, A. (1989). *Transplant. Int.,* **2,** 2.
15. Ahern, A. T., Artruc, S. B., and Della Pelle, P. (1982). *Transplantation,* **33,** 103.
16. Falk, J. A., Cardella, C. J., Halloran, P. J., Bear, R. A., and Arhus, G. S. (1987). *Transplant. Proc.,* **19,** 720.
17. Taylor, C. J., Chapman, J. R., Ting, A., and Morris, P. J. (1989). *Transplantation,* **48,** 953.
18. Roelen, D. L., Datema, G., Van Bree, S., Zhang, L., Van Rood, J. J., and Claas, F. H. J. (1992). *Transplantation,* **53,** 899.
19. Gjertson, D. W. and Terasaki, P. I. (1990). *Organ Transplantation,* (ed. G. M. Abouna, M. S. A. Kumar and A. G. White), p. 203. Kluwer Academic Publishers, Dordrecht.
20. Matas, A. J., Frey, D. J., Gillingham, K. J., Noresen, H. J., Reinsmoen, N. L., Payne, W. D., Dunn, D. L., Sutherland, D. E., and Najarian, J. S. (1990). *Transplantation,* **50,** 599.
21. Opelz, G. (1988). *Lancet,* **2,** 61.
22. Thorogood, J., Persijn, G. G., Schreuder, G. M. Th., D'Amaro, J., Zantvoort, F. A., Van Houwelingen, J. C., and Van Rood, J. J. (1990). *Transplantation,* **50,** 146.
23. Gilks, W. R., Gore, S. M., and Bradley, B. A. (1990). *Transplantation,* **50,** 141.
24. Takemoto, S., Carnahan, E., and Terasaki, P. I. (1990). *Clinical Transplants,* (ed. P. I. Terasaki), p. 485. UCLA Tissue Typing Laboratory, Los Angeles, CA.
25. Opelz, G., Mytilineos, J., Scherer, S., Dunckley, H., Trejant, J., Chapman, J., Middleton, D., Savage, D., Fisher, D., Bignon, J. D., Bensa, J. C., Albert, E., and Noreen, H. (1991). *Lancet,* **338,** 461.
26. Claas, F. H. J., De Waal, L. P., Beelen, J., Reekers, P., Van de Berg-Loonen, P., De Gast, E., D'Amaro, J., Persijn, G. G., Zantvoort, F., and Van Rood, J. J. (1989). *Clin. Transplants,* (ed. P. I. Terasaki), p. 185; UCLA Tissue Typing Laboratory, Los Angeles, CA.
27. Claas, F. H. J. (1993). *Immunology of heart and heart-lung transplantation,* (ed. M. Rose and M. Yacoub). Edward Arnold.
28. Welsh, K. I. (1989). *Curr. Opin. Immunol.,* **1,** 1178.
29. Kaminski, E., Hans, J., Man, S., Brookes, P., Mackinnan, S., Hughes, T., Avakian, O., and Goldman, J. M. (1989). *Transplantation,* **48,** 608.

5

Molecular methods

DAVID SAVAGE, LEE ANN BAXTER-LOWE, JACK GORSKI, and DEREK MIDDLETON

1. Introduction

The application of molecular techniques to tissue typing has led to a fundamental change in methodology for HLA class II typing. The first part of this chapter describes the use of Southern blot (1) techniques which permit the identification of restriction fragment length polymorphisms (RFLPs) which correlate with known serological (HLA-DR/DQ) and cellular (HLA-Dw) defined specificities (2). The latter part of the chapter describes the enzymatic amplification of the second exon of HLA-DR/DQ class II genes using the polymerase chain reaction (PCR) (3) and analysis of the product with allele- and sequence-specific oligonucleotides (4).

Compared to serotyping (see Chapter 3), DNA typing techniques provide more information on genetic variation. Other advantages include:

- cell type, cell viability, or cell surface expression are generally unimportant
- homozygosity can be confirmed
- DNA probes can be re-generated, whereas continuous screening for antisera is necessary for serotyping (see Chapter 2)
- DNA samples can be easily transported from one location to another
- DNA typing lends itself to batch analysis
- there is less subjectivity involved in the interpretation of data

2. Application of restriction fragment length polymorphism typing

The most convenient and economical approach to HLA-DR/DQ RFLP typing utilizes a single restriction enzyme (*Taq*I) and short, (i.e. subregion or exon-specific) complementary DNA (cDNA) probes to HLA-DRB, -DQB, and -DQA loci (2). Typing by DNA RFLP reveals polymorphisms in both coding and non-coding regions, however most of the polymorphism generated is the result of differences in non-coding regions. Due to linkage between the non-coding regions and the coding regions these polymorphisms can be used to identify the alleles.

The major drawback to DNA RFLP typing compared to serotyping is that it cannot be applied prospectively to matching cadaver organ donors, because of the length of time involved in obtaining a result. A further limitation is that certain HLA-DR specificities give rise to identical DRB RFLP patterns, (e.g. DR1 and DR103). In addition there are a few indistinguishable *Taq*I/DRB heterozygote combination patterns, [e.g. DR17 (Dw24)/DR14 (Dw9), and DR17 (Dw25)/DR13 (Dw18, Dw24)]. Due to the strong linkage disequilibrium between HLA-DR and -DQ loci, certain indistinguishable DRB RFLP patterns can be inferred from the DQ RFLP type in European Caucasoid populations (2).

A recent report has shown that there is a 20% difference in results between DNA RFLP typing and serotyping for HLA-DR (5). This is principally because of the lack of monospecific antisera to certain specificities, notably the alleles associated with HLA-DR52. More recently a retrospective study using DNA RFLP typing has shown that if typing mistakes are eliminated from the determination of HLA-DR antigens, the success rate of HLA matched cadaver transplants is almost as good as that of HLA-matched sibling transplants (6).

3. HLA class II cDNA probes

3.1 Source of cDNA probes

The cDNA probes used for HLA-DRB, -DQB, and -DQA *Taq*I RFLP typing are pRTV1 (7), pII-β-I (8), and pDCH1 (9) respectively. Short cDNA probes are used in preference to full-length cDNA probes because they give rise to less inter-locus cross-hybridization resulting in fewer bands and easier analysis of RFLP patterns. The probes are usually obtained as bacterial stab cultures from the laboratory of origin. These stab cultures contain bacteria, (e.g. *Escherichia coli* HB101) that are derived from a single clone which has been transformed with plasmid DNA, (e.g. pBR322) carrying the appropriate probe DNA.

The stab cultures have been prepared in the laboratory of origin by inserting the probe DNA into the plasmid. This is performed by cleaving the plasmid with a restriction enzyme that recognizes a single plasmid restriction site, followed by ligation of the probe to the plasmid using DNA ligase. The plasmid is used to transform competent bacterial cells and the resulting transformants selected on agar plates containing the appropriate antibiotic. The competent cells can be prepared and transformed according to standard techniques (10) or may be purchased from commercial companies, (e.g. Gibco–BRL). Further confirmation that the clone contains plasmid/probe DNA involves growing the clone in broth containing appropriate antibiotic, isolation and lysis of the bacterial cells, purification of plasmid DNA, and restriction enzyme digestion of plasmid DNA, followed by agarose gel electrophoresis to check the fragment sizes obtained.

3.2 Preparation of plasmid DNA

Preparation of plasmid containing probe DNA is performed by growing a sample of bacterial stab culture in broth containing appropriate antibiotic. The bacterial cells are lysed, the plasmid DNA purified, and resuspended in buffer solution.

In the case of the pRTV1 clone, the probe is inserted into a polycloning site, (i.e. synthetic sequence containing closely arranged series of restriction sites) within the plasmid pSP64, and selection is carried out in ampicillin. For pII-β-I and pDCH1 probes each are inserted into the ampicillin gene (insertional inactivation) within the plasmid pBR322, and selection is carried out using tetracycline.

Various methods are available for large scale preparation of plasmid DNA, e.g. caesium chloride-ethidium bromide density gradient or alkali-lysis. However the alkali-lysis method (*Protocol 1*) is preferred because of convenience, and as it does not require ultracentrifuge facilities.

Protocol 1. Large scale preparation of plasmid DNA

Good microbiological practice should be used throughout when preparing plasmid DNA.

(a) L-broth—dissolve 2.5 g bactotryptone (Difco 0123-01), 2.5 g NaCl, and 1.25 g yeast extract (Difco 1880-17) in a total volume of 250 ml deionized distilled water (ddH_2O), and autoclave.

(b) SET buffer—dissolve 20 g sucrose in 50 ml ddH_2O, add 5 ml 1 M Tris-HCl pH 7.6, and 10 ml 0.5 M Na_2 EDTA pH 8.0. Make up to 100 ml with ddH_2O and autoclave. To 30 ml SET buffer add 600 μl lysozyme solution (50 mg/ml) (Sigma 6876) just prior to use.

(c) RNase solution—dissolve 0.1 g RNase (Sigma R4875) in 7 ml ddH_2O, add 37.5 μl 4 M NaCl, and 100 μl 1 M Tris-HCl pH 7.6, and make up to 10 ml with ddH_2O. Boil RNase solution for 15 min to inactivate deoxyribonucleases, cool to room temperature, and store at −20 °C.

(d) Supplement all media with ampicillin (Sigma A9393; 50 mg/litre final concentration) for the pRTV1 (DRB) probe, or with tetracycline hydrochloride (Sigma T3383; 12.5 mg/litre final concentration) for the pII-β-I (DQB) and pDCH1 (DQA) probes. (*Note.* Do not autoclave antibiotic stock solutions or media containing antibiotics.)

Note. There are no speed controls on the microcentrifuge (MSE microcentaur) used in this laboratory. Therefore the microcentrifuge can only be

Protocol 1. *Continued*

set for a period of time and no *g* value is given in the protocols. The maximum *g* value for the 1.5 ml microcentrifuge tubes (Nycomed 1012-01-0) is 11 600 *g* and for the 0.6 ml microcentrifuge tubes (Nycomed 1048-01-0) 10 500 *g*. However these speeds are not necessary and the centrifugations can be performed at lower *g* values.

1. Plate out a sample of bacterial stab culture on an agar plate [1.5% (w/v) bactoagar (Difco 0140-01) in L-broth] containing appropriate antibiotic. Grow overnight for 16–20 h in a 37°C incubator.
2. Select a single bacterial colony and inoculate 10 ml sterile L-broth containing appropriate antibiotic in a 20 ml sterile Universal tube. Grow overnight for 16–20 h at 37°C in an orbital shaking incubator (Gallenkamp INR-200).
3. Inoculate 200 ml sterile L-broth containing appropriate antibiotic in a 500 ml conical flask with 1 ml of culture from step **2** and incubate overnight for 16–20 h at 37°C in an orbital shaking incubator. Take the remainder of the culture from step **2** and prepare 1 ml aliquots of bacterial suspensions in 16% (v/v) sterile glycerol and store at −80°C for future use.
4. Pour contents of the flask into a 250 ml Nalgene bottle (Techmate 3120-0250). Balance bottles using scales and centrifuge for 20 min at 2000 *g* (MSE Hi-Spin 21 centrifuge).
5. Carefully pour off supernatant without disturbing the pellet. Sterilize the supernatant prior to disposal.
6. Resuspend the cells in 30 ml SET buffer (containing lysozyme) and leave at room temperature for 5 min.
7. Add 65 ml of freshly prepared lysis mix [1% (w/v) sodium dodecyl sulphate (SDS), 0.2 M NaOH], mix well, and incubate bottles for 30 min at 50°C in a shaking water-bath.
8. Remove bottles and place on ice for 25 min.
9. Slowly add 20 ml 5 M potassium acetate (pH 4.8) with swirling. Return bottles to ice for a further 40 min.
10. Centrifuge bottles at 4°C for 30 min at 2000 *g*. Filter the supernatant through a Kleenex tissue in a glass funnel, into a 250 ml Nalgene bottle.
11. Add 200 μl RNase solution and incubate for 20 min at 40°C in a water-bath.
12. Add 25 ml Tris-saturated phenol, pH 7.9 (Camlab AD/0945-52) and 25 ml choroform:isoamyl alcohol (24:1). Shake well and centrifuge for 20 min at 2000 *g*.
13. Without touching the interface, carefully transfer upper aqueous phase into a 250 ml Nalgene bottle and add an equal volume of isopropanol

(propan-2-ol). Mix, leave at room temperature for 15 min, and centrifuge for 30 min at 3000 *g*.

14. Carefully remove all fluid and resuspend plasmid DNA pellet in 3 ml Tris-EDTA buffer (TE) (10 mM Tris-HCl pH 7.6, 1 mM EDTA pH 8.0). Add 120 µl 5 M NaCl followed by 3 ml isopropanol, and transfer into a 10 ml polycarbonate centrifuge tube (Techmate 3117-0120). Centrifuge for 10 min at 10 000 *g* (MSE Hi-Spin 21).

15. Carefully remove and discard the supernatant. Drain the tube and dry the pellet thoroughly for 30–60 min in a 37°C incubator.

16. Re-dissolve the DNA pellet in 500 µl TE buffer, transfer into a 1.5 ml microcentrifuge tube, and store at 4°C overnight.

17. Add 500 µl Tris-saturated phenol (pH 7.9), vortex for 15 sec, and centrifuge for 1 min in a microcentrifuge, Without touching the interface, carefully remove upper aqueous phase into a 1.5 ml microcentrifuge tube.

18. Repeat step **17** above.

19. Add 500 µl chloroform:isoamyl alcohol (24:1), vortex for 15 sec, centrifuge for 1 min in a microcentrifuge, and carefully transfer the upper aqueous phase into a 1.5 ml microcentrifuge tube.

20. Add 50 µl 4 M sodium acetate (pH 6.0) and 1 ml ice-cold ethanol, mix, and place the tube on ice for 5 min.

21. Centrifuge for 10 min in a microcentrifuge. Without disturbing the DNA pellet, carefully remove and discard the supernatant. Wash the pellet with 1 ml ice-cold ethanol and centrifuge for 2 min in a microcentrifuge.

22. Carefully remove and discard the ethanol. Dry pellet thoroughly for 30–60 min in a 37°C incubator, and re-dissolve pellet in 200 µl TE buffer. Store overnight at 4°C, and assess concentration and purity of DNA sample (see *Protocol 3*). Adjust plasmid concentration to 1 µg/µl, aliquot in 45 µl amounts, and store at −20°C.

3.3 Removal of probe DNA from plasmid

For pRTV1 the restriction enzyme *Pst*I is used to cleave the plasmid, giving rise to two fragments at 3054 bp (pSP64) and 517 bp (DRB insert). For pII-β-I the restriction enzyme *Ava*I is used, giving rise to three fragments at 2634 bp (pBR322), 2228 bp (pBR322), and 620 bp (DQB insert). For pDCH1 the restriction enzyme *Pst*I is used, giving rise to two fragments at 4362 bp (pBR322) and 797 bp (DQA insert). The fragments are electrophoretically separated in a low melting point agarose gel and the probe fragment excised (*Protocol 2*).

Protocol 2. Restriction enzyme digestion of plasmid/probe DNA and preparative electrophoresis

1. Digest 40 μg of plasmid containing probe DNA in a total volume of 60 μl with appropriate restriction enzyme (3 U/μg) (see Section 3.3) according to the manufacturer's instructions.
2. Add 15 μl of 5 × gel loading buffer (GLB) [50 mM Tris-HCl pH 7.6, 50 mM Na_2 EDTA pH 8.0, 0.5% (w/v) SDS, 40% (w/v) sucrose, 0.1% (w/v) bromophenol blue] to stop the reaction. Gel loading buffer is used to increase the density of the DNA when submerged in the gel slots, and also as a tracker dye to indicate how far the DNA has travelled.
3. Electrophorese samples using a mini-gel apparatus at 50 V (constant voltage) in a 1% (w/v) low melting point agarose (BRL 5517UA) gel in 1 × Tris-acetate EDTA (TAE) (40 mM Tris-acetate, 1 mM EDTA) buffer containing 0.5 μg/ml ethidium bromide (10 mg/ml, Sigma E1510) for 3–4 h (see *Protocol 5*). (*Caution*. Wear gloves, ethidium bromide is mutagenic.) Also run 0.7 μg of a *Hin*dIII digest of lambda (λ) bacteriophage (λ/*Hin*dIII: BRL 520-5612SA) to check the sizes of the fragments obtained, and to check that the electrophoresis has proceeded with no problems. (*Note*. To create a slot large enough to accommodate the relatively large volume of plasmid/probe digest, use three teeth of a slot-forming comb together.)
4. Place gel/template on a UV transillumінator (UV Products, Model TM20) (λ_{302nm}) and check if all the appropriate fragments are present, (i.e. digestion is complete). *Caution*. Wear UV eye protection. Photograph the gel with a camera (Polaroid MP4 loaded with 667 film) fitted with a Wratten 23A filter (Kodak 14955589). Fragments appear red/orange because of ethidium bromide fluorescence.
5. Under UV light excise the band corresponding to the probe with a sterile scalpel blade and trim off as much excess agarose as possible from the slice. Do not expose the DNA to UV light for any longer than is necessary.
6. Place the agarose slice into a pre-weighed 1.5 ml microcentrifuge tube. Re-weigh the tube and deduce the weight of the slice.
7. Add 1.5 ml ddH_2O for each gram of agarose slice and boil the sample for 7 min. Centrifuge for 5 sec in a microcentrifuge (see *Protocol 1*), and incubate for 10 min at 37°C in a water-bath.
8. Determine the new volume by aspirating the molten agarose into a pipette tip attached to a Gilson P1000 pipette. Estimate the concentration of probe using the equation below: Concentration of probe (ng/μl) =

$$\frac{\text{insert (bp)} \times \text{plasmid + probe (ng)/plasmid (bp) + insert (bp)}}{\text{new volume } (\mu\text{l})}$$

For example:

$$\text{Concentration of pRTV1} = \frac{517 \times 40\,000/(3054 + 517)}{500} = 11.6\ \text{ng}/\mu\text{l}$$

9. Aliquot the molten agarose containing probe in 50 ng or 100 ng amounts into 0.6 ml microcentrifuge tubes. Allow samples to solidify at room temperature, then store at −20°C.

4. Extraction, assay, and storage of genomic DNA

Genomic DNA can be extracted from any nucleated cell. The salting-out method (11) is preferred over methods employing organic deproteinizers, (e.g. phenol), since it is rapid, safe, and economical (*Protocol 3*). In those laboratories where the facilities are available DNA can be extracted using an automated DNA extractor, (e.g. Applied Biosystems). When resuspended in buffer solution the DNA can be stored for daily use at 4°C, or long-term at −80°C. The salting-out method is used as distinct from the method in *Protocol 10* as it is essential that the DNA be pure, homogeneous, not degraded, and assayed prior to use for DNA RFLP analysis.

Protocol 3. DNA extraction from nucleated cells

(a) Red cell lysis buffer (RCL)—0.32 M sucrose, 10 mM Tris-HCl pH 7.5, 5 mM $MgCl_2$, 1% Triton X-100 (Sigma X100). RCL buffer is filtered through a 0.45 μm filter (Fisher Biotech/Nalgene 09-40-22G) into a sterile container and stored at 4°C for a maximum of ten days.

(b) Nuclei lysis buffer (NLB)—50 mM KCl, 1.5 mM $MgCl_2$, 0.1 mg/ml gelatin (EIA grade, Bio Rad 170-6537), 0.45% (v/v) Nonidet P-40 (Sigma P1379), 0.45% (v/v) Tween 20 (Calbiochem 492015), 10 mM Tris-HCl pH 8.3. Since heating may be required to dissolve gelatin, it is useful to prepare a 1% gelatin (w/v) stock solution incubated at 70°C until the gelatin is completely dissolved.

1. Collect 10 ml blood into 2 ml 5% (w/v) Na_2 EDTA in a 15 ml conical Falcon tube (Becton Dickinson 2097).
2. Centrifuge blood for 10 min at 1300 *g*.
3. Remove plasma and transfer buffy coat into a 15 ml conical tube.
4. Add 10 ml RCL, mix, and stand for 10 min at room temperature mixing intermittently. Centrifuge for 10 min at 1300 *g* and remove red cell lysate.
5. Repeat step **4** above.
6. Resuspend white cell pellet in 3 ml NLB. Add 600 μl 1 × proteinase K solution [2 mg/ml proteinase K in ddH_2O, (stored at −20°C in aliquots)

Protocol 3. *Continued*

Sigma P0390] and 200 μl 10% (w/v) SDS, mix well but gently, and incubate for 16–18 h at 37°C in a water-bath. Alternatively a one day extraction can be performed using 600 μl 10 mg/ml proteinase K solution instead of 600 μl 2 mg/ml proteinase K solution, keeping all other components the same, and the sample incubated for 3 h at 55°C. After incubation proceed directly to step **7**. However in the interests of economy the overnight method is preferred.

7. Add 1 ml 6 M NaCl solution and shake vigorously for 15 sec to precipitate protein.
8. Centrifuge for 15 min at 1500 *g* and transfer supernatant into a 15 ml conical tube.
9. Repeat step **8** above.
10. Add 8 ml absolute ethanol, cap the tube, and mix gently by inversion to precipitate the DNA.
11. Remove the DNA by wrapping it around the end of a heat-sealed glass Pasteur pipette, squeezing the DNA gently against the inside wall of the tube to remove excess ethanol.
12. Dissolve the DNA in 100–200 μl TE buffer in a 1.5 ml microcentrifuge tube. Try to achieve a DNA concentration of approximately 1 μ/μl. Heat samples for 2 h at 50°C in a water-bath to aid dissolution.
13. Store DNA samples at 4°C for three days, vortexing once each day prior to determination of DNA purity and concentration.
14. To assay the DNA dilute 5 μl of stock DNA solution in 495 μl TE buffer and record optical density (OD) readings at 260 nm (λ_{max} DNA), and 280 nm (λ_{max} protein), against a TE blank using optically matched quartz micro-cuvettes (Hellma 6042).
15. To determine the concentration of the DNA (μg/μl) multiply optical density reading at 260 nm by a factor of five. To assess the purity of the DNA calculate the ratio of OD 260 nm to OD 280 nm which should be 1.7–2.0 (before use).
16. For DNA extraction from small numbers of previously isolated cells, (e.g. frozen cells), scale down all reagents used above by a factor of ten, wash cells twice in phosphate-buffered saline (PBS), and perform extraction in 1.5 ml microcentrifuge tubes.

5. Restriction endonuclease digestion of genomic DNA

Since the human genome comprises 3×10^9 bp it is necessary to cleave the DNA into smaller fragments to permit any kind of analysis. This is performed using restriction enzymes, which cleave the DNA at specific points within DNA recognition sequences. There are many restriction enzymes, which are produced naturally by bacteria, each of which recognize unique base sequences. Restriction enzymes can be purchased from a number of commercial companies, [e.g. Northumbria Biologicals Ltd (NBL), Bethesda Research Laboratories (BRL), Boehringer].

The restriction enzyme *Taq*I recognizes the DNA sequence 5′-TCGA-3′ and reveals a most extensive polymorphism which correlates with HLA-DR and -DQ specificities. This is possibly because *Taq*I has a DNA recognition sequence containing the dimer CpG, and it is known that CpG dimers tend to undergo transition mutation to TpG dimers (12), thus destroying *Taq*I restriction sites when mutation occurs. Most restriction enzymes operate at 37°C, however *Taq*I differs in that its optimum efficiency is at 65°C.

For RFLP analysis it is recommended to digest and process control DNAs, (i.e. samples of known HLA-DR/DQ type) alongside unknown samples. The control samples should cover the range of polymorphic fragments usually encountered so as to aid interpretation of data (see Section 11). The controls can either be DNA from homozygous typing cells or DNA from heterozygotes, the latter being preferred since fewer controls are required.

Protocol 4. Restriction endonuclease digestion of genomic DNA

1. Combine the following in a 1.5 ml microcentrifuge tube:

ddH_2O	(to a final volume of 20 μl)
DNA	8 μg
10 × reaction buffer[a]	2 μl
*Taq*I restriction enzyme[b] (NBL 011106)	40 U

2. Vortex briefly, then centrifuge for 5 sec in a microcentrifuge (see *Protocol 1*).

3. Incubate for a total of 3 h at 65°C in a water-bath. Two drops of mineral oil may be added to *Taq*I digests to reduce evaporation, but this is not essential.

[a] Supplied by manufacturer with *Taq*I restriction enzyme.
[b] Add the enzyme in two aliquots of 20 U at 1.5 h intervals.

---

6. Agarose gel electrophoresis

In order to analyse DNA fragments created by restriction enzyme digestion it is necessary to separate the fragments using submerged horizontal agarose gel electrophoresis (*Protocol 5*). Electrophoretic separation is based on the fact that DNA molecules have an overall negative charge. The rate of migration of DNA molecules is based on molecular size, (i.e. migration is inversely proportional to the $\log_{10}$ of the number of base pairs) and the percentage concentration of agarose used. Since a large number of DNA fragments of various lengths are created following restriction enzyme digestion, completely digested DNA appears as a continuous red/orange fluorescent streak when stained with ethidium bromide and examined under UV light following agarose gel electrophoresis.

Electrophoresis equipment can be purchased from a number of commercial companies, (e.g. BRL, Pharmacia). Two sizes of gel are generally used to fractionate digested DNA. Mini-gels are used to monitor an aliquot of the digested DNA to see if digestion is completed, and if approximately equal amounts of DNA have been digested for each sample. Transfer gels are used to fractionate the main digest prior to Southern blotting (see *Protocol 6*). For clinically non-urgent samples the monitoring of digested DNA can be omitted because incomplete digestion with *Taq*I or loading of inappropriate amounts of DNA are rare, and if incomplete digestion or incorrect loading of the amount of DNA occurs the digest can be repeated. For urgent samples it is recommended that the digest is monitored using a mini-gel, so as any problems can be rectified prior to transfer gel electrophoresis, thus saving time.

Protocol 5. Electrophoresis of digested DNA

A. *Mini-gel electrophoresis*

The following description is based on BRL H5 horizontal electrophoresis equipment.

1. Prepare a mini-gel by adding 0.7 g agarose (BRL 5510UA) to 100 ml 1 × TAE buffer. Boil, then cool molten agarose to 65°C.
2. Add 5 μl ethidium bromide (10 mg/ml), mix gently, and pour into a level gel template (14 cm × 11 cm) which has been sealed at both ends with tape. (*Caution*. Wear gloves, ethidium bromide is mutagenic.)
3. Remove air bubbles, insert slot-forming 14 well combs, and allow gel to set for 1 h.
4. Remove combs and tape. Immerse the gel in electrophoresis buffer (1 × TAE containing 0.5 μg/ml ethidium bromide) to a depth of 2–3 mm above the level of the gel.
5. Combine 2.5 μl of digested DNA (*Protocol 4*) with 5.5 μl TE and 2 μl 5 × GLB from each individual sample. Load samples into the slots of the gel and electrophorese at 50–100 V (constant voltage) for 2–3 h.
6. Place gel/template on a UV transilluminator and check if digestion is complete and that the amount of DNA in all tracks is approximately equal, then photograph the gel (see *Protocol 2*).

B. *Transfer gel electrophoresis*

The following description is based on BRL H4 horizontal electrophoresis equipment.

1. Prepare a 0.7% (w/v) agarose gel as previously described, but use 2.1 g agarose in 300 ml 1 × TAE buffer. Add 15 μl ethidium bromide (10 mg/ml) to the molten agarose once it has cooled to 65 °C, and pour into a level gel template (25 cm × 20 cm) which has been sealed at both ends with tape.
2. Insert a 30 slot comb (Labosystems, 10th International Histocompatibility Workshop design) into the template at one end. Allow the gel to set for 1 h.
3. Remove comb and tape. Immerse the gel in electrophoresis buffer to a depth of 2–3 mm above the gel.
4. Add 4 μl of 5 × GLB to the digested DNA and load samples (20 μl) into the slots of the gel, as well as 0.7 μg λ/*Hin*dIII size markers into slot number two. *Note*. Do not use end slots as these are distorted. Run samples slowly into the gel at 25 V (constant voltage) for a minimum of 30 min, then electrophorese at 50 V (constant voltage) for 18–20 h, with re-circulation of electrophoresis buffer using a peristaltic pump (Philip Harris Scientific, P85-230). (*Note*. Ensure that the vertical part of the electrode wire is insulated, because current from the vertical wire can cause electrophoretic distortion of DNA in the gel.)
5. After 18–20 h check that the λ/*Hin*dIII 2.0 kb marker has moved approximately 16 cm from the zero position of the slots by placing the gel/template on to a UV transilluminator, then stop electrophoresis. (*Note*. Do not expose DNA to UV light for any longer than is necessary.)
6. Align a transparent ruler alongside the λ/*Hin*dIII markers so that the zero mark is positioned at the level of the slots and photograph the gel/template under UV light (see *Protocol 2*).

---

7. Southern blotting

It is more convenient to hybridize radiolabelled probe to DNA fragments fixed to a nylon membrane than directly to DNA fragments in an agarose gel. Consequently, DNA fragments are replica-transferred to a nylon membrane using the technique of Southern blotting (1), either by capillary action (*Protocol 6*) or by vacuum transfer using a commercial vacuum blotter, (e.g. Pharmacia), prior to hybridization with radiolabelled probe.

Prior to transfer, the gel is treated with acid, (partial depurination) followed by alkali, (hydrolysis at points of depurination) in order to cleave high molecular weight DNA (>10 kb) to permit complete transfer, and to denature the DNA prior to hybridization. The DNA is then transferred from the gel to a charged nylon membrane and fixed *in situ* using alkali (13) (*Protocol 6*).

Protocol 6. Southern blotting using charged nylon membranes

A. *Pre-treatment of gel*

1. After electrophoresis slide the gel from the template on to a piece of perspex. Cut the transfer gel at the level of the slots. Measure 20 cm from the position of the slots and cut off the bottom of the gel, then cut off the bottom corner of the gel corresponding to slot number one. Place a second piece of perspex on top of the gel and carefully invert the gel. *Note*. The gel is inverted because when performing Southern blotting the DNA has less distance to travel through the gel when being transferred to the membrane.
2. Immerse gel in 500 ml 0.25 M HCl and shake for 10 min at room temperature using a platform shaker (Luckham, Reciproshake 30).
3. Pour off HCl and rinse twice for 30 sec in ddH_2O.
4. Immerse gel in 500 ml 0.4 M NaOH and shake for 30 min at room temperature using a platform shaker.

B. *Southern transfer*

1. Place a perspex support (25 cm × 22.5 cm × 5 cm) inside a tray. Drape three sheets of Whatman 3MM paper (35 cm × 22.5 cm) over the support.
2. Pour approximately 1750 ml 0.4 M NaOH over the Whatman sheets (wicking sheets), and remove air bubbles using a roller.
3. Using a spent X-ray film lift the gel from step **A4** and slide on to the wicking sheets. Remove air bubbles with the back of a gloved finger, and seal the outside of the gel with cling film.
4. Number the bottom left hand corner of a piece of Hybond N+ membrane (Amersham RPN 203B, 20 cm × 20 cm) and place it exactly on top of the gel so that the numbered corner corresponds to position number one on the gel. Remove air bubbles with a roller.
5. Place three sheets of pre-wet, (i.e. in 0.4 M NaOH) Whatman 3MM paper (20 cm × 20 cm) exactly on top of the membrane. Remove air bubbles between each sheet of paper with a roller.
6. Place a 10 cm stack of absorbent tissues on top of the Whatman paper, followed by a piece of perspex, and a 1 kg weight. Transfer overnight for 16–18 h.

7. After transfer, break the blot down and immerse membrane in 2 × saline, sodium phosphate, EDTA (SSPE: 0.3 M NaCl, 0.02 M NaH_2PO_4, 2 mM Na_2 EDTA pH 7.4) for 10 min with shaking, to neutralize the membrane and remove adhering agarose. Blot off excess fluid from the membrane and either wrap moist membrane in Saran Wrap (Genetic Research SW1), seal in a plastic bag, and store at 4°C, or proceed directly to *Protocol 8*.
8. Re-stain the gel in 1 × TAE buffer containing 0.5 μg/ml ethidium bromide for 1 h and check under UV light that transfer has occurred.

---

8. Radiolabelling of probe DNA

The most efficient method for labelling cDNA probes to high specific activity is the random hexamer, or oligolabelling method (14). The technique is extremely convenient as the probe can be labelled directly in low melting point agarose (*Protocol 7*) and purification, (i.e. removal of agarose) can be performed by simple filtration (see *Protocol 8*). In addition oligolabelling kits can be purchased from a number of commercial companies, (e.g. Pharmacia, Amersham).

Protocol 7. Radiolabelling of cDNA probes using random hexamer method

Guidelines for handling, storage, and disposal of radioactive [α^{32}P] deoxycytidine triphosphate (dCTP) should be obtained from the local radiation protection adviser prior to use.

(a) Solution A—combine 1 ml 1.25 M Tris-HCl pH 8.0, 0.125 M $MgCl_2$, 18 μl 2-mercaptoethanol (Sigma M6250), 5 μl deoxyadenosine triphosphate (100 μmoles/ml: Pharmacia 27-2050-01), 5 μl deoxythymidine triphosphate (100 μmoles/ml: Pharmacia 27-2080-01); 5 μl deoxyguanosine triphosphate (100 μmoles/ml: Pharmacia 27-2070-01).

(b) Solution B—dissolve 45.66 g Hepes (Boehringer 223778) in approximately 80 ml ddH_2O. Adjust pH to 6.6 with 4 M NaOH and make up to 100 ml with ddH_2O.

(c) Solution C—dissolve 50 optical density units of hexanucleotide (Pharmacia 27-2166-01) in 555 μl 3 mM Tris-HCl pH 7.6, 0.2 mM Na_2 EDTA pH 8.0.

(d) 5 × oligolabelling buffer (OLB)—combine 367 μl solution A, 918 μl solution B, and 550 μl solution C. Store at −20°C.

1. Tape the cap of a microcentrifuge tube containing probe DNA in low melting point agarose (from *Protocol 2*) and boil for 4 min. Centrifuge for 5 sec in a microcentrifuge (see *Protocol 1*) and incubate for 10 min at 37°C in a water-bath prior to labelling.

Protocol 7. *Continued*

2. For hybridization of one membrane[a] combine the following in a 0.6 ml microcentrifuge tube:

probe DNA	50 ng
ddH_2O	(to a final volume of 50 μl)
5 × OLB	10 μl
[α^{32}P] dCTP (Amersham PB10205, 3000 Ci/mmole)	4 μl
DNA polymerase I Klenow fragment (Pharmacia 27-0928-02)	5 U

3. Vortex briefly and centrifuge for 5 sec in a microcentrifuge.
4. Place the tube in a lead pot containing a little water and incubate for 5–16 h at 37°C in a water-bath.

[a] For two membranes use 100 ng of probe, keeping 5 × OLB, [α^{32}P] dCTP, and enzyme amounts the same in a total volume of 50 μl.

9. Pre-hybridization and hybridization of membrane

The main objective when applying hybridization techniques is to obtain high signal to background ratio. In order to reduce non-specific binding of probe to the membrane a pre-hybridization step is performed. This involves incubating the membrane in hybridization buffer to which herring sperm DNA has been added. The herring sperm DNA and Denhardt's reagent are blocking reagents which block out non-specific binding sites on the membrane.

Hybridization should be performed in as small a volume as possible either in sealed plastic bags in water-baths, or in special bottles in hybridization incubators. The latter is recommended because of safety, convenience, efficiency, and economy, (i.e. less radiolabelled probe is required because of the greater contact of probe with the membrane) (*Protocol 8*).

Protocol 8. Pre-hybridization and hybridization of membrane

All procedures below are carried out in a Robbins Scientific Hybridization Incubator (Model 310) using Robbins Scientific hybridization bottles (1040-01-0).

(a) Herring sperm DNA (10 mg/ml)—dissolve 1 g herring sperm DNA (Boehringer 223646) in a total volume of 100 ml 10 mM Tris-HCl pH 7.6, 1 mM Na_2 EDTA pH 8.0, 0.1 M NaCl. Sonicate for 2 min (MSE Soniprep 150), boil for 10 min, and chill on ice for 5 min. Store in aliquots at −20°C. Prior to use boil for 10 min and chill on ice for 5 min.

(b) 50 × Denhardt's reagent—add 2 g of bovine serum albumin (BSA, Sigma A7906) to approximately 80 ml ddH_2O and allow to dissolve without mixing (do not autoclave BSA). Adjust pH to 3.0 with 1 M HCl, boil for 15 min, then chill on ice for 10 min. Re-adjust pH to 7.5 with 4 M NaOH and make up to 100 ml with ddH_2O. Dissolve 2 g of polyvinylpyrrolidone (PVP, Sigma PVP40) and 2 g Ficoll 400 (Sigma F4375) in a total volume of 100 ml ddH_2O and autoclave. Combine 100 ml BSA solution with 100 ml
PVP/Ficoll solution, aliquot into 20 ml amounts, and store at −20°C.

(c) Hybridization buffer—combine 5 ml 50 × Denhardt's reagent, 12.5 ml 20 × SSPE, 5 ml 50% (w/v) dextran sulphate (Sigma D6001), and 5 ml 10% (w/v) SDS, and make up to 50 ml with ddH_2O.

A. *Denaturation and purification of probe DNA*

1. Add 250 μl herring sperm DNA to the tube containing radiolabelled probe (see *Protocol 7*), seal the cap of the tube with tape, and boil mixture for 10 min in a fume cupboard with extractor on. Centrifuge for 5 sec in a microcentrifuge (see *Protocol 1*) and chill on ice for 5 min.
2. Aspirate 5 ml pre-warmed (67 °C) hybridization buffer, from a total of 10 ml hybridization buffer in a sterile Universal tube, into a 10 ml disposable syringe. Attach a blunt end needle (Sherwood Medical Supplies Gauge 18) on to the syringe, and suck up denatured probe/herring sperm DNA into the syringe.
3. Detach needle and carefully remove air from the syringe, then filter contents through a 0.45 μm filter (Millipore SLHA 025BS) back into the remaining 5 ml of hybridization buffer. Cap the tube and gently mix contents. (*Note*. If 100 ng of labelled probe is used this can be filtered using 5 ml of hybridization buffer into a total of 20 ml hybridization buffer, and the sample split into 2 × 10 ml amounts for two membranes).

B. *Pre-hybridization and hybridization*

1. Add 500 μl herring sperm DNA to 15 ml pre-warmed (67 °C) hybridization buffer.
2. Add mixture to a hybridization bottle containing a membrane, screw cap on tightly then clamp the bottle to the rotisserie inside the hybridization incubator, and incubate for 4 h at 67 °C.
3. Discard hybridization buffer from the bottle. Carefully unravel the membrane around the inside of the tube, and add hybridization fluid (10 ml) containing denatured probe/herring sperm DNA (see A above) to the bottle. Re-cap the bottle and return it to the hybridization incubator and incubate for 24 h at 67 °C.

10. Post-hybridization procedures

10.1 Stringency washes

In order to reduce inter-locus cross-hybridization of the probe the membrane is stringently washed following hybridization in a series of salt solutions ranging from high salt concentration, (i.e. low stringency) to low salt concentration, (i.e. high stringency) (*Protocol 9*). In addition, temperature, (i.e. 65–68 °C) is also used to reduce probe cross-hybridization (*Protocol 9*).

10.2 Autoradiography

Positive signals corresponding to specifically bound radiolabelled probe can be detected using autoradiography (*Protocol 9*). This involves applying the membrane to an X-ray film(s), which is then placed inside an X-ray cassette with intensifying screens. The film is exposed at −80 °C because low temperature stabilizes the silver ions that form the latent image of the radioactive source. The signals appear as black bands on the X-ray film (autoradiograph) when processed by hand or using an X-ray processor.

10.3 Dehybridization

The bonding between transferred genomic DNA and the membrane is covalent, whereas DNA–DNA hybrids are held together by hydrogen bonding. Thus DNA denaturation can remove hybridized probe, whilst membrane-bound DNA is retained (*Protocol 9*). Consequently the dehybridized membrane can be re-hybridized. Although during each dehybridization a little membrane-bound DNA is removed resulting in decreased signal with subsequent hybridizations, this does not cause any problem when re-hybridizing with HLA-DQB and -DQA probes.

Protocol 9. Stringency washes, autoradiography, and dehybridization

All stringency washes and dehybridization are carried out in a Robbins Scientific Hybridization Incubator using Robbins Scientific hybridization bottles

Note. It is preferable to use two hybridization incubators, one set at 67 °C, and the other set at 20°C, so that bottles can be quickly transferred from one temperature to the other during stringency washes. Alternatively if a laboratory has only one Robbins incubator, washes can be performed in trays at room temperature on a platform shaker.

1. Carefully discard hybridization fluid containing radiolabelled probe into a designated sink, and wash as follows using approximately 180 ml of solution for each wash:
   - three washes in 2 × SSPE/0.1% (w/v) SDS for 20 min each at 20°C

- one wash in 1 × SSPE/0.1% (w/v) SDS for 20 min at 67°C
- one wash in 0.2 × SSPE/0.1% (w/v) SDS for 20 min at 67°C

2. Remove membrane, blot off excess fluid (but do not let membrane dry out), and seal in Saran Wrap. Check radioactivity level on the membrane using a Geiger counter.
3. In a darkroom tape the membrane to Kodak X Omat AR5 X-ray film (Kodak 1651454) by aligning the side of the membrane, corresponding to the position of the slots on the gel, along one side of the film. Align a second film on top of the membrane, place film/membrane/film sandwich inside an X-ray cassette with intensifying screens (Dupont, Cronex), and expose for a minimum of 24 h at −80°C.
4. Process one of the X-ray films after 24 h in an X-ray processor (Agfa, Gevamatic 60). Process the second film depending on the signal to background ratio of the first film.
5. To dehybridize the membrane place it inside a bottle, add approximately 180 ml boiling 0.5% (w/v) SDS to the bottle, and incubate for 1 h at 20°C.
6. Check that the membrane is dehybridized by exposing it to one X-ray film as described in step **3**.
7. Store the membrane flat in a sealed plastic bag at 4°C until further use.

---

11. Interpretation of RFLP data and trouble-shooting

11.1 Regime for interpretation

(a) Identification of fragments should be made by visual comparison with control fragments. If sizing is required plot a semi-logarithmic curve [$\log_{10}$ bp versus electrophoretic mobility (mm)], calibrated using λ/*Hind*III size markers, and read off fragment size (bp) knowing the distance travelled by the fragment (mm).

(b) Look first for fragments that are unique to a particular specificity, then locate other bands associated with that specificity.

(c) In the case of heterozygotes, assign one specificity, see what fragments are left, then assign the second specificity. Check that there is no other possible combination type(s).

(d) Make use of the fact that different polymorphic fragments within a specificity reveal different signal intensities to identify RFLP types.

(e) Identify fragments which are very close in size by comparison with fragments in nearby tracks, since DNA samples in different parts of the gel may run at different rates, (e.g. DNA samples in the middle of the gel tend to run faster than DNA samples at the ends of the gel).

11.2 *Taq*I/DRB RFLPs

Generally unique patterns of *Taq*I/DRB RFLP fragments correlate with serological and cellular defined specificities (*Figure 1*).

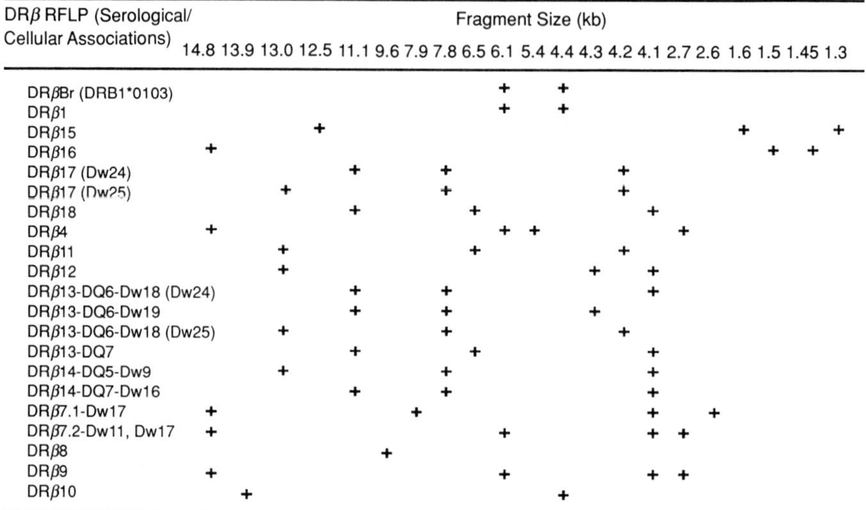

DRβ RFLP (Serological/ Cellular Associations)	Fragment Size (kb)																				
	14.8	13.9	13.0	12.5	11.1	9.6	7.9	7.8	6.5	6.1	5.4	4.4	4.3	4.2	4.1	2.7	2.6	1.6	1.5	1.45	1.3
DRβBr (DRB1*0103)										+		+									
DRβ1										+		+									
DRβ15				+														+			+
DRβ16	+																		+	+	
DRβ17 (Dw24)					+			+						+							
DRβ17 (Dw25)			+					+						+							
DRβ18					+				+						+						
DRβ4	+									+	+					+					
DRβ11			+						+					+							
DRβ12			+										+		+						
DRβ13-DQ6-Dw18 (Dw24)					+			+							+						
DRβ13-DQ6-Dw19					+			+					+								
DRβ13-DQ6-Dw18 (Dw25)			+					+						+							
DRβ13-DQ7					+				+						+						
DRβ14-DQ5-Dw9			+					+							+						
DRβ14-DQ7-Dw16					+			+							+						
DRβ7.1-Dw17	+						+								+		+				
DRβ7.2-Dw11, Dw17	+									+					+	+					
DRβ8						+															
DRβ9	+									+					+	+					
DRβ10		+										+									

Figure 1. *Taq*I/DRB RFLPs and associated serological/cellular specificities.

11.3 *Taq*I/DQB RFLPs

*Taq*I/DQB RFLPs reveal at least seven informative RFLPs, designated 1a (5.5 kb), 1b (2.8 kb), x (4.55 kb), 2a (4.6 kb, 2.7 kb, 1.6 kb), 2b (7.2 kb, 2.7 kb, 1.6 kb), 3a (2.4 kb, 1.9 kb), and 3b (4.6 kb, 2.5 kb). These patterns correlate strongly with serologically determined HLA-DQ specificities (2).

11.4 *Taq*I/DQA RFLPs

*Taq*I/DQA RFLPs reveal five informative DQA1 associated fragments under standard conditions above, designated 1a (2.7 kb), 1b (6.7 kb), 1c (7.4 kb), 2 (4.8 kb), and 3 (5.8 kb). These fragments are not as strongly associated with HLA-DQ specificities as are *Taq*I/DQB RFLP fragments (2). The DQA probe also identifies two allelic forms of the DQA2 gene, designated DQA2U (2.1 kb) and DQA2L (2.0 kb).

11.5 Typical DR/DQ associations

In European Caucasoids certain *Taq*I/DRB, DQB, and DQA RFLP combinations are found to be strongly associated together because of the strong linkage disequilibrium between HLA-DR and -DQ (2). These associations can be used to infer the HLA-DR type of an individual in the following occasions where identical *Taq*I/DRB RFLP patterns are encountered:

- 17 (Dw25) and 13-DQ6-Dw18 (Dw25)
- 18 and 13-DQ7 (Dw 'HAG')
- 13-DQ6-Dw18 (Dw24) and 14-DQ7-Dw16

These inferences should not be made in populations where the RFLP correlations to serological specificities has not been established.

11.6 Novel RFLPs and atypical DR/DQ associations

Occasionally (<1%) novel RFLPs are observed in European Caucasoid populations. Novel RFLPs should be repeated, and if possible family studies performed to confirm their existence. Atypical DR/DQ associations also occur, presumably as a result of recombination events, but these are rare in European Caucasoid populations (2).

11.7 Trouble-shooting

Table 1 highlights some of the common technical problems and their causes encountered when performing RFLP analysis, and gives remedies to solve or avoid these problems.

12. Theory and application of oligotyping

The second generation of molecular techniques for tissue typing, oligotyping, is based upon the ability of oligonucleotides to hybridize to target DNA sequences with exquisite specificity. This sequence-specific hybridization of oligonucleotides is first utilized for oligonucleotides which serve as primers to initiate selective amplification of HLA genes using the PCR. The genes targeted by the HLA-specific oligonucleotides can be amplified a million times. Therefore, the ratio of nucleic acids encoding HLA to non-target nucleic acids becomes so large that non-target sequences are not detected in subsequent steps. The second step of oligotyping uses sequence-specific hybridization of oligonucleotide probes to detect polymorphic sequences in the amplified DNA. Sequence-specific hybridization is achieved using conditions that prevent formation of hybrids when a single base pair difference exists between the probe and target. This step is often referred to as sequence-specific oligonucleotide probe hybridization (SSOPH).

One fundamental difference between oligotyping and other HLA typing methods is that the nucleotide squences of the alleles must be known in order to synthesize primers and probes. All known nucleotide sequences of HLA genes (see Chapter 7) (15) are aligned to identify the polymorphic sequences that distinguish a particular allele or group of alleles from all others. Sequences

Table 1. Trouble-shooting guide for RFLP analysis

Problem	Common cause	Remedy
OD 260/280 <1.7	DNA contaminated with protein	Re-extract with phenol/chloroform
Tracks distorted on gel	Buffer not re-circulated	Follow protocol exactly
	Electrophoresis buffer or gel wrongly prepared	Follow protocol exactly
Poor or no signals on autoradiograph	Poor transfer of DNA	Check re-stained gel
	DNA not properly fixed	Replace batch of membrane
	Probe not properly labelled	Follow protocol exactly
	Probe not properly denatured	Follow protocol exactly
	Strigency washes too high	Correct the concentration of salt solutions
Generalized background on autoradiograph	Stringency washes too low	Correct the concentration of salt solutions and/or temperature of washes
	Ineffective blocking of non-specific sites on membrane	Correct the amount of herring sperm DNA
	Probe concentration too high	Reduce probe DNA concentration
Extra bands on autoradiograph	Partial digestion of DNA	Add more enzyme and incubate for longer, or re-purify stock DNA, and re-digest
Bands curved/distorted on autoradiograph	DNA overloaded on gel	Check monitor gel and reduce amount of DNA loaded
	Voltage too high	Reduce voltage
No dehybridization	Solutions incorrectly prepared	Follow protocol exactly

that are present in all of the target HLA alleles, (e.g. all HLA-DRB alleles) are selected as sites for primers which are used to amplify the target genes. The presence or absence of certain polymorphic sequences in amplified products is used to assign an oligotype.

One of the primary advantages of oligotyping is that the method can be customized for each application. Oligotypes can be defined to correspond to serological specificities, cellular specificities (HLA-Dw), or alleles. For example, approximation of HLA-DR serological specificities requires detection of polymorphic sequences which distinguish alleles of one specificity from all others (*Figure 2*). This is often referred to as 'low resolution' or 'generic' oligotyping. Alternatively, oligotyping can resolve all known alleles or small groups of alleles such as those correlated with cellular specificities. This is often referred to as 'high resolution oligotyping'. An example of high resolution oligotyping is the detection of 12 oligotypes corresponding to the 12 alleles which are serologically defined as HLA-DR4. High resolution oligotyping usually requires selective amplification of a group of related alleles. For the previous example, all HLA-DRB1*04 alleles are specifically amplified and the DNA is hybridized with a panel of probes which distinguish each HLA-DRB1-04 allele (16–18). Assignment of oligotypes is usually based upon HLA Nomenclature from the World Health Organization (15), substituting '-' for the '*'. For example, detection of polymorphic sequences found in the allele HLA-DRB1*0401 would result in the assignment of an HLA-DRB1-0401 oligotype.

13. Amplification using the PCR

The PCR consists of three steps carried out in succession.

(a) Denaturation—double stranded DNA is incubated at high temperatures to dissociate the two strands.

(b) Annealing—the temperature is lowered to permit annealing of an oligonucleotide primer to each target strand at the 3′ end of the region to be amplified. Formation of primer:target hybrids is favoured over target:target hybrids by using excess primer.

(c) Primer extension—the temperature is raised to allow DNA polymerase to initiate DNA synthesis from the primer to yield a complementary strand of DNA.

These three steps constitute one cycle of the PCR and result in a theoretical doubling of the number of target molecules. Amplification is exponential, yielding 2^n copies of a target in n cycles. After many cycles, the rate of synthesis declines due to several factors including reannealing of target molecules and lack of substrates (3).

Thermal cycling is performed using a programmable thermal cycler. Thermal cycler variation can be a major cause of difficulty. This includes instrument to instrument variation, (e.g. rate of temperature change, efficiency of heat exchange, block temperature), cycle to cycle variation for a single instrument, and well to well variation for each cycle. For many thermal cyclers, temperature settings pertain to the block which holds the tubes rather than the temperature inside the tubes. Therefore variation in the efficiency of heat transfer can be a significant problem. If appropriate temperatures are not realized, amplification can fail for all or a few samples. Trouble-shooting may require monitoring the temperature of the solution inside the sample tubes. Temperatures inside a tube can be monitored by inserting a thermocouple probe (Omega, Type K, Model HH72K2) through a small hole in a tube which contains the same amount of solution as samples. The hole should be sealed during cycling. Some thermal cyclers are equipped with thermocouples to measure the temperature of the reactions.

Many factors influence the specificity of the amplification including sequence of the primer, (i.e. length, GC content), free Mg^{2+} concentration, ratio of primer to target, and polymerase concentration (3). Conditions must be optimized for each primer pair. Minor modification of the conditions can significantly alter the specificity of an amplification. For example, the specificity of many primer pairs is influenced by the concentration of free Mg^{2+}. Since dNTPs bind Mg^{2+}, changing the concentration of dNTPs can influence the specificity of an amplification by changing the concentration of free Mg^{2+}.

A method for low resolution HLA-DRB and -DQB1 oligotyping is provided in *Protocols 10–17. Protocol 10* describes isolation of DNA from whole blood. This method can be used to extract DNA from whole blood (fresh or frozen), nucleated blood cells (fresh or frozen), or buccal mucosa. Blood collected using any anti-coagulant can be used as a source of DNA, but EDTA is the anti-coagulant of choice for many investigators. If frozen cell suspensions are used:

(a) thaw specimens at room temperature
(b) transfer to a 1.5 ml microcentrifuge tube
(c) wash cells at least three times as described in steps **2–4** of *Protocol 10*

Washing removes inhibitors of amplification which are frequently present in cryopreserved cells.

Protocol 10. Extraction of genomic DNA for amplification

1. Add 125 μl anti-coagulated blood and 1.0 ml RCL buffer to a 1.5 ml microcentrifuge tube and vortex briefly.
2. Microcentrifuge for 30 sec.
3. Discard the supernatant using a pipette.
4. Resuspend the pellet in 1 ml RCL buffer.

5. Repeat steps **2–4** until all indication of haemoglobin coloration is lost.
6. After the final wash cycle, microcentrifuge for 30 sec, and remove the supernatant using a micropipettor.
7. Resuspend the final pellet in 120 μl NLB.
8. Add 30 μl proteinase K solution.
9. Incubate at 63–65 °C for 1.5 h, then 95 °C for 10 min.
10. Microcentrifuge for 1 min.
11. Store the supernatant at 4 °C. Samples can be stored for at least one year.

Protocol 11. Amplification of HLA-DRB and -DQB1

It is extremely important to take precautions to prevent contamination of samples with previously amplified DNA (19). Contamination with less than 10^{-6} μl of a previous amplification can yield an incorrect oligotype. Contamination of reagents can be minimized by preparing solutions in facilities which have not been exposed to amplified products and aliquoting reagents for single use. It is beneficial to store some uncontaminated water because a general water supply can become contaminated with the amplification product.

- 5 × PCR buffer—250 mM KCl, 7.5 mM $MgCl_2$, 0.05% gelatin, 50 mM Tris-HCl pH 8.3
- dNTP stock—25 mM each dATP, dCTP, dGTP, dTTP, pH 7.0 (Boehringer Mannheim, 1051440, 1051458, 1051466, and 1051482)
- *Taq* polymerase—AmpliTaq (Perkin Elmer Cetus, N801-0060) dilute in ddH_2O immediately before use
- Primer 1—intron-RFL6: 5′-GTCCCCACAGCACGTTTCTTG
- Primer 2—ESFTVQ87: 5′-CGCCGCTGCACTGTGAAGCTCTC
- Primer 3—MCYFTNGT14: 5′-TGTGCTACTTCACCAACGGGAC
- Primer 4—AEYWNSQKE58: 5′-TCCTTCTGGCTGTTCCAGTACTCGGC

Custom oligonucleotides can be synthesized using an oligonucleotide synthesizer (PCR-MATE, Model 391, Applied Biosystems Inc.), or purchased from one of several companies which provide oligonucleotides within a few days.

1. To amplify HLA-DRB alleles, combine the following in a 0.6 ml microcentrifuge tube and vortex.

5 × PCR buffer	30 μl
dNTP stock	1.2 μl
Primer 1	1.0 μg
Primer 2	1.1 μg

Protocol 11. *Continued*

DNA	10 μl (alternatively 0.3–1.5 μg DNA from *Protocol 3* may be used)
ddH_2O	to adjust to a total volume to 150 μl

One or more reagent controls using ddH_2O rather than DNA should be included.

2. To amplify HLA-DQB1 alleles combine the following in a 0.6 ml microcentrifuge tube and vortex.

5 × PCR buffer	20 μl
dNTP stock	0.8 μl
Primer 3	0.71 μg
Primer 4	0.84 μg
DNA	10 μl (alternatively 0.2–1.0 μg DNA from *Protocol 3* may be used)
ddH_2O	to adjust to a total volume to 100 μl

One or more reagent controls using ddH_2O rather than DNA should be included.

3. Add one drop of mineral oil to each tube. Omit this step if the thermal cycler is specifically designed to eliminate evaporation during thermal cycling.

4. Incubate the samples for 3 min in a heat block, set at 100–105°C.

5. Rapidly cool the samples in an ice/water bath.

6. Add 1 U/100 μl of reaction volume of *Taq* polymerase in a volume of 1–2 μl, close the tube, vortex, and return to the ice bath.

7. Transfer all tubes to a thermal cycler preheated to 94°C (DNA thermal Cycler 480, Perkin Elmer Cetus) programmed for 30 cycles.
   (a) 98°C for 30 sec (tube temperature about 90°C).
   (b) 55°C for 60 sec (tube temperature about 55°C).
   (c) 72°C for 105 sec (tube temperature about 67°C).
   (d) Incubate 72°C for 7 min after completion of the 30th cycle.

The specificity and quantity of amplified product are monitored by hybridization to consensus probes (*Protocol 15*). Alternatively, the quantity and quality of the products of the amplification can be evaluated prior to hybridization using mini-gel electrophoresis as described in *Protocol 12*.

Protocol 12. Electrophoresis of amplified DNA

- 1 × TBE—10.8 g Tris base, 5.5 g boric acid, 4 ml 0.5 M EDTA, pH 8.0, adjust to 1 litre with ddH_2O
- dye—0.25% (w/v) bromophenol blue, 0.25% (w/v) xylene cyanol, 30% (v/v) glycerol in ddH_2O

The following description is based on the use of a horizontal mini-gel apparatus from FisherBiotech (FB-MSV-965), but could be performed using the apparatus described in *Protocol 5*.

1. Mix the following:
   1 g NuSieve Agarose (FMC, 50082)
   1 g LE Agarose (FMC, 50004)
   100 ml TBE
   Heat the mixture in a boiling water bath or microwave until the agarose is dissolved. Cool to 65°C, add ethidium bromide to 0.5 μg/ml and pour on to the gel plate.
2. Insert a comb (16 well combs, Fisher Biotech) in each gel.
3. Cool until gel is solid (10–20 min).
4. Fill the electrophoresis chamber with 1 × TBE, containing 0.5 μg/ml ethidium bromide.
5. Carefully remove the comb and place the gel in the electrophoresis chamber.
6. Mix 7 μl amplified sample and 1 μl dye.
7. Add one sample to each well of the gel and add a molecular weight marker (*Msp*I digest of pBR322, New England Biolabs, 303-2C) to at least one well.
8. Replace the cover, connect to the power source, and run at 120 V for approximately 20 min. The DNA bands migrate between the purple and blue dye bands. The purple dye should not run off the end of the gel.
9. View the DNA in the gel using a UV transilluminator. The product should be a single band of 278 bp for HLA-DR and 156 bp for HLA-DQ. Photograph the gel if a permanent record is desired (*Protocol 2*).

14. Sequence-specific oligonucleotide probe hybridization

Since many HLA alleles have similar sequences, sequence-specific hybridization is essential for accurate oligotyping. SSOPH requires precise experimental conditions and comprehensive controls are required to detect subtle changes in conditions which can alter the specificity of hybridization.

14.1 Controls

It is important to include controls to monitor the amount of amplified DNA bound to the membrane and the hybridization specificity of each probe. The amount of amplified DNA which is bound to the membrane is monitored using a consensus probe. Poor amplification and/or variation in binding the DNA to the membrane are detected using this control. Hybridization specificity

Table 2. Cell lines for controls

Name	Number	HLA-DR Oligotype	HLA-DQ Oligotype
WT100BIS	9006	1	1
AMAI	9010	2 (15)	1
WJRO76	9012	2 (16)	1
STEINLIN	9087	3,52	2
WT51	9029	4,53	8
JHAF	9030	4,53	7
SWEIG	9037	11,52	7
TISI	9042	11,52	7
TEM	9057	13,52	1
AMALA	9064	14,52	7
LBUF	9048	7,53	2
MADURA	9069	8	4
DKB	9075	9	9
BM16	9038	12	7
RAJI	—	3,10,52	1,2

is monitored using a panel of controls including at least one positive control for each probe and negative controls for all sequences which are sufficiently similar to the probes for potential cross-hybridization. Subtle variation in conditions can alter the hybridization to one negative control but not another, even in assays that have historically been highly reliable. The importance of using a comprehensive panel of negative hybridization controls in every experiment cannot be over emphasized. A panel of cell lines which can be used for controls for low resolution HLA-DR and -DQ oligotyping is listed in *Table 2*. Comprehensive controls should be included on every membrane. If separate strips are used for controls and unknown samples, a hybridization error may be missed due to membrane exchange or subtle differences in washing conditions.

14.2 Preparation of dot blots

Prepare at least one membrane for each probe in the assay. Extra membranes are useful for repeating unsatisfactory hybridizations or future studies to test additional probes. The preparation of dot blots using a 96 well manifold is described in *Protocol 13*. This approach is advantageous because it minimizes sample to sample variation in loading and provides a relatively large surface area to aid in evaluation of ambiguous dots. If dot blot manifolds are not available, DNA can be directly spotted on to membranes (20) or transferred from gels using a Southern transfer technique (1, 17).

One factor which is critical to the success of non-radioactive assays is compatibility between the membrane and the detection system. Assays can be affected by subtle differences among membranes, including lot to lot variation from a single manufacturer. Several lots should be tested before purchasing membranes and a long term supply should be obtained when a suitable batch is identified. Rarely, a small area of one membrane will be defective. This can be detected if several dots are partially stained clearly demonstrating the existence of a defective area. This type of defect can be routinely detected by re-hybridizing every membrane with a consensus probe. *Protocol 13* describes a method using a Magnagraph membrane and digoxigenin-based detection of probes which has proved reliable in thousands of assays.

Protocol 13. Preparation of dot blots

Note. Always wear gloves for handling membranes and minimize contact. Whenever possible handle membranes in the corners using membrane forceps.

- denaturation buffer—0.5 M NaOH, 3.0 M NaCl
- neutralization buffer—3 M NaCl, 2 M Tris-HCl, pH 7.4
- india ink solution (optional)—1 μl india ink (Koh-I-Noor) in 2 ml ethanol

1. Determine the number of membranes needed. If the number of controls and unknown samples exceeds 48, one membrane must be used for each probe. If the total number is less than 48, one manifold can be used to prepare membranes for two or more probes by cutting sections containing a complete set of controls and unknown samples at step **11**.
2. For each membrane needed, cut one Magnagraph membrane (Micron Separations Incorporated) and two filters (Millipore MBBD 81150) to 8.5 × 11 cm.
3. Wet each membrane individually in ddH_2O by curving the membrane to touch the centre of the membrane to the water and gradually lowering the ends to wet the entire membrane. Soak at room temperature for 5 min.
4. Assemble the dot blot manifold by placing in order:
   (a) the base of the manifold (Millipore Milliblot MBBD BP000) (plenum, filter support, and dot duct)
   (b) two wet filters (remove air bubbles)
   (c) wet membrane (remove air bubbles)
   (d) the top of the manifold

   Slide the side clamps in place. If it is difficult to put the clamps in place, a C-clamp (any hardware store) can be used to hold the apparatus in place while sliding the side clamps on the assembly. Alternative dot blot apparatus can be purchased from BioRad (170–6545).

Protocol 13. *Continued*

5. Denature DNA samples by adding to cubetube (USA Scientific Plastics 14200012).
   $(n + 1) \times 5$ µl amplified DNA
   $(n + 1) \times 12.5$ µl denaturation solution
   $(n + 1) \times 7.5$ µl ddH_2O
   where n is the number of probes in the assay.
   Incubate at room temperature for 5 min.
6. Add $(n + 1) \times 75$ µl neutralization buffer to each tube. Apply 100 µl of each sample in the same position on each membrane, recording the position of each sample. An 8 or 12 channel programmable micropipette is recommended for adding samples to membranes. Incubate at room temperature for 10 min.
7. The membrane can be marked for orientation by placing 5 µl/well india ink solution in one or several unused wells at this time. Alternatively membranes can be marked after step **9** by numbering with india ink or cutting one corner.
8. Draw the liquid through the membrane using a vacuum aspirator.
9. Remove clamps and the top of the manifold.
10. Place the membrane in a DNA transfer lamp (Spectrolinker, Fisher Biotech, UV-XL-1500) with DNA facing up, and expose the membrane to UV light for 5 min.
11. If one membrane contains two or more sets of dots, cut the membrane into sections containing a complete set of controls and unknown samples. The membranes can be used immediately for hybridization or stored at room temperature indefinitely. For storage, place the membranes on filter paper, dry in air, cover with filter paper, and store at room temperature.

14.3 SSOPH

Many alternative methods for SSOPH have been reported in the literature. The major differences between these methods are:

- length and sequence of oligonucleotide probes
- solvent for wash steps
- reporter molecule and its detection

Protocols 14 and *15* utilize fairly short oligonucleotide probes (12–14 bases) to minimize the stability of the mismatched target:probe hybrids. This improves discrimination between similar sequences, (e.g. single base pair mismatches). Non-specifically bound oligonucleotides are removed using washes containing tetramethylammonium chloride (TMAC) which reduces the effect of GC content on stability of hybrids. TMAC can reduce the number of different wash temperatures which are required for a panel of probes. Success requires the correct concentration of TMAC which can be determined using the refractive index (*Protocol 15*). The probes are labelled with digoxigenin which is detected using a high affinity antibody conjugated to alkaline phosphatase. The digoxigenin can be attached to the end of oligonucleotides using an efficient chemical method described in *Protocol 14*. The digoxigenin moiety is added to 5′-amino oligonucleotides by incubating with a digoxigenin ester under mild alkaline conditions. Alternatively, a kit is available for 3′-labelling using terminal deoxynucleotidyl transferase (Boehringer Mannheim, 1362-372). The efficiency of the enzymatic addition of digoxigenin-11-ddUTP is fairly low. Although the efficiency of enzymatic addition can be increased by adding other deoxynucleotides to the reaction, the specificity of the probes can be affected. The use of labelled digoxigenin probes for low resolution HLA-DRB and -DQB1 oligotyping is described in *Protocol 15*. Optimal hybridization and wash conditions have been empirically determined for each digoxigenin-labelled probe. Changing any component, (e.g. probe sequence, labelling method) can alter the specificity of the hybridization. An alternative method for detection using chemiluminescence is described in *Protocol 16*.

Protocol 14. Labelling of oligonucleotides with digoxigenin-3-O-methylcarbonyl-ε-aminocaproic acid-N-hydroxy-succinimide ester (DigMAAHS ester)

- 3 M sodium acetate, pH 8.5
- 3 M sodium borate, pH 8.5
- DigMAAHS ester—dissolve 5 mg of the DigMAAHS ester (Boehringer Mannheim 1333054) in 200 μl N,N-dimethylformamide (EM Science DX1726-3)
- oligonucleotides—5′-amino substituted oligonucleotide prepared using Aminolink (Applied Biosystems Inc.) for the 5′ base

High quality reagents and solvents are required for successful labelling.

1. Dissolve 0.2 μmol dried 5′-amino substituted oligonucleotide in 300 μl ddH_2O and transfer to a 1.5 ml microcentrifuge tube.
2. Add 30 μl 3 M sodium acetate, pH 8.5 and 900 μl 100% ethanol.
3. Mix and store at −20°C for at least 2 h.
4. Microcentrifuge for 15 min. Carefully decant supernatant and drain tube.
5. Wash the pellet by adding 500 μl 70% ethanol and vortex. Microcentrifuge for 10 min, carefully decant supernatant, and drain the tube.
6. Dry the oligonucleotide pellet in a Speed Vac (Savant Instruments) at the high temperature setting.
7. Dissolve the dried oligonucleotide in 50 μl 3 M sodium borate, pH 8.5.

Protocol 14. *Continued*

8. Add 50 μl DigMAAHS ester to the oligonucleotide and incubate at room temperature overnight.
9. Dry the mixture in a Speed Vac (usually requires 1–2 h using the high setting).
10. Resuspend the pellet in 500 μl ddH_2O. This oligonucleotide can be used directly, or purified using standard methods such as HPLC or electrophoresis. The precise concentration of oligonucleotides cannot be determined from A_{260} because digoxigenin contributes to the absorbance. Purification often improves the performance of unsatisfactory probes.

Protocol 15. Hybridization and interpretation of data

- hybridization solution—10 × Denhardt's, 5 × saline sodium citrate (SSC 0.75 M NaCl, 75 mM Na_3 citrate pH 7.0), 100 μg/ml herring sperm DNA, 5 mM EDTA, 7% (w/v) SDS, 20 mM Na phosphate
- TMAC solution—3 M TMAC, 1% (w/v) SDS, 2 mM EDTA pH 8.0, 50 mM Tris-HCl pH 8.0, pre-warmed to wash temperature
  The concentration of TMAC stock is determined using the formula: Concentration (M) = (N − 1.333)/0.018, where N is the refractive index.
- 6 × SSC
- buffer 1—0.15 M NaCl, 0.1 M Tris-HCl pH 7.4
- buffer 2—0.3% (w/v) non-fat dry milk (Carnation) in buffer 1
- buffer 3—0.1 M NaCl, 50 mM $MgCl_2$, 0.1 M Tris-HCl pH 9.5
- buffer 4—1 mM EDTA, 10 mM Tris-HCl pH 8.0
- antibody conjugate—20 μl anti-digoxigenin [Fab]-alkaline phosphatase (Boehringer Mannheim 1093 274) in 100 ml buffer 1 (prepare immediately before use)
- NBT solution—0.38 g 4-nitroblue tetrazolium chloride (NBT) (Boehringer Mannheim 1087479), 13.5 ml dimethylformamide, 1.5 ml ddH_2O (store in the dark at 4°C)
- X-phosphate solution—0.5 g 5-bromo-4-chloro-3-indolyl-phosphate (Boehringer Mannheim 760994), 10 ml dimethylformamide (store in the dark at 4°C)
- coloration solution—30 μl NBT solution, 30 μl X-phosphate solution, 10 ml buffer 3

1. Prepare one 50 ml conical plastic centrifuge tube (disposable) for each probe in the assay by adding the following to each tube (label the tubes with the name of each probe):
   3 ml hybridization solution
   1–10 μl probe (80–400 ng)
   The optimum amount of probe is empirically determined for each lot.

2. Use gloves and filter forceps to roll each membrane with the DNA side facing inward and place each membrane into appropriate tubes. Remove any air bubbles located between the membrane and the side of the tube.
3. Place the tube into the hybridization oven at the correct temperature (*Figures 2* and *3*) for 2 h. Hybridization times ranging from 2 h to overnight can be used.
4. Transfer all membranes to one plastic dish (Rubbermaid) containing enough 6 × SSC to cover all membranes and shake the container for 1 min. Pour the liquid from the container. Repeat two times. Be certain to remove all of the solution after the last wash.
5. Submerge the membranes in TMAC solution and shake the container for 1 min to ensure that all SSC is removed.
6. Transfer membranes to a plastic dish for each wash temperature. Add enough TMAC solution to submerge membranes, cover the dishes, and incubate in a shaking water-bath (Hot Shaker, Bellco) at the indicated temperature (*Figures 2* and *3*) for 20 min. Pour the solution from the container, add fresh TMAC solution, and repeat the incubation. Pour the solution from the container.
7. Rinse the membranes three times in buffer 1 at room temperature. Pour the solution from the container.
8. Add enough buffer 2 to submerge the membranes, place on an orbital shaker, and shake at room temperature for 35 min. Prepare the antibody conjugate during this incubation.
9. Pour the buffer from the container and rinse in buffer 1.
10. Submerge membranes in antibody conjugate solution, place on an orbital shaker, and shake at room temperature for 35 min.
11. Pour the antibody solution from the container and rinse the membranes for 1 min in buffer 1. Repeat the wash step two times. Pour all solution from the container.
12. Submerge membranes in buffer 1, place on an orbital shaker, and shake gently at room temperature for 5 min. Pour the buffer from the container.
13. Submerge membranes in buffer 3, place on an orbital shaker, and shake gently at room temperature for 5 min. Pour the buffer from the container.
14. Transfer membranes to a small glass container and add a small amount of coloration solution (200 μl per membrane). Incubate at room temperature in a dark place, (e.g. a drawer). Check the membranes every 30 sec for the first 5 min, and every 5 min thereafter until the positive controls are clearly visible as blue-purple spots. Most membranes are completed between 30 sec and 25 min. If conditions are not optimal, colour development can be continued for several hours. After the coloration is complete, place each membrane in buffer 4 to stop the reaction. If membranes are left in the coloration solution too long negative samples

Protocol 15. *Continued*

will become coloured, making it difficult to discriminate between positive and negative dots.

15. Incubate the membranes in buffer 4 for at least 5 min. Wash one time in fresh buffer 4. Place membranes on clean filter paper to air-dry. The background of the membranes will gradually increase during storage and this process is accelerated with light exposure. Membranes can be stored in a dark place for several months without significantly affecting the interpretation. If a permanent record is desired, a photocopy machine or a camera can be used.

16. Score the colour intensity for each spot on the membrane.

1 No hybridization.
2 Very light hybridization, probably negative.
3 Ambiguous.
4 Light positive.
5 Clear positive.
0 Technical failure.

A good membrane will have all dots scored entirely as 1 or 5, with the exception of samples that have an atypical amount of DNA bound to the membrane. Atypical amounts of DNA on the membrane are detected by comparing the intensity of staining for all samples using the consensus control probe. If the amount of DNA is less than usual, positive probes may be scored as 3 or 4. If the amount of DNA is more than usual, negative probes may be scored as 2 or 3. It is best to repeat these samples. If a membrane has scores of 2, 3, or 4 which cannot be attributed to loading of the DNA, it is best to repeat the hybridization and wash steps using an extra membrane. Alternatively, the membranes can be stripped (*Protocol 17*) and reused, but the result is usually inferior to that observed with fresh membranes.

17. Compare the scores for the controls for each probe with the expected pattern. Membranes which do not have the expected pattern must be repeated.

18. Score the dots from the unknown samples and compare the hybridization patterns with that expected for known alleles (*Figures 2* and *3*) to assign the oligotypes of each sample. For example, a sample scored as 5 for WNSQ61, H33, ERCI28, and N26I28 is assigned oligotypes 1, 4, and 53. For certain heterozygotes, a single hybridization pattern will be consistent with multiple combinations of oligotypes due to potential masking of some HLA-DRB1*13 and *14 alleles. This circumstance can be overcome by performing high resolution oligotyping for ambiguous samples. If desired, the number of ambiguous typings can be reduced by adding more probes to the panel.

PROBE	B1 1	B1 2	B1 10	B1 4	B1 7	B1 9	B1 3	B1 11	B1 8	B1 12	B1 1301/2/4	B1 1303	B1 1305/6	B1 1401/4	B1 1402	B1 1403	B1 1405	B1 1406	B3 52	B4 53	B5 2	HYB (°C)	WASH (°C)	SEQUENCE (5'→ 3')
WNSQ61	■	■	■	■	■	■	■	■	■	■	■	■	■	■	■	■	■	■	■	■	■	30	30	T CTG GCT GTT CC
ERCI28	■																					30	30	AT GCA TCT TTC C
R13		■																				30	30	CTC CCT CTT AGG C
R74N77							■															35	32	A GTT GTC CAC CCG
H33				■																		30	30	C CTC TTG GTG AT
E58								■														30	30	CTC CTC ATC AGG
LL37										■												30	32	CG CAG GAG CTC C
Y16									■	■				⁞								32	30	C TCT ACG GGT GA
YST10							■	■			■	■	■	⁞	■	■	■	■				32	30	TCT ACG TCT GAG
N37							■				⁞		■		■	■		■				35	30	C GTT CTC CTC CT
F37					■									■			■					35	35	CG CAC GAA CTC C
DE70	⁞			⁞				⁞			■											30	32	CCG CTC GTC TTC
RRAA	⁞			⁞											■							30	40	AGG CGG GCC GCG
R71								⁞					■									30	48	CGC GGC CCG CCT GT
A57														■								30	30	TG CTC CGC AGC AG
DK70												■										30	32	CTT GTC TTC CAG
E28L30					■																	30	32	A GAA GAG TCT TTC
L74									⁞							■						30	32	CAG GGC CCG CCT
YLH26						■																30	30	GTG CAG ATA CCG
ERV28			■																			30	32	AC GCG TCT TTC C
N77.2																			■			30	30	GCA GTA ATT GTC C
N26I28																				■		30	32	T GAT CAG GTT CC
D11																					■	30	30	TAT CCT GCT GCA

Figure 2. Probes for low resolution HLA-DRB oligotyping. Probe names (*left*) are listed along with nucleotide sequences (*right*), hybridization temperatures (HYB), and wash temperatures (WASH). The oligotypes defined at the top correspond to HLA-DR alleles from HLA-DRB1, B3, B4, and B5 loci, as indicated. Expected hybridization patterns with this panel of probes; hybridization to all alleles in the group is shown as ■, hybridization to some, but not all members of the group is shown as ⁞. No entry indicates no hybridization.

PROBE	DQB1 OLIGOTYPE 1	2	4	7	8	9	HYB (°C)	WASH (°C)	SEQUENCE (5'-3')
RFDS39	■	■	■	■	■	■	33	33	CGC TTC GAC AGC
QGRP53	■						33	33	GGC CGC CCC TGC
GEFR46		■					33	33	CGG AAC TCC CCC
RLDA56			■				33	33	GCG TCA AGC CGC
EVYR45				■			33	33	CGG TAC ACC TCC
PPAA57					■		35	42	GGC GGC AGG CGG
PPDA57				■		■	33	33	CGG CGT CAG GCG
GVYR47	▒		■		■	■	33	33	CGA TAC ACC CCC
LGPP55				■	■	■	33	33	AGG CGG CCC CAG

Figure 3. Probes for low resolution HLA-DQB1 oligotyping. Probe names (*left*) are listed along with nucleotide sequences (*right*), hybridization temperatures (HYB), and wash temperatures (WASH). The oligotypes listed at the top are assigned according to the indicated hybridization patterns. Hybridization to all alleles for an oligotype is shown as ■, and hybridization to some, but not all members of the group is shown as ▒. No entry indicates no hybridization.

Protocol 16. Detection by chemiluminescence

The procedure below is based on the Boehringer Digoxigenin Luminescent Detection Kit (Boehringer 1363514) for one membrane.

- buffer 1—0.1 M maleic acid (Sigma M0375), 0.15 M NaCl pH 7.5
- buffer 2—1% (w/v) blocking reagent in buffer 1
- buffer 3—0.1 M Tris-HCl, 0.1 M NaCl, 50 mM $MgCl_2$, pH 9.5
- washing buffer—0.3% (v/v) Tween 20 in buffer 1

Note. (a) Steps **1–6** below are performed at room temperature using a platform shaker.
(b) Separate boxes with lids, (e.g. Boehringer 800058) should be used for each buffer solution.

1. Take membrane (from *Protocol 15*, step **6**) and immerse in 100 ml of washing buffer and shake for 5 min.
2. Transfer membrane into 100 ml of buffer 2 and shake for 30 min.
3. Discard buffer 2 and add 80 ml of anti-digoxigenin-alkaline phosphatase conjugate in buffer 2 (8 μl of conjugate [750 U/ml] in 80 ml of buffer 2), and shake for 30 min.
4. Transfer the membrane to 100 ml of washing buffer and shake for 15 min. Discard washing buffer and repeat.

5. Transfer membrane to 100 ml of buffer 3 and shake for 5 min.
6. Transfer the membrane to a plastic bag and add 10 ml 3-(2′-spiroadamantane)-4-methoxy-4-(3″-phosphoryloxy)-phenyl-1,2-dioxetane (AMPPD) in buffer 3 (100 μl AMPPD [10 mg/ml] in 10 ml of buffer 3). Remove air bubbles and seal the bag. Place the bag inside a light-tight box and shake for 5 min.
7. Remove the bag, cut a corner of the bag and pour off the fluid. (*Note*. The solution may be reused several times.) Remove the membrane from the bag, blot off excess liquid, and wrap in Saran Wrap.
8. Expose the membrane to an X-ray film (see *Protocol 9*) at room temperature for 1 min.
9. Dehybridize membrane by adding 100 ml 0.5% (w/v) SDS to the membrane in a tray and incubating at 65°C for one hour.

Protocol 17. Stripping membranes after *Protocol 15*

Caution dimethylformamide is a liver toxin. Wear gloves and use a fume hood.

- dimethylformamide (DMF, Aldrich)
- 0.2 M NaOH, 0.1% (w/v) SDS

1. Pour 100 ml dimethylformamide into a 200 ml beaker and place the beaker into a larger beaker (500 ml) filled with water.
2. Heat the DMF to 65°C. (*Note*. The flash point for DMF is 70°C.)
3. Use membrane forceps to immerse the membrane into the DMF for 30 sec. Swirl until the colour disappears.
4. Rinse the membranes with ddH_2O.
5. Incubate the membranes in 0.2 M NaOH and 0.1% (w/v) SDS for 30 min at 37°C.
6. Rinse the membranes several times with ddH_2O.
7. Place the membranes on clean filter paper. Air-dry and store at room temperature until used for hybridization.

References

1. Southern, E. M. (1975). *J. Mol. Biol.,* **98,** 503.
2. Bidwell, J. L., Bidwell, E. A., Savage, D. A., Middleton, D., Klouda, P. T., and Bradley, B. A. (1988). *Transplantation,* **45,** 640.

3. Innis, M. A. and Gelfand, D. H. (1990). In *PCR Protocols* (ed. M. A. Innis, D. H. Gelfand, J. J. Sninsky, and T. J. White), pp. 3–12. Academic Press, London.
4. Baxter-Lowe, L., Hunter, J., Casper, J., and Gorski, J. (1989). *J. Clin. Invest.*, **84,** 613.
5. Middleton, D., Savage, D. A., Cullen, C., and Martin, J. (1988). *Transplant. Int.,* **1,** 161.
6. Opelz, G., Mytilineos, J., Scherer, S., Dunckley, H., Trejaut, J., Chapman, J., Middleton, D., Savage, D., Fischer, G., Bignon, J. D., Bensa, J. C., Albert, E., and Noreen, H. (1991). *Lancet,* **338,** 461.
7. Bidwell, J. L. and Jarrold, E. A. (1986). *Mol. Immunol.,* **23,** 1111.
8. Larhammar, D., Schenning, L., Gustafsson, K., Wiman, K., Claesson, L., Rask, L., and Peterson, P. (1982). *Proc. Natl Acad. Sci. USA,* **79,** 3687.
9. Auffrey, C., Lillie, J. W., Arnot, D., Grossberger, D., Kappes, D., and Strominger, J. L. (1984). *Nature,* **308,** 327.
10. Sambrook, J., Fritsch, E. F., and Maniatis, T. (ed.) (1989). *Molecular Cloning, A Laboratory Manual.* Cold Spring Harbor Press, Cold Spring Harbor, NY.
11. Miller, S., Dykes, D. D., and Polesky, H. (1988). *Nucleic Acids Res.,* **16,** 1215.
12. Barker, D., Schafer, M., and White, R. (1984). *Cell,* **36,** 131.
13. Reed, K. C. and Mann, D. A. (1985). *Nucleic Acids Res.,* **13,** 7207.
14. Feinberg, A. and Vogelstein, B. (1983). *Anal. Biochem.,* **132,** 6.
15. Marsh, S. G. E. and Bodmer, J. G. (1991). *Hum. Immunol.,* **31,** 207.
16. Gao, X., Fernandez-Vina, M., Shumway, W., and Stastny, P. (1990). *Hum. Immunol.,* **27,** 40.
17. Lanchbury, J. S. S., Hall, M. A., Welsh, K. I., and Panayi, G. S. (1990). *Hum. Immunol.,* **27,** 136.
18. Petersdorf, E. W., Smith, A. G., Mickelson, E. M., Martin, P. J., and Hansen, J. A. (1991). *Immunogenetics,* **33,** 267.
19. Higuchi, R. and Kwok, S. (1989). *Nature,* **339,** 237.
20. Kafatos, F. C., Jones, W. C., and Efstratiadis, A. (1979). *Nucleic Acids Res.,* **7,** 1541.

6

Cellular methods

NANCY L. REINSMOEN

1. Mixed lymphocyte culture

1.1 Introduction

The mixed lymphocyte culture (MLC) is perhaps the most widely used of cellular assays, and represents a functional assay of cellular response to stimulatory determinants associated predominantly with HLA class II molecules including HLA-DR, and -DQ, and to a lesser extent HLA-DP molecules. The first descriptions of this assay as a measurement of cellular immunity (1, 2), together with the development of a one-way method of stimulation, allowed the correlation of proliferative responses between siblings and the conclusion that a single genetic locus or region, now known as HLA, controlled the MLC reactivity. The recognition of disparate HLA class II molecules and the resulting T cell activation as measured in MLC are thought to represent an *in vitro* model of the afferent arm of the *in vivo* allograft reaction.

The MLC reaction is an *in vitro* test of lymphocytes responding to stimulation by disparate HLA class II molecules predominantly expressed on B cells and monocytes of the stimulator cell population. Proliferative reactivity to HLA class I molecules has been reported, but plays a minimal role in the overall bulk MLC response. In the MLC assay stimulator cells have been inactivated, usually by X-irradiation, and can no longer divide. The resulting proliferation of responding cells involves the logarithmic expansion of multiple clones of alloactivated T cells. This response can be measured by incorporation of the radioisotope tritiated thymidine into replicating DNA during the logarithmic phase of cellular expansion, usually on the fifth day of culture. The amount of thymidine incorporated into cellular DNA is then assayed by liquid scintillation spectrophotometry. Exogenous tritiated thymidine added to *in vitro* cultures is incorporated during DNA replication via the salvage pathway, in which free purine bases are formed by hydrolytic degradation of nucleic acid and nucleotides. Exogenous tritiated thymidine is added to cultures for a period of time that is longer than the 'S' phase of the cell cycle but shorter than the cell cycle itself, usually 18 hours.

The degree of reactivity observed correlates with the degree of antigenic disparity between responding and stimulating cells. On average, sibling combinations disparate for both HLA haplotypes demonstrate reactivity comparable to that observed in combinations of cells from unrelated individuals, while cells from HLA identical sibling combinations demonstrate essentially no reactivity in MLC. Although MLC is extremely useful in assessing the overall proliferative response to stimulation by the combined disparate stimulatory determinants, this methodology cannot be used to define cellular specificities or to distinguish the reactivity to individual gene products. With the advent of the DNA based methodologies for HLA typing (see Chapter 5), and the testing of unrelated donors for bone marrow transplantation, it is now possible to correlate the degree of MLC reactivity with a given disparity. That is, isolated disparities for HLA-DQ, DR52 subgroups, and HLA-DP all elicit a low degree of MLC reactivity in the range of 5–12% relative response (RR), while disparity for DRB1 αβ molecules elicits a greater degree of reactivity in the range of greater than 18% RR. MLC has been used clinically for donor selection, predominantly for bone marrow transplantation. With the more recent application of DNA based HLA typing methods, the use of MLC as a method to identify HLA class II disparities is controversial. Whether MLC can be used to identify acceptable mismatches relevant to transplant outcome remains to be investigated. The use of MLC and other cellular methodologies in other areas of transplantation have vital applications and are discussed in Section 7.

Protocol 1. Mixed lymphocyte culture

- lymphocyte separation medium (LSM), (Pharmacia Biotechnologies 17084003)
- culture medium: RPMI 1640 with Hepes buffer supplemented with 100 U/ml penicillin, 100 U/ml streptomycin (Grand Island Biological 380-2400AJ), 10 U/ml preservative free heparin (Monoparin heparin, Accurate Chemical & Scientific Corp. A6500), 2 mM L-glutamine, and 10–20% pooled human sera
- pooled human sera: serum from 10–20 healthy non-transfused male donors, heat inactivated at 56°C for 30 min (see technical considerations for additional specifications)
- radiation source (usually gamma emitting radiation source): alternatively, mitomycin-C (Sigma, MO503) can be used
- [^{3}H]thymidine: a specific activity of 6.7 Ci/mM, and 0.5–1.0 mCi/well are commonly used (New England Nuclear, NEN-027)
- Hanks' balanced salt solution (HBSS) (Grand Island Biological, 310-4170PJ)
- laminar flow hoods, Baker 60, Sanford, Maine

- liquid scintillation counter, 1205 Betaplate, Pharmacia LKB
- rate freezer CryoMed model 70014
- liquid nitrogen refrigeration unit, Model CAIIIL CryoMed

1. Isolate mononuclear cells by centrifugation of peripheral blood diluted 1:2 with HBSS over LSM (see also Chapters 2 and 3).
2. Remove peripheral blood mononuclear cells (PBMC) from the LSM interface and dilute with HBSS, centrifuge at 500 *g* for 10 min. Decant and repeat wash steps two additional times.
3. Resuspend cells in an exact quantity of complete culture medium.
4. Perform white cell count and determine viability via dye exclusion.
5. Dilute the cell suspension to a final concentration of 5×10^5 PBMC/ml using the culture medium.
6. Inactivate stimulator cells by either irradiation at 1500–3000 Rads or incubation with mitomycin-C according to the manufacturers' instructions.
7. Using a repeating microlitre pipette, add stimulator and responder cells in triplicate to round bottom microtitre plates (ICN 76-042-05), such that each well receives 100 μl of stimulator cells (5×10^4 PBMC) and 100 μl of responding cells (5×10^4 PBMC). A complete culture set up includes:
   (a) allogeneic cultures containing all possible combination of responder and stimulator cells, including cells from three control cell donors of a known HLA phenotype
   (b) autologous cultures containing the responder and stimulator cells from the same cell donor
   (c) control wells containing either responder or stimulator cells alone with an equal volume of complete culture medium
   (d) double irradiation control cultures containing stimulator cells from two different cell donors.
8. Incubate at 37°C in a humidified atmosphere of 5% CO_2 in air for five days.
9. Add 0.5–1.0 mCi of tritiated thymidine to each well, and incubate the cultures for an additional 18 h.
10. The culture plates can then be harvested immediately, or sealed with pressure sensitive film and placed in the refrigerator until harvesting can be performed.
11. There are a number of different harvesting machines and counting systems available, ranging from harvesting the cells on to filter disc sheets, to counting the samples in vials, or cassettes, or directly without the need for scintillation fluid. Consult the manufacturer's instruction manual for the appropriate procedures to follow.

Protocol 1. *Continued*

Technical considerations

(a) Collection of specimens. Care must be taken throughout the procedure to ensure a sterile specimen is obtained. The specimen may be saved overnight but should be processed within 24 hours of the phlebotomy. The specimen should be maintained at room temperature even if being shipped by overnight carrier. Poor cell yields may result from either too cold or too warm temperature conditions.

(b) Drugs. If a patient is receiving one of the following drugs, the proliferative response may be compromised: prednisone, myaluran, hydroxyurea, cytoxan, conamycin, L-asparaginase.

(c) Serum. One of the most common sources of technical problems which occurs in any of the cellular procedures is a poor serum source. Each individual lot of a serum source, or preferably each individual serum unit comprising the lot, should be screened for growth support capabilities and possible HLA antibodies. The screen should include a control response to a pool of allogeneic cells to measure maximum response, and an autologous control to ensure low backgrounds. If sporadic high backgrounds are observed an endotoxin test may be advisable.

(d) Tritiated thymidine. If low c.p.m. are observed, the scintillation counter and the shelf life of the tritiated thymidine should be checked. The half-life of the tritium is 12.3 years, but the shelf life of the thymidine is considerably shorter.

(e) Frozen cells. Cells to be used as responder cells in the cell cultures can be frozen prior to use (see Chapter 2). However, better viability and cell recovery are experienced if the cells are rate frozen and stored in the vapour phase of a liquid nitrogen storage unit.

The ASHI Procedure Manual (3) is an excellent source for additional information and details regarding cellular methods.

1.2 Results and interpretation

The results are usually expressed as raw counts per minute (c.p.m.) tritiated thymidine incorporation. The data may be reduced to allow for easier interpretation and comparability from one test to another. The two most common forms of data reduction include the stimulation index and the relative response (RR). The stimulation index is a simple ratio of the c.p.m. from an experimental MLC combination divided by the c.p.m. of the autologous control. The relative response (RR) is the ratio of the net (after subtraction of autologous control c.p.m.) c.p.m. of an allogeneic MLC combination, and the net c.p.m. in a maximally stimulated or control MLC combination (usually the response to a pool of allogeneic cells), multiplied by 100 to obtain a percentage.

2. HLA-Dw typing using homozygous typing cells (HTC)

With the application of DNA methodologies for typing HLA class II specificities, the HTC approach to typing for HLA-D region identity is no longer commonly used. Historically, cells homozygous for the major stimulatory determinants associated predominantly with the HLA-DR and HLA-DQ molecules were used as stimulator cells or typing cells in the MLC. Theoretically, cells from individuals who are homozygous for these HLA-D region determinants can be used to identify responder cells possessing these HLA-D region determinants, since a weak MLC reaction would be expected from these combinations compared to responder cells that do not share HLA-D region determinants with the HTC. Responder cells showing a weak response to a particular HTC, or set of HTCs defining a given specificity, are assumed to express the specificity that is defined by that particular HTC. The HTC-defined HLA-Dw specificities are actually haplotype designations and represent clusters of antigenic determinants associated with the HLA class II molecules. Thus, the response to a given HTC represents the aggregate reactions of multiple clones recognizing determinants associated with the HLA-DR, HLA-DQ, and HLA-DP molecules. The 23 HLA-Dw specificities defined by HTCs identify subgroups of the serologically-defined DR antigens but are not attributable to a single HLA-D region locus. The counts per minute tritiated thymidine incorporation are normalized for variation in the responding capabilities of the various responding cells and for the different stimulatory capabilities of the various HTCs (4). One future application of this test is the measurement of the change of a response with time. For example, the development of donor antigen-specific hyporeactivity has been assessed post-transplant for a number of kidney transplant recipients by measuring the change in response to HTCs defining a given donor's specificities (see Section 7).

Protocol 2. HLA-Dw typing using HTCs

The HLA-Dw typing technique utilizes the basic MLC procedure (see *Protocol 1*) using HTCs of well-defined specificity as stimulator cells. In contrast to the MLC assay, the HLA-Dw typing assay is usually set up using frozen stimulator and responder cells. Typically, three to four HTCs per Dw specificity, and three pooled stimulating reagent cells (three unrelated cells per pool, selected to include no duplication of DR/Dw specificities) are required.

1. Thaw cells according to the usual standard procedure (see Chapter 2).
2. Use the autologous response and responding cells, or stimulator cells cultured alone in the culture medium as negative controls.

Protocol 2. *Continued*

3. Perform cell viabilities before plating the cells. In round bottom microtitre plates add to each well 50 000 responding cells and 50 000 stimulating cells (irradiated at 3000 Rads) in a total volume of 0.2 ml.
4. Incubate, label, harvest, and count the cultures as usual according to the standard MLC procedure (*Protocol 1*).

2.1 Results and interpretation

The results are expressed as double normalized values (DNVs) calculated according to the standard methods established by the International Histocompatibility Workshops (4). Briefly, the individual responses to the various HTCs are normalized by dividing the median c.p.m. of the triplicate value by the 75th percentile ranked response of the c.p.m. medians, and multiplying by 100 to obtain the responder normalized values (RNV). The DNVs are obtained by ranking the RNV for each stimulating cell, dividing each RNV by the 75th percentile values, and multiplying by 100.

A positive typing response is assigned when the majority of the responses to HTCs defining a given specificity result in DNV <30 or when DNV results of 31–50 are reproducible. A possible typing result is assigned when DNV results of 31–50 are observed for at least 2 of the HTCs defining a given specificity. No typing response is assigned when all DNV results are greater than 50. Typings are repeated when there is a discrepancy with the DR type obtained by another method, or when DNV of 31–50 are obtained.

3. Primed lymphocyte test (PLT)

The principle of the PLT assay is to generate responder cells primed against disparities expressed by the stimulator cell by incubating the cells together for a period of ten days. These primed cells, presumably memory cells, will respond in an accelerated or secondary response when re-stimulated by cells from the original stimulator, or other cells which share stimulatory determinants with the sensitizing cell. The expansion and cloning of these primed cells provided valuable reagents to identify the cellularly-defined determinants/epitopes associated with the Mhc molecules and enhanced our understanding of the allogeneic response. Stimulatory determinants in PLT, which presumably reflect those determinants capable of stimulation in the primary MLC, have been reported as being associated with HLA-DR, -DQ, -DP, the HLA-A, -B chromosomal segment and, in addition, even determinants not linked to HLA. By careful selection of responder and stimulator cells which share some of the HLA specificities and differ for others, highly specific and sensitive reagents can be generated against the disparate stimulatory determinants. For example, the HLA-DP segregant series was defined by using unrelated priming cells phenotypically identical for HLA-A, -B, -C, -DR, and -DQ but disparate for one HLA-DP specificity. The PLT can be used in monitoring transplant recipients to assess if primed cells reactive to donor antigens are present in the allograft. When the PLT is used in this manner to assess the alloproliferative response, all disparate molecules can potentially elicit a response. T cell cloning may be necessary to differentiate the response to the individual HLA molecules.

Protocol 3. Primed lymphocyte test

A. *PLT testing for determination of a HLA specificity*

1. The cells are obtained in the same manner as described in *Protocol 1*.
2. Culture 10^7 responding and 10^7 stimulating cells (X-irradiated, 3000 Rads) together in a final volume of 15 ml culture medium for a period of 10–12 days, adding 2 ml culture medium on alternate days.
3. After 10–12 days, the primed cells may be used immediately, frozen for future use, or cloned.
4. The primed cells may be tested at concentrations ranging from 10^4–10^3 cells/well with 5×10^5 stimulating cells.
5. Incubate for 48 h in a humidified 37°C, 5% CO_2 incubator and add 2 mCi/well [^{3}H]thymidine (6.7 specific activity) for 18 h.
6. Terminate, harvest, and count cultures as described in *Protocol 1*.

B. *Expansion of primed bulk T cell populations*

1. Dilute reagent cells to $0.3–0.5 \times 10^6$ cells/ml with culture medium.
2. Incubate cultures at 37°C in a 5% CO_2 humidified environment.
3. On the third day, add undiluted rIL-2 (recombinant interleukin 2, Sandoz Research Institute) to obtain an optimal concentration of 20 U/ml.
4. Continue to adjust the volume such that the cell concentration does not exceed 0.5×10^6 cell/ml.
5. The T cell population may be expanded by adding the feeder cells (see technical considerations) at weekly intervals, and keeping the cell culture diluted to the appropriate cell concentration.

C. *T cell cloning*

Bulk primed cells may be cloned by the following procedure.

1. Centrifuge the primed cells over LSM as described in *Protocol 1*, wash, and count.
2. Adjust the cell concentration with culture medium containing 20 U/ml rIL-2, such that 0.3 cells are delivered per well in 5 μl.

Protocol 3. *Continued*

3. Adjust the feeder cells in culture medium containing rIL-2 such that 2 × 10^3 LCLs or 1 × 10^4 PBL are delivered per well in 5 μl.
4. Combine equal volumes of primed cells and feeder cells and plate 10 μl/well in sterile Terasaki trays (Robbins Scientific, 1006-01-0) to obtain the desired 0.3 primed cells/well with 10^4 PBL feeder cells/well, or 2 × 10^3 LCL feeder cells/well.
5. Wrap plates in aluminium foil and place in humidified 37°C, 5% CO_2 environment. Add 5 μl culture medium containing rIL-2 after 3–4 days of incubation.
6. After 7–10 days, score wells for positive growth by viewing under a phase contrast microscope.
7. Transfer positive wells to 96 well flat well plates (ICN 76-032-05) containing 10^5 feeder cells in 0.2 ml rIL-2 culture medium.
8. After a week, transfer positive wells to 24 well plates (ICN 76-033-05) containing 10^6 feeder cells in 2 ml rIL-2 culture medium. The T cell clones may be maintained in a manner similar to that for the T cell bulk populations; however, the clones must be assayed periodically to make certain function is maintained.

Technical considerations

(a) Variation in cell numbers. If lymphoblastoid cell lines (LCLs) are used as stimulator cells, the cell concentration should be lowered to 10^4 LCLs per well. Cloned T cell reagents are usually tested at concentration of 10^4 cells per well or lower.

(b) Label times. Label times of eight hours have been described especially when assaying cloned cells.

(c) Controls including:

i. Cells of the original stimulating cell used to generate the primed reagent.

ii. Cells of the original responding cell.

iii. Medium controls.

iv. Pool primed PLTs. Pool primed PLTs are often used to control for the varying stimulatory capabilities of the various stimulating cells. Cells from two donors are used as responding cells, and a pool of irradiated cells from at least three HLA-DR, -DQ, -DP disparate donors are used as stimulating cells. The responding cells are primed separately against the stimulator pool as previously described. The primed reagents are then combined to generate the pool primed reagent. These cells should be primed against a broad array of HLA antigens such that most stimulator cells used will re-stimulate this reagent if the cells are functioning effectively.

(d) Feeder cells. For long term cultures, feeder cells are necessary to maintain the T cell cultures. The choice of feeder cells depends upon the specificity and the use of the reagent and, therefore, must be left to the discretion of the investigator. PMNC that share the same specificity as the sensitizing antigen are irradiated and added to the culture system at 2–5 fold the number of the reagent cells. LCLs (irradiated at 10 000 Rads) that are the same specificity as the stimulator cells can be used; however, usually only equal numbers of feeder cells and reagent cells are used so as not to add excessive amounts of antigen to the T cell cultures. Autologous feeder cells may be used in conjunction with the antigen-specific PLT or LCL feeder cells at ratio of 4:1 (auto:specific).

3.1 Results and interpretation

The results may be expressed as a per cent relative response calculated in the same manner as in the MLC assay. The positive versus negative re-stimulation values are determined by a cluster analysis programme (5).

4. T cell precursors

The T cell precursor analysis is performed by limiting dilution assays, whereby limiting numbers of responder cells are cultured with a constant number of stimulator cells and assayed for reactivity (cytotoxic, proliferative, or cytokine release) against additional stimulator cells. The frequency of responding cells is determined by a maximum likelihood estimation using a computer programme. Limiting dilution assays imply that the lysis is a single-hit process such that the presence of a single killer precursor cell will initiate the sequence of events which lead, for example, to the eventual lysis of the cell in the case of cytotoxic precursors. Also imperative is the clear distinction between a response and a lack of that response, which is made more difficult in the cytotoxicity assay where an arbitrary threshold is set separating the spontaneous release of ^{51}Cr from the release from the lysed targets.

Protocol 4. Limiting dilution assay

- phytohaemagglutinin (PHA) (Burroughs Wellcome, HA-17), 28.75 U/ml RPMI 1640

1. Culture limiting numbers of PBMCs (5×10^4–0.125×10^4) in round bottom microtire plates with constant numbers (5×10^4) irradiated (3000 Rads) stimulator cells.
2. Multiple wells per dilution are necessary to ensure an accurate assessment of the frequency. Usually 30 wells per dilution are set up. The culture medium is the same as used in *Protocol 1*.

Protocol 4. *Continued*

3. Add IL-2 and additional culture medium on days three and six such that the final concentration is 5 U/ml.
4. Prepare target cells on day seven using unirradiated stimulator cells from the initial priming combination. Adjust cell concentration to 10^6 cells/ml and place 1 ml of the cells with 1 ml PHA in 4–6 wells of a 24 well microtitre plate. Incubate at 37°C in a humidified atmosphere of 5% CO_2 for three days.
5. On day ten assay the cultures for cytotoxicity (or other functional assays) against ^{51}Cr labelled target cells. Label 2×10^6 PHA target cells with 0.5 mCi ^{51}Cr for 1.5 h, wash, and adjust cell concentration to 2×10^4 cells/ml. Replace 100 μl of the media in the priming cultures with 100 μl of labelled target cells, spin plates at 100 *g* for 5 min, and incubate for 4 h.
6. Supernatants are harvested and tested for released ^{51}Cr, either using a gamma counter to detect the gamma rays or a beta counter to detect the auger rays released.
7. Wells are scored as positive by determining that the reactivity observed is significantly [>3 SD (standard deviation)] greater than that of the control background (wells containing irradiated stimulator cells incubated alone without responder cells are used as background controls).
8. The frequency of responding cells is determined by a maximum likelihood estimation using a computer programme, and the variance by the use of 95% confidence intervals. Regression analysis is used to generate a straight line and chi-square analysis is used to show that the data obtained are in accordance with single-hit kinetics. A programme which will perform all the necessary calculations for this analysis is listed in reference 5.

---

5. Epstein–Barr virus (EBV) transformed lymphoblastoid cell lines

Long-term culture of cells can consist of either normal cells or transformed cells. Examples of long-term culture of normal cells will be addressed in Section 6. Normal B cells can be infected with EBV through the CD21 receptor. EBV transformed lymphoblastoid cell lines are often called lymphoblastoid cell lines (LCLs). An LCL panel on bank is a valuable source of reference DNA, reference reagents in tissue typing, and cellular studies of epitopes recognized by antibody or T cells. Since the LCLs have been transformed, they do not depend upon the addition of exogenous lymphokines for their continual growth. These cell lines are easy to grow, but critical attention must be given to the sterility and growth conditions of the lines so that mycoplasma infection does not overtake the cultures.

Mycoplasma is an organism which lacks cell walls and thus is not susceptible to many antibiotics and is not readily visible by light microscopy. However, the effects of mycoplasma infection are readily evident and severe. Mycoplasma contains the thymine kinase enzyme which interferes with nucleic and amino acid metabolism, thereby interfering with DNA replication of the contaminated cell culture. Thus the growth rate of these cells is affected and they eventually die. Meticulous care must be taken since mycoplasma infection can easily spread to other cultures in the laboratory. Mycoplasma infection should be suspected in short term cultures if values for tritiated thymidine incorporation have decreased or are very low.

The single greatest method of prevention of mycoplasma contamination is the aseptic culture technique. The cultures must never be allowed to overgrow the culture medium to the point where the medium is a yellow colour. As outlined below, LCLs must be checked and split a minimum of every second day. A second common source of mycoplasma contamination is the water-bath. When a frozen vial of LCLs is thawed in a water-bath, the outside of the vial should be wiped with a 2% bleach solution before it is opened. One of the first signs of mycoplasma infection in LCLs is a dirty or gritty appearance of the cultures when viewed under the microscope. In addition, LCL generally will not divide at the usual rapid rate. LCLs must be checked regularly for mycoplasma infection.

There are a variety of methods available including Hoechst staining, or biochemical methods such as nucleic acid hybridization using DNA probes. Simple kits are now available for the detection of mycoplasma which rely on DNA hybridization techniques (Gen-Probe, 1591). The kit consists of small tritium labelled DNA probes which are complementary to known specific sequences of mycoplasma RNA. These probes are added to the supernatant from cell cultures. If mycoplasma is present, the DNA probe will bind to the mycoplasma RNA forming hybrids which then are detected in a scintillation counter.

Protocol 5. Generation and maintenance of LCLs

The laminar flow hood must be cleaned daily with a 2% bleach solution, and the filters checked routinely. Incubators must be cleaned monthly, including washing with a 2% bleach solution, autoclaving all of the racks, and preferably keeping autoclaved water in the bottom of the incubator.

(a) LCL culture medium RPMI 1640 without Hepes buffer, supplemented with 100 U/ml penicillin, 100 U/ml streptomycin, 0.05 mg/ml Gentamicin, 2 mM L-glutamine, and 10% fetal calf serum.

(b) EBV supernatant. The supernatant used to transform PBMC can be obtained by the culture of an EBV shedding line, such as B95-8 cultured

Protocol 5. *Continued*

in LCL media. The line is cultured for a period of two weeks until the medium is a yellow colour, at which time the culture is centrifuged at 1500 *g* for 10 min, and the supernatant filtered through a 0.4 mm filter.

(c) Cyclosporin A medium. CSA is stored as a stock 50 mg/ml at 4°C, and is diluted in LCL culture medium to a final concentration of 0.5 μg/ml (also stored at 4°C).

A. *Transformation*

1. Suspend 2×10^6 PBMC in 1 ml CSA culture medium, add 1 ml EBV supernatant, and incubate for a period of 21 days. The culture is shedding virus during this time so care should be taken to keep the cultures isolated from other cultures.
2. Add CSA culture medium during this period when the culture medium turns yellow.
3. After 21 days remove the cultures from CSA culture medium and culture in regular LCL culture medium.

B. *Maintaining LCLs*

1. Inspect the cultures for contamination, mycoplasma, and cell density using an inverted phase microscope every other day.
2. Split the cultures by diluting the cells to approximately 0.2×10^6 cells/ml.
3. The LCLs can be rate frozen (usually at $5–10 \times 10^6$ LCLs/vial) and stored in vapour phase liquid nitrogen.

6. Propagation of graft infiltrating cells

Although immunohistochemical studies provide information identifying different types of infiltrating T cells by cell surface markers (see Chapter 8) there is little information regarding the function and specificity of these cells and their role in allograft rejection and tolerance. Other investigators have shown that monitoring of peripheral blood population has a limited value. However, a more direct approach to the study of infiltrating T cells is to isolate cells from the graft and perform various studies, including the cellular assays outlined in this chapter as well as additional molecular assays. This section outlines the basic approach used to propagate cells from biopsies and lung bronchoalveolar lavages (BAL). Other approaches such as the use of donor or allogeneic cells as feeder cells is at the discretion of the investigator. Since the ultimate goal is to expand the cell population infiltrating the allograft, and not to prime *in vitro* any cells obtained from the graft, this section will deal with methods which minimize the possibility of priming against donor antigens *in vitro*.

Protocol 6. Propagation of cells from biopsies and BAL

- tissue culture medium is as previously described with the addition of antibiotics and pooled human sera, plus rIL-2 added to a final concentration of 10 U/ml

1. Prepare autologous feeder cells from recipient blood by LSM centrifugation (*Protocol 1*), and irradiate the cells with 3000–5000 Rads.
2. Place the biopsies in flat bottom 96 well microtitre plates with IL-2 culture medium and 50 000 autologous feeder cells. Incubate BAL derived cells (0.5×10^6) with 1.0×10^6 autologous feeder cells in 24 well plates.
3. Incubate the biopsies/BAL derived cells as usual, and regularly observe for cell growth.
4. Once cell growth is observed split the cultures and expand to successively larger well culture plates, and eventually to flasks.
5. The cultures should be fed with medium containing IL-2 every other day once the cultures are growing rapidly, and fed with the autologous feeder cells once a week.
6. Discard all biopsies/BAL derived cells lacking cell growth after three weeks.

Technical considerations

When culturing cells from BAL, fine needle aspirates, or biopsies, addition of anti-CD3 MoAb may help in the expansion of the activated cell population without compromising the specificity.

---

7. Clinical applications

The current use of MLC and HLA-Dw typing in the context of the clinical laboratory has changed with the advent of DNA typing for HLA polymorphisms (see Chapter 5). These tests are no longer used exclusively for the assessment of HLA class II identity but can serve as useful adjuncts to other assays. Although DNA technologies provide more exact information regarding the HLA class II polymorphisms, the question still remains whether cellular assays such as MLC can provide information concerning acceptable mismatches, that is polymorphisms which are not recognized as being different by the effector T cells. Currently, cellular assays are being used to define donor antigen-specific hypoactivity, to investigate the function and specificity of graft infiltrating cells, to determine T cell precursor frequencies, and to identify T cell recognized epitopes on Mhc molecules.

MLC, HLA-Dw typing, and cell mediated lympholysis techniques are being used currently to investigate the development of donor antigen-specific hypoactivity following transplantation (6). These assays may be useful in identifying those patients who are good candidates for having their immunosuppression withdrawn or tapered, based on their apparent immunoregulation of response to disparate antigens. The development of donor antigen-specific hypoactivity, as measured by MLC and HLA-Dw typing assays, correlates with improved late renal transplant outcome as evidenced by fewer late rejection episodes and fewer graft losses.

PLT has been used to detect donor antigen-specific reactivity of BAL lymphocytes associated with acute lung rejection and obliterative bronchiolitis (OB), as well as cells infiltrating transplanted renal allografts (7). Our previous studies (8) demonstrated a predominant CD8+ cell population mediating class I donor antigen-specific reactivity correlating with OB in 3/3 recipients tested, and a predominant CD4+ cell population mediating class II donor antigen-specific reactivity correlating with acute rejection episodes in 13/15 recipients tested. These studies are of importance not only in monitoring recipients, but also in investigating the immunological basis of pulmonary disease. Taken together, these results suggest that distinct immunopathogenetic events may be occurring during acute lung rejection and OB.

Propagation of T lymphocytes from renal, cardiac, and hepatic allografts demonstrates a strong correlation between long-term T cell growth and the clinical presence of acute cellular rejection. PLT has been used to investigate the specificity of these graft infiltrating cells (9). This technique has been used to demonstrate functional characteristics of graft infiltrating cells, and to provide information regarding the activation state of the T cell infiltrate.

Previous studies have suggested a correlation between cytotoxic T cell precursor (CTLp) frequency and the clinical grade of graft-versus-host (GVHD) in patients receiving marrow from unrelated donors (10, 11). Those patients with CTLp frequencies higher than 1:100 000 tended to have more severe GVHD than those patients with lower CTLp frequencies. This approach may be useful in predicting GVHD in unrelated donor bone marrow transplant recipients.

In conclusion, the cellular assays described in this chapter have been used historically for the determination of class II identity and for the investigation of T cell recognized epitopes. These techniques remain useful for assessing T cell recognized epitopes and investigating immune regulation, and will undoubtedly provide valuable information in evaluating the immune status of transplant recipients.

References

1. Bain, B., Vas, M., and Lowenstein, L. (1964). *Blood,* **23,** 108.
2. Bach, R. H. and Hirschhorn, K. (1964). *Science,* **143,** 813.
3. Reinsmoen, N. L. and Mickelson, E. M. (1990). In *ASHI Laboratory Manual,* (ed. A. A. Zachary and G. Teresi), pp. 333–424. American Society for Histocompatibiity and Immunogenetics, Lenexa, Kansas.
4. Ryder, L. P., Thomsen, M., and Platz, P. (1975). In *Histocompatibility Testing 1975*, (ed. F. Kissmeyer-Neilson), p. 528. Munksgaard, Copenhagen.
5. Carroll, P. G., DeWolf, W. C., Mehta, C. R., Rohan, J. E., and Yunic, E. J. (1979). *Transplant. Proc.,* **11,** 1809.
6. Reinsmoen, N. L., Kaufman, D., Sutherland, D., Matas, A. J., and Bach, F. H. (1990). *Transplantation,* **50,** 1472.
7. Robinowich, H., Zeevi, A., Paradis, I. L., Yousen, S. A., Dauber, J. H., Kormos, R., Hardesty, R. L., Griffith, B. P., and Duquesnoy, R. V. (1990). *Transplantation,* **49,** 115.
8. Reinsmoen, N. L., Bolman, R. M., Savik, K., Butters, K., and Hertz, M. (1992). *Transplantation,* **53,** 181.
9. Miceli, C., Barry, T. S., and Finn, O. J. (1988). *Hum. Immunol.,* **22,** 185.
10. Kaminski, E., Hows, J., Man, S., Brookes, P., Maxkinnon, S., Houghes, T., Avakian, O., Goldman, J. M., and Batchelor, J. R. (1989). *Transplanttion,* **48,** 608.
11. Kaminski, E., Hows, J., Brookes, P., Mackinnon, S., Houghes, T., Avakian, O., Sharrock, C., Goldman, J. M., and Batchelor, J. R. (1989). *Transplant. Proc.,* **21,** 2986.

7

HLA class I gene and protein sequence polymorphisms

ANN-MARGARET LITTLE and PETER PARHAM

1. Introduction

The unique feature of the HLA-A, -B, -C class I proteins is their polymorphism. Differences between class I proteins can be detected by alloantisera in the classical serological tissue typing assay, and it is this method which is most widely used to define and distinguish class I polymorphism (see Chapter 3).

Comparison of class I sequences, together with the 3D structure of HLA-A2 (1) has shown the majority of polymorphic residues are located within and around the peptide binding cleft (2). It is likely that certain residues found deep within this cleft are not directly accessible to allo-recognition mechanisms and go undetected by alloantisera.

Additional techniques have been applied to define class I polymorphisms undetected by serology, these include analysis of variants recognized by cytotoxic T lymphocytes (CTL), electrophoretic heterogeneity of immunoprecipitated class I molecules, and molecular structure determination via nucleotide sequencing. Collectively these techniques show that many more class I alleles exist than are detected at present by serology.

Both alloreactive and self-restricted CTLs seem sensitive to substitutions in the peptide binding cleft and can discriminate almost all serologically indistinguishable variants (3). Indeed, CTLs may be so sensitive that they can distinguish between identical alleles which present different polymorphic self peptides in different target cells (4). Thus not all CTL defined variants need be allelic variants of the class I heavy chain.

The first non-serologically detected HLA 'variant' or 'subtype' to be defined was M7 (HLA-A*0202) which was shown to differ from the common serologically defined HLA-A2 (A*0201) by CTL recognition (5). At present 12 alleles of HLA-A2 have been identified, however this group of alleles can only be split into two groups using antisera. One group is represented by the A*0203 allele, the other embraces the remaining 11 alleles.

Analysis of HLA class I molecules by isoelectric focusing (IEF) demonstrated structural heterogeneity for many serological specificities, thus promoting interest in the characterization of those subtypes which fail to be specifically recognized by serology. Such characterization has resulted in the determination of the nucleotide and protein sequences of over 100 HLA class I alleles.

1.1 Application

Accurate definition of HLA class II polymorphisms that are ill-defined by serology has been shown to be advantageous in the matching of organs and bone marrow for transplantation (6). As bone marrow transplantation has become increasingly dependent on unrelated donors, accurate class I, in addition to class II, typing is certainly a pre-requisite. Due to the immunological competence of the tissue being transplanted, the donor's immune system is particularly sensitive to HLA mismatches. Indeed mismatching between two variants of HLA-B44 which are not detected by serology and differ by one amino acid, has been associated with bone marrow rejection (7).

Conversely it has been argued that it is not necessary to define HLA class I polymorphisms precisely for successful organ transplantation (8), particularly with the success of immunosuppressive therapy. In contrast molecular analysis of serologically related alleles has shown that class I proteins which are related by serology may not necessarily be as closely related in their molecular structure. Sequencing analysis of alleles from the A19 family: HLA-A29, 30, 31, 32, and 33, showed HLA-A30 to be structurally more related with members of the A1/A3/A11 family than with the A19 family (9). The serological similarity between HLA-A30 and -A31 can be mapped structurally to one residue, arginine 56, and similarity with the rest of the A19 family can be mapped to residues 144 and 151 in the α2 domain, which also form the 4E epitope shared by all HLA-B and -C locus and A29, 30, 31, 32, and 33 molecules (10).

Many diseases are associated with HLA specificities, ankylosing spondylitis (AS) with HLA-B27 being the strongest (11). Sequence comparison of seven B27 subtypes identifies a negatively charged B pocket as a common feature correlating with disease susceptibility. Speculation that HLA-B*2703 is not associated with AS (12) supports the hypothesis that unique aspects of peptide binding by B27 plays a role in the diseased state as B*2703 differs by a unique substitution in the peptide binding cleft. Definition of B27 polymorphism may thus have provided insight important for the elucidation of mechanisms predisposing B27 positive individuals to disease.

The extensive polymorphism of the HLA-A, -B, -C genes makes them attractive markers for human population genetics and for the study of the evolution and migration of human populations. Serological analysis has revealed many such differences including for example, A43, B42, B46, and B54 which are highly specific to particular ethnic groups. From molecular analysis of A2 and B27 it was found that a significant proportion of the subtypes were specific to particular populations, and similar results are emerging for other antigens. For example, three tribes of South American Indians have been found to express different subtypes of HLA-B35 (13, 14). Thus the molecular analysis of class I HLA alleles reveals they have a hitherto unexploited potential for the analysis of human populations.

1.2 Non-serological techniques which define polymorphisms

HLA class I polymorphism can be detected by electrophoretic analysis whereby immunoprecipitated class I proteins are separated on IEF gels and charge variants identified. This technique is most useful for the identification of class I alleles when there is confusion over the serological typing of the cell. Like serology it has limitations and allelic variants that show no difference in overall charge cannot be resolved. For example only five subtypes of HLA-A2 can be defined by IEF.

The ultimate definition of polymorphism is through the characterization of the amino acid sequence of the class I molecule. Determination of the primary structure of class I molecules via DNA sequencing has resulted in the complete definition of ill-defined serological types and serological 'blanks', in addition to characterization of variants of common serological specificities.

2. Immunoprecipitation of HLA-A, -B molecules and electrophoretic characterization

2.1 Background

The heavy chains (HC) of the class I molecules are separated according to their isoelectric points (pI), therefore class I 'subtypes' or 'variants' which differ in the number of their charged amino acids can be identified (15, 16). IEF is most useful for defining HLA-A, -B types when there is confusion over the serological typing for a particular individual:

- cells which do not give clear patterns by serology
- cells which are typed as 'blank'
- those which come from ethnic minorities

Since the pIs of many class I alleles are similar it is not possible to assign the HLA type of a cell without serological data.

2.2 Methodology

Class I molecules are immunoprecipitated from metabolically radiolabelled cell lysates with HLA specific monoclonal antibodies (MoAb). [^{35}S]methionine is incorporated into both HC and β_2-microglobulin (β_2-m) as the polypeptides are synthesized. The time of incubation with the radioisotope depends on the nature of the cells being labelled, since lymphoblastoid cell lines (LCLs) express higher levels of class I molecules a shorter labelling time can be used, e.g. two to six hours. With peripheral blood lymphocytes

(PBL) and the same quantity of isotope, a longer labelling time, 6 to 15 hours is required. Radiolabelled cell lysates are pre-cleared with *Staphylococcus aureus* (SA) cells which possess an immunoglobulin binding molecule, protein A on their cell surface. This binds immunoglobulin Fc domains but not the Fab fragment which participates in specific antibody-antigen interactions. Rabbit anti-mouse serum is also added. Pre-clearing is necessary to remove radiolabelled proteins which non-specifically bind to the SA cells and to immunoglobulin molecules.

Interpretation of the banding patterns obtained on the gels is simplified by performing two sequential immunoprecipitations. For the first, the MoAb 4E (17) can be used. W6/32, a MoAb which reacts with a determinant common to all class I HC in association with β_2-m (18), is used for the second precipitation where HLA-A, -B, and -C locus molecules which are not recognized by 4E are precipitated. Class I HC possess between zero to three negatively charged sialic acid residues on their carbohydrate chain further complicating interpretation of the gels as four bands may be present for each allele. These sialic acid residues can be removed by neuraminidase digestion prior to electrophoresis. Although this is not 100% efficient, completely desialyated HC are represented by the major bands while incompletely desialyated molecules appear fainter. Monomorphic class I monoclonal antibodies can also be used, e.g. ME1 which reacts with B27 molecules and MA2.1 which recognizes A2 molecules.

The immunoprecipitated desialyated class I molecules are separated on IEF gels according to their pI. This is defined by the composition of charged amino acids (aspartic acid, glutamic acid, lysine, histidine, and arginine) in the protein structure. The non-polymorphic β_2-m is easily detected as an invariant band for all cells studied, and provides a useful reference band. Proteins migrate through the polyacrylamide gel which contains 'ampholytes'; a mixture of multicharged low molecular weight mixed polymers of aliphatic amino acids and either carboxylic or sulphuric acids. These ampholytes cover a wide range of pIs and upon application of an electric field to the IEF gel the ampholytes migrate and establish the pH gradient. Thus the protein under study will migrate through this pH gradient until it reaches a point at which its overall charge is neutral. The pH of the gradient at this position is the pI of the protein. The pI range for HLA class I HC ranges from pH 5.0 to pH 7.0 (19).

Protocol 1. Metabolic radiolabelling, and immunoprecipitation of HLA class I molecules

A. *Metabolic radiolabelling of class I molecules in cultured cells and cell lysis*

- TE buffer: 50 mM Tris (Calbiochem, 648311) pH 7.4–7.6, 5 mM EDTA (Sigma, E4884)
- lysis buffer: 0.5% Nonidet P-40 (NP-40) (Sigma, N3516) in TE buffer

1. Use 3–10 × 10^6 fresh or cryopreserved viable PBLs isolated from whole blood, or 3–5 × 10^6 LCLs. Pellet cells by centrifugation at 800 *g* for 5 min. Resuspend the cells in 1 ml methionine-free RPMI 1640 (Irvine Scientific, 9160) or appropriate culture medium for cell lines; supplement with 10% dialysed fetal calf serum (FCS, Irvine Scientific, 3000), glutamine (Gibco, 320-5030AG), and penicillin and streptomycin (Gibco, 600-5140AG). Incubate in 24 well tissue culture plates at 37°C in 5% CO_2 incubator for 45 min.
2. Add 50 μCi L-[^{35}S] methionine (Amersham International, SJ1015) to the cultures and continue incubations for 4 h for cell lines, and for 10 h for PBLs.
3. Transfer cell cultures to microfuge tubes (in which all additional steps will be performed); microfuge for 5 min, and lyse by resuspending in 1 ml lysis buffer. Add 10 μl 100 mM PMSF (Sigma, P7626), (dissolved in 100% ethanol). Incubate on ice for 30 min.
4. Microcentrifuge for 10 min to remove insoluble nuclear and cytoskeletal debris, and transfer the supernatant to a clean tube.

B. *Pre-clearing and immunoprecipitation*

- TNEN: 20 mM Tris-HCl pH 7.6, 10 mM EDTA, 0.1 M NaCl, 0.5% NP-40
- wash solution 1: 0.1% sodium dodecyl sulphate (SDS) (BioRad, 161-0301), 0.5% sodium deoxycholate (Sigma, D6750), 1% bovine serum albumin (BSA) (Sigma, A2153) in TNEN
- wash solution 2: 0.5 M NaCl, 10% TNEN

1. Use 350 μl SA cells (Calbiochem, 10% Pansorbin) per sample for two immunoprecipitations. Wash the SA cells three times in PBS, and resuspend in wash solution 1.
2. Add 5 μl rabbit anti-mouse serum (ICN Immunobiologicals, 65-125-1) or normal mouse serum to each cell lysate, mix the contents of the tube, and add 200 μl SA cells. Mix and incubate on ice, or rotate at 4°C for a minimum of 30 min to a maximum of 15 h.
3. Microfuge to pellet the SA cells, transfer the supernatant to a clean tube, and add 50 μl SA cells. Repeat previous incubation (step **2**), and keep the supernatant.
4. Add 5 μg purified 4E IgG or 5 μl of 4E ascites to each pre-cleared cell lysate, mix, add 50 μl SA cells, and incubate on ice or rotate at 4°C for 30 min.
5. Pellet SA cells as before, remove supernatant to a clean tube, and keep the pellet. Add 5 μg W6/32 to the supernatant and repeat incubation as for 4E, again keep the pellet. The supernatant can be disposed of at this point or kept for further precipitations (it takes at least three rounds of immunoprecipitation to remove all class I molecules which react with W6/32).
6. Wash both 4E and W6/32 pellets in wash solution 1 and in wash solution 2.

Protocol 1. *Continued*

C. *Removal of sialic acid residues by neuraminidase treatment*

- IEF loading buffer: 14.26 g urea (ICN, 821527), 5 ml 10% NP-40, 1.25 ml ampholine pH 3.5–10.0 (LKB, 80-1125-87), 0.38 g dithiothreitol (Sigma, D0632), 0.25 ml 1% bromophenol blue (Sigma, B6896), distilled H_2O to a final volume of 25 ml

1. Resuspend washed pellets carrying immunoprecipitated class I molecules in 20 μl neuraminidase type VI (Sigma, N3001) or type VIII (Sigma, N5631) 10 U/ml dissolved in 0.05 M EDTA pH 6.5–7.0, and incubate for a minimum of 3 h to a maximum of 15 h at 37°C, shaking to avoid sedimentation of the SA cells.
2. Wash the pellets in 1 ml TNEN and resuspend in 40 μl IEF loading buffer. Elute class I molecules from the SA pellets at room temperature for 15 to 30 min. Pellet the SA cells and load gel with 25 μl from the supernatant.

Protocol 2. Preparing and running IEF gels

- gel solution for one 16 × 18 × 0.075 cm gel: 11.4 g urea, 3 ml 30% acrylamide (BioRad, 1610101)/bisacrylamide (BioRad, 161-0201) stock (29:1), 4 ml 10% NP-40, 5 ml dH_2O
- ampholines: 0.4 ml pH 3.5–10.0 (LKB 80-1125-87), 0.3 ml pH 5.0–7.0 (LKB, 80-1125-91), 0.3 ml pH 6.0–8.0 (LKB, 80-1125-93)
- vertical electrophoresis tank apparatus, (e.g. Hoefer SE600 and LKB 2001 vertical electrophoresis systems)

1. Wash glass plates with scouring powder, (e.g. 'Comet', 'Vim', 'Ajax') dH_2O and then with alcohol. Prepare gel casting mould according to manufacturer's instructions.
2. Dissolve the urea in the gel solution by warming. Cool to room temperature and add ampholines. Use fresh, high quality electrophoretic reagents, (e.g. BioRad, electrophoresis purity reagents) to avoid de-gassing the acrylamide solution.
3. Add 50 μl 10% ammonium persulphate (APS) (BioRad, 161-0700) and 20 μl TEMED (BioRad, 161-0801). Mix and pour the gel immediately into mould and add comb.
4. After polymerization of the gel (1 h), remove the comb, and rinse the wells with dH_2O. Load 25 μl from each sample supernatant (see *Protocol 1*) in each well. Add 25 μl IEF loading buffer to any empty wells. Each well is overlayed with IEF loading buffer diluted 1:2 with dH_2O.
5. Attach the gel sandwich to the upper and lower buffer tanks. Add the lower anode buffer (15 mM H_3PO_4, 1 ml concentrated H_3PO_4/litre) and

the upper cathode buffer (20 mM NaOH, 0.8 g/litre). Use an apparatus which allows the lower tank buffer to completely cover the gel giving an equal distribution of the heat generated. Add a stir bar to the lower gel tank and place the whole apparatus on top of a magnetic stirrer.

6. Initiate focusing at 200 V by setting a current of 10 mA and limit to 400 V. The voltage increases to 400 V with decreasing current over a period of two hours, then it remains constant. Run larger gels (280 × 180 × 1 mm) at 1000 V. Focus for 13–16 h and increase the voltage to 800 V for the last hour of the smaller gel run.
7. Fix gels in 15% acetic acid for 30 min and treat with fluorographic reagent, e.g. 'Amplify' (Amersham International, NAMP100) for 15–30 min. Dry the gel at 60°C under vacuum in a slab dryer (BioRad, 483) using dry-ice ethanol traps. Expose to X-ray film in cassettes with intensifying screens at −70°C for 6 to 24 h.

2.3 Variations in methodology

2.3.1 Chymotrypsin and carboxypeptidase B treatment

Some HLA-A and -B locus molecules have identical focusing positions, e.g. HLA-A3 and -B7, and may be difficult to distinguish if the sequential immunoprecipitation is not performed. Treating neuraminidase digested immune complexes bound to SA (see *Protocol 1*) with chymotrypsin which cleaves the carboxy side of tyrosine, phenylalanine, and tryptophan residues, changes the focusing positions of both HLA-A and -B locus heavy chains. Both HLA-A and -B locus molecules have a tyrosine residue at position 320 and cleavage at this position removes the carboxy terminal of the heavy chain which, for both HLA-A and -B locus molecules contains two negatively charged amino acids. In addition, HLA-A locus molecules have a positively charged lysine residue at position 340 which is absent from -B locus molecules; thus the overall change in position for HLA-A locus molecules is less acidic than that for -B locus molecules, and this is reflected by a differential change in banding positions after IEF (20).

Carboxypeptidase B treatment changes the pI of HLA-A locus molecules by cleaving at lysine 340 in the cytoplasmic domain, causing an acidic shift in their pI without affecting the pI of HLA-B locus molecules (21).

Protocol 3. Use of chymotrypsin and carboxypeptidase

A. *Chymotrypsin treatment*

1. Wash SA immune complexes in 2 mM Tris pH 7.6, 1 mM EDTA, 0.5 M NaCl, and 0.05% NP-40.
2. Resuspend SA pellets in 17 μl PBS and add 3 μl (0.4 U) α-chymotrypsin (Sigma, 4129) dissolved in PBS.

Protocol 3. *Continued*

3. Incubate 30 min at room temperature, wash 1 × 1 ml TNEN. Add IEF loading buffer as before.

B. *Carboxypeptidase B treatment*

1. Wash as above and resuspend in 20 μl PBS.
2. Add 3.5 U carboxypeptidase B (Sigma, C7261) and incubate for 1 h at room temperature.
3. Wash as above and resuspend in IEF loading buffer.

2.3.2 Western blotting analysis

Western blotting analysis of IEF gels has the advantage of not requiring metabolic radiolabelling of the cells. Concentrated detergent lysates are loaded directly on to the IEF gels. Following electrophoresis the proteins are transferred from the gel to either a nitrocellulose or nylon membrane by electroblotting. Proteins immobilized as such can be detected by incubating the membrane with specific monoclonal or polyclonal antibodies (22, 23) which recognize denatured HC. This technique has been useful for the identification of soluble class I molecules in serum (24).

Protocol 4. Western blotting of class I molecules

A. *Pre-condensation of Triton X-114 lysis buffer*

1. Dilute 0.25 ml Triton X-114 (Sigma) to 50 ml with TE in a 50 ml Falcon tube (VWR Scientific). Leave overnight at 4°C.
2. Warm the tube in a 37°C water-bath until the solution is turbid (5 min).
3. Centrifuge for 3 min at 300 *g*.
4. Aspirate the upper aqueous phase and discard. Add ice-cold TE to increase the volume to 50 ml.
5. Repeat steps **2–4**.

B. *Cell lysis and neuraminidase treatment*

1. Resuspend 3–5 × 10^6 cells in 1 ml pre-condensed Triton X-114 lysis buffer. Add 10 μl 100 mM PMSF and incubate on ice for 30 min.
2. Microfuge lysates for 10 min to pellet condensed chromatin and cytoskeletal material. Transfer the supernatant to a clean tube and incubate at 37°C in a water-bath for 5 min until the solution turns turbid.
3. Centrifuge for 3 min at 300 *g* to separate aqueous and detergent phases. Integral membrane proteins including HLA molecules are found in the detergent phase.

4. Add 50 μl neuraminidase (2 U/ml) to the detergent phase and incubate with shaking at 37 °C for a minimum of 3 h to a maximum of 15 h. Repeat the phase separation in step **3** and discard the aqueous phase. Add 30 μl IEF loading buffer to 30 μl detergent phase, mix, and load 30 μl on the gel. Perform IEF as *Protocol 2*.

C. *Electroblotting*

- use either a semidry (e.g. BioRad Trans-blot semidry transfer cell, 170-3940; Hoefer TE70 Semi Phor semidry transfer unit) or tank electroblotting apparatus, (e.g. Hoefer, TE42) according to manufacturer's instructions
- transfer buffer—for nitrocellulose membranes: 25 mM Tris, 192 mM glycine, 20% methanol; for nylon membranes omit the methanol
- Tris-buffered saline (TBS): 0.02 M Tris, 0.5 M NaCl, pH 7.5
- TTBS: TBS with 0.2% Tween 20

1. After IEF dismantle the apparatus and place the gel in a container of transfer buffer; incubate at room temperature for 20 min with gentle shaking. Alternatively if using nitrocellulose filters wash the gel in 50% methanol, 1% SDS, 5 mM Tris-HCl, pH 8.0 prior to blotting. This reduces the NP-40 concentration which may inhibit the binding of proteins to nitrocellulose.
2. Prepare blotting 'sandwich' as described in manufacturer's instructions, placing the membrane on the anode side of the gel. Electroblot for a semidry blotter at 200 mA for 1 h, and tank blot 400 mA for 2 h.

D. *Immunoprobing of filter*

All incubations and washings are performed at room temperature with gentle shaking.

1. Block unbound sites on the membrane by incubating for 1 h in TBS with 5% dried skimmed milk added.
2. Wash the membrane in TTBS for 20 min.
3. Add specific monoclonals, e.g. HC-10 and HC-A2 at 1:1000 dilution of ascites in TTBS with 1% dried skimmed milk. Incubate for 1 h.
4. Wash membrane twice in TTBS for 10 min.
5. Add secondary antibody; goat anti-mouse IgG alkaline phosphatase conjugate (Sigma A4656) diluted 1:1000 in TTBS with 1% dried skimmed milk added. Incubate for 1 h.
6. Wash membrane twice in TTBS for 10 min and once in TBS for 15 min.

E. *Staining membrane*

- alkaline phosphatase buffer: 100 mM Tris, pH 9.3, 100 mM NaCl, 5 mM $MgCl_2$
- nitroblue tetrazolium (NBT) solution: dissolve 250 mg NBT (Sigma, N-6876) in 5 ml, 70% dimethylformamide

Protocol 3. *Continued*

1. For 20 ml substrate solution, dissolve 2 mg 3-indoxyl phosphate (Sigma, I-5505) in 20 ml alkaline phosphatase buffer.
2. Add substrate solution to container containing membrane and add 40 μl NBT solution.
3. Stop the reaction when bands appear by washing in dH_2O.
4. Dry by blotting between filter paper.

2.4 Analysis of HLA class I IEF banding patterns

To simplify assignment of HLA types to bands run control samples on the same gel which share at least one of the HLA specificities of the sample under investigation. The B-LCL cell panel which was characterized for both HLA class I and class II specificities as part of the 10th IHW are ideal control cells to use (25).

In B-LCL the cell surface expression of HLA-C locus molecules is about ten times less than that of HLA-A and -B locus molecules; those that are expressed are recognized by the monoclonal antibodies 4E and W6/32 and can be seen on IEF gels as faint bands. A further complication to interpretation of HLA-C locus bands is that they give two to three bands per allele. These extra bands do not correlate with carbohydrate heterogeneity and their cause is unknown. HLA-C molecules can be identified by both radiolabelling/immunoprecipitation and immunoblotting (26).

IEF has certainly been most useful for defining HLA class I variants, however there are limitations to its application. Class I molecules which differ in their composition of uncharged amino acids or which have equal differences in numbers of charged amino acids will not be detected by this technique. IEF could not detect the elusive HLA-A9.3 variant which is serologically related to A24 but indistinguishable from HLA-A24 by IEF. Subsequent sequencing analysis has shown that the serological heterogeneity is indeed due to A9.3 being a different allele from A24. These alleles differ by two amino acid substitutions which do not affect the pI of the proteins (27).

It is also possible to confuse subtypes of common serological specificities, e.g. the B22 group has been serologically split into three subtypes HLA-B54, 55, and 56. Similarly there are three IEF banding positions observed for cells typed as HLA-B22. Sequence analysis has shown that this group is more heterogeneous and so far variants of both HLA-B55 and -B56 have been detected (28). Of these four alleles two share the same IEF position, thus the three bands observed cannot be assigned to B54, B55, and B56 (see *Figure 1*).

Two alleles have been characterized for each of the serological specificities HLA-A34 and -A66. Both these specificities belong to the A10 family and they are not easily distinguished by serology. IEF analysis of these alleles

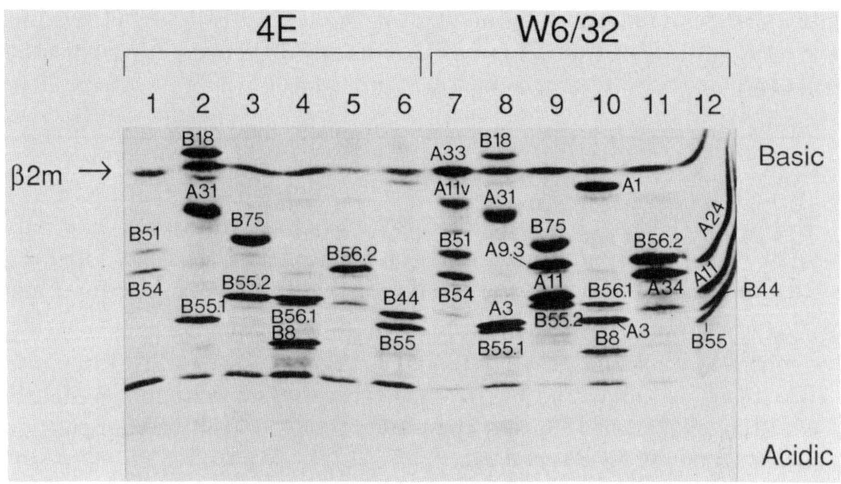

Figure 1. IEF gel showing five subtypes of the HLA-B22 group. Class I molecules immunoprecipitated with the MoAb 4E are present in lanes 1 to 6. Those not precipitated by 4E are precipitated with W6/32 and are shown in lanes 7 to 12. Cell lines used are as follows. Lanes 1 and 7—'TTL': HLA-A11, 33 B51, 54. Lanes 2 and 8—'VEN': HLA-A3, 31 B18, 55.1. Lanes 3 and 9—'APA': HLA-A9.3, 11 B55.2, 75. Lanes 4 and 10—'VOO': HLA-A1, 3 B8, 56.1. Lanes 5 and 11—'ENA': HLA-A34, B56.2. Lanes 6 and 12—'RW': HLA-A11, 24 B44, 55. The subtype of B55 in RW has not yet been defined but it is most likely to be B55.1. TTL expresses a variant of A11 which has yet to be characterized by nucleotide sequencing.

does not aid in the differentiation of A34 from A66 as A*6601 focuses at the same position as A*3401 and A*6602 focuses at the same position as A*3402.

For these examples additional methods for defining class I polymorphism were necessary, and indeed characterization of the nucleotide sequence of these class I alleles has proved to be the most effective and accurate method for defining HLA class I protein polymorphism.

3. Nucleotide sequencing of HLA class I genes

Identification of class I polymorphisms by sequencing is a straightforward technique which any laboratory with minimal molecular biology equipment can perform. Expect to spend four weeks to obtain a sequence.

3.1 Amplification of cDNA by the polymerase chain reaction (PCR)

The methodology for cloning HLA genes follows standard molecular techniques for which Maniatis (29) is an excellent reference. Several methods can be employed to prepare mRNA and cDNA. We prepare mRNA from 100 ml cultures of cells following the method of Bragado *et al.* (30). First strand cDNA is synthesized from 20 μg mRNA using AMV (avian myeloblastosis virus) reverse transcriptase as described in *Protocol 5*.

Protocol 5. First strand cDNA synthesis

1. Resuspend 20 μg mRNA in 16 μl RNase free dH_2O, vortex, and pulse spin. Add 4 μl 1 mg/ml oligo dT (Collaborative Research Inc., d(pT)12–18). Incubate at 55 °C for 3 min followed by incubation on ice for 3 min.
2. Add 0.55 μl RNasin (Promega, N2111), vortex, and pulse spin. Add 8 μl 5 mM dNTPs (Promega, U1240), 4 μl 0.1 M DTT, 4 μl 10 × RTase buffer (Promega, M5301), and 1.5 μl AMV reverse transcriptase (Promega, M5101). Incubate at 42 °C for 45 min, add 1.5 μl AMV reverse transcriptase, and incubate at 42 °C for 45 min.
3. Clean cDNA preparation by adding 60 μl dH_2O to cDNA. Add 100 μl phenol/chloroform, vortex to mix, and leave at room temperature for 4 min, microcentrifuge for 2 min.
4. Transfer aqueous phase to a clean tube and save. Add 100 μl dH_2O to the organic phase, mix, spin as before, and add aqueous phase to that already saved.
5. In a fume hood, add 200 μl dH_2O saturated ether to the aqueous phase, mix by vortexing and microfuge for 2 min, aspirate the upper ether layer with a finely drawn glass pipette, and repeat extraction with another 200 μl dH_2O saturated ether. Heat at 55 °C for 2 min to remove traces of ether.
6. Add 185 μl 4 M ammonium acetate, 850 μl 100% ethanol, and precipitate DNA by freezing at −70 °C for 30 min or −20 °C for 2 h, pellet DNA by microcentrifugation at 4 °C for 15 min, and dry DNA pellet under vacuum for 5–10 min. The DNA pellet is very difficult to see after this precipitation. Resuspend in 50 μl dH_2O.

HLA-A, -B, and -C locus genes are specifically amplified from cDNA (see *Protocol 6*) using pairs of locus specific oligonucleotide primers (see *Figure 2*) which recognize sequences unique to the untranslated region of each locus. The universal 5′ amplification primer (5UT) is from the 5′ region immediately after the initiation codon (31), and the antisense 3′ universal primer (3UT) is from the sequence immediately downstream of the termination codon. The three locus specific 3′ primers are from the middle of the 3′ untranslated region (32). These primers are not always 100% specific and can amplify class I genes from other loci, in particular B locus genes are amplified when using the A and C locus primers.

5UT	5'- GGGCGTCGACGGACTCAGAATCTCCCCAGACGCCGAG -3'	HLA class I universal 5' oligo
3UT	5'- CCGCAAGCTTTCTCAGTCCCTCACAAGGCAGCTGTC -3'	HLA class I universal 3'oligo
3UTA	5'- CCGCAAGCTTTTGGGGAGGGAGCACAGGTCAGCGTGGGAAG -3'	HLA-A locus 3' oligo
3UTB	5'- CCGCAAGCTTCTGGGGAGGAAACACAGGTCAGCATGGGAAC -3'	HLA-B locus 3' oligo
3UTC	5'- CCGCAAGCTTTCGGGGAGGGAACACAGGTCAGTGTGGGGAC -3'	HLA-C locus 3' oligo

Figure 2. PCR primers used to amplify HLA class I genes. A *Sal*I site has been engineered into the 5UT primer and *Hin*dIII sites have been incorporated into the 3′ primers. These sites are underlined. The 3UTA, B, and C primers are individually used with the universal 5UT primer.

Protocol 6. Amplification of HLA class I genes by PCR

Use sterile tips and tubes throughout. Controls should include reactions without primers and reactions without added DNA.

1. To each PCR tube, including controls, add in the following order: sterile double distilled H_2O (calculate volume required so that final volume in tube is 100 μl), 10 μl 10 × reaction buffer (Perkin Elmer, Cetus, N808-0006), 16 μl nucleotide mix (Perkin Elmer, Cetus N808-0007), 3–6 μl cDNA, 300 ng each 5UT and 3UT primers. Mix the contents of each tube by vortexing.
2. Denature the DNA by heating to 94°C in a thermal cycler for 4 min, remove tubes and place on ice, pulse spin to concentrate solution from the sides of the tubes.
3. Add 1 μl *Taq* polymerase (Perkin Elmer, Cetus, N801-0060), mix by vortexing, and add two drops mineral oil (Sigma, M-3516).
4. Replace tubes in thermal cycler and programme as follows:
   24 cycles: denature for 1 min at 94°C, anneal for 45 sec at 65°C, and extend for 2 min at 72°C. Final cycle: 94°C 1 min, 65°C 45 sec, and 72°C 10 min. This programme is effective for amplification with all the primers described.
5. During the PCR run prepare a 1% agarose gel with TBE (0.02 M EDTA pH 8.0, 0.89 M boric acid, 0.89 M Tris).
6. Remove 10 μl from each PCR reaction mixture, add gel loading buffer, and run on agarose gel with size markers, e.g. *Hae*III digest of ϕX174 DNA (United States Biochemicals (USB), 70063). Run 10 μl of previously similarly amplified cDNA as a positive control for comparison.
7. Clean remaining amplified DNA (90 μl). Add 10 μl TE (10 mM Tris-H Cl, 1 mM EDTA, pH 7.6) and 100 μl phenol/chloroform, vortex, and leave at room temperature for 4 min. Microfuge for 2 min and save aqueous phase.

Protocol 6. *Continued*

Re-extract organic phase with 100 μl TE as before, and combine aqueous phases in a clean tube. Add 200 μl chloroform/isoamyl alcohol (24:1), mix, and microfuge for 2 min. Add 20 μl 3 M NaAc pH 5.2 and 400 μl 100% ethanol. Precipitate DNA as in *Protocol 5*.

3.2 Cloning amplified HLA genes into M13

3.2.1 M13, the sequencing vector

The vector commonly used for cloning DNA to be sequenced is the filamentous *Escherichia coli* bacteriophage M13 (33). The replicative form of M13 is a circular double stranded DNA molecule which allows cloning by standard restriction endonuclease methods. M13 mp18 and mp19 are derivatives of M13 constructed specifically to facilitate cloning and sequencing of DNA fragments. They are both 7250 nucleotides long and differ only in the orientation of the 54 base polylinker they carry thus facilitating asymmetric cloning. This polylinker contains restriction enzyme recognition sites including *Sal*I and *Hin*dIII (see *Figure 2*). Insertion of DNA into these sites disrupts the gene encoding β-galactosidase and results in colourless plaques when plated in the presence of IPTG (isopropyl-β-D thiogalactopyranoside) and X-gal (5-bromo-4-chloro-3-indolyl-β-D-galactoside). Using M13 mp18 and mp19, both sense (mp19) and non-sense (mp18) strands of the HLA class I genes are sequenced in opposite directions. Non-recombinant M13 gives blue plaques.

3.2.2 Restriction enzyme digestion

Digest both vector (M13 mp18 and M13 mp19) and HLA class I DNA using an excess of enzyme, but do not allow the volume of enzyme to exceed 0.1 × reaction volume. Enzymes differ in the salt concentration required in their reaction buffers which enable optimum digestion, *Hin*dIII requires a medium salt buffer and *Sal*I a high salt buffer. The optimal salt concentrations vary for enzymes obtained from different manufacturers. Efficiency of digestion is greater when the DNA is phenol extracted and precipitated between digestions (see *Protocol 7*).

Protocol 7. Phenol extraction of DNA

1. Add an equal volume of phenol:chloroform:isoamyl alcohol (25:24:1) to the reaction mix, centrifuge (500 *g*), remove aqueous phase, and save.
2. Re-extract organic phase by adding an equal volume of TE, vortex, centrifuge (500 *g*), and pool aqueous phases.
3. Add 1 × volume chloroform:isoamyl alcohol (24:1), mix, centrifuge (500 *g*), and precipitate DNA from the aqueous phase with 0.1 × volume 3 M sodium acetate pH 5.2, and 2.0 × volumes 100% ice-cold ethanol.

3.2.3 Purification of digested DNA

Double digested DNA samples must be cleaned to prevent re-annealing of digested fragments. Cut DNA can be cleaned using glass beads prepared in the laboratory or bought commercially (Bio 101 Inc., 'Geneclean'). As an extra precaution to prevent self-ligation of the M13 DNA, although *Hin*dIII and *Sal*I sites are not compatible and should not re-ligate, the 5′-phosphate groups are removed (29) with the enzyme calf intestinal alkaline phosphatase (Boehringer, 713023). Re-ligation can only occur when a DNA fragment with intact 5′-phosphate groups is inserted, thus the only circular molecules formed will be recombinants.

3.2.4 Ligation, transformation, and single stranded recombinant M13 preparation

Clean digested M13 and HLA class I DNA are ligated in 100 ng reactions (although higher amounts will work). The molar ratio of M13 (7.25 kb) to HLA (1–1.5 kb) in the ligation reaction is approximately 1:1, therefore for 17 ng HLA DNA, 85 ng M13 is required. Ligation is performed with one unit T4 DNA ligase in a total volume of 20 μl, either at 15 °C overnight or two hours at room temperature. Include controls with cut M13 and uncut M13. Transformation is performed as in *Protocol 8*.

Protocol 8. Transformation

B Broth: 5 g tryptone, 4 g NaCl, in a total volume of 500 ml ddH_2O. To 400 ml add 6 g bactoagar (bottom) and to remaining 100 ml add 0.7 g bactoagar (top). Autoclave and cool to 50 °C, pour 'bottom' agar into sterile Petri dishes, when the agar has set, invert and dry at 37 °C. Maintain 'top' agar at 50 °C until needed.
Also use controls from ligations.

1. Prepare competent *Escherichia coli* (JM109 or DH5F′α) cells according to the Standard Transformation Protocol of Hanahan (34).
2. Mix 200 μl competent cells with 10 μl from the ligation mix and incubate on ice for 40 min.
3. Heat shock the cells by placing in a 42 °C water-bath for 90 sec and immediately return to ice.
4. Add 3 ml top agar to 15 ml polypropylene tubes (two tubes per sample), maintain at 42 °C in a water-bath.
5. To each of these tubes add 25 μl X-gal (20 mg/ml), 25 μl IPTG (20 mg/ml), 200 μl exponentially grown JM109 cells (not made competent), and one of two aliquots of transformed cells, either 40 μl or 170 μl. Mix the contents and immediately pour on top of previously prepared agar plates.

Protocol 8. *Continued*

6. Allow top agar to solidify at room temperature, invert, and incubate at 37°C until plaques appear, usually 12–15 h.
7. Store plates at 4°C.

Single stranded recombinant M13 are prepared from colourless plaques formed by infected JM109 cells by following *Protocol 9*.

Protocol 9. 'Mini-preparation' of single stranded M13 template

- PEG/NaCl: 20% carbowax, polyethylene glycol 8000 (Fisher Scientific, P155-500), 2.5 M NaCl
- 2 × YT: 16 g bactotryptone (Difco, 0123-01-1), 10 g yeast extract (Difco, 0127-01-7), 5 g NaCl, dH_2O to a final volume of 1 litre

1. Pick a colony of JM109 from a minimal media plate and grow overnight shaking at 37°C in 2 ml 2 × YT. This can be set up on the day of the transformation.
2. The next day, pick single colourless recombinant M13 plaques from the transformation plates by pressing the end of a sterile yellow pipette tip into the colony. Drop pipette tip into a 15 ml centrifuge tube containing 1.5 ml 2 × YT. Add 20 μl from the JM109 culture prepared in step **1** and incubate 10 to 15 h shaking at 37°C.
3. Decant culture into sterile microcentrifuge tubes and pellet bacteria by microfugation for 5 min. Transfer 1.2 ml supernatant to a clean tube, reserve the remainder in a second tube, and store at −70°C. This is the stock supernatant which will be used for future M13 preparations (see *Protocol 13*).
4. Add 300 μl PEG/NaCl solution to 1.2 ml supernatant, mix by vortexing thoroughly for 7 sec. Precipitate the phage particles at room temperature for 15 min.
5. Pellet the phage by microfugation for 10 min. The pellet should be visible but will be small. Discard supernatant and microfuge for 1 min. Discard remaining supernatant by aspirating through a finely drawn glass pipette.
6. Add 100 μl TE and dissolve pellet.
7. Add 50 μl Tris-saturated phenol, mix by vortexing, leave at room temperature for 4 min, and microfuge for 4 min.

8. Transfer 85 μl of the aqueous phase to a clean tube, taking care not to remove any of the organic phase.
9. Add 200 μl dH_2O saturated ether, vortex, and microfuge for 2 min. Aspirate upper ether phase and repeat three times.
10. Add 7 μl 3 M sodium acetate pH 7.5 and 200 μl 100% ethanol. Precipitate DNA as in *Protocol 5*.
11. Resuspend DNA in 18 μl TE.

To verify that the extracted M13 DNA does contain insert, 3 μl of single stranded M13 DNA is run on a 1% agarose gel in TBE with single stranded M13 as a control. Recombinant M13 are identified by their reduced mobility compared to single stranded M13, and these clones are characterized further by sequencing their first 200 base pairs. This also allows identification of the insert as either HLA-A, -B, or -C locus by identifying locus specific nucleotides (see *Figure 3*). Alleles from the same locus can be differentiated by comparing the sequences to published data for closely related alleles (35).

3.3 Annealing and sequencing reactions

3.3.1 Sequencing primers

To completely sequence a HLA class I gene five sequencing primers are used for each clone; one is the universal M13 primer (USB) and the other four are HLA class I specific primers (see *Figure 4*). These primers are used at 0.5 pmol/μl (300 ng). Gels are run for 'short' and 'long' time periods with the same reactions in order to get overlapping sequences for the five primers, and ten independent fragments of sequence will be obtained per clone. From the 'short' gel, sequences up to 200 bp from the primer can be read, whereas for the 'long' gel sequences begin approximately 70 bp after the primer and continue for another 150–200 bp.

The four HLA specific primers correspond to relatively non-polymorphic regions in exons two, three, and four. The sequencing protocol is designed to obtain overlap of all fragments sequenced and therefore any polymorphism present in these areas will be detected. In order to sequence the first 200 bp and to identify locus and allelic specificity of clones, any of the primers for exons two and three, where the majority of class I polymorphisms are located, may be used. Generally we use the universal M13 primer for mp19 clones and the 2N HLA specific primer for mp18 clones and sequence within exons one and two. Follow *Protocol 10* for the annealing reaction.

Nucleotide	4	5	6	36	61	175	198	200	206	226	228	270	277	283
HLA-A	G	C	C	A	C	C	C	A/G	T	A	A	T/A	G	T
HLA-B	C	G/T	G	G	G	A	T	C	A/T/C	A	A	C*	A/G*	A/G
HLA-C	C	G	G	G	G	C	T	C	G	G	G	G/C	C	G

Nucleotide	299	345	468	485	616	621	627	637	744	756	774	777	783	787
HLA-A	T/A/C	T	T	T	A	C	C	A	G	G	G	T	G	G
HLA-B	A*	G	C	C	G	C	A/T	G	A	T/C	A	A	A	A
HLA-C	T	G	C	C	G	A	A	G	A	C/T	G	A	A	G

Nucleotide	831	846	855	861	867	873	874	910	933	934	947	949	954	956
HLA-A	G	C	G	T	T	C	A	C	C	A	T	C	T	G
HLA-B	A	A	A	T	G	G/A	A	T	T	G	C	G	A	C
HLA-C	A	G	G	C	G	G/A	G	C	C	G	C	G	G	C/T

Nucleotide	957	959	960	961	962	963	965	970	972	999	1016	1018	1034
HLA-A	A	-	-	-	C	T	T	A/G	T	G	A	A	A
HLA-B	A	-	-	-	T	T	T	A	C	T	G	G	A
HLA-C	T	T	C	C	T	A/G	C	C	A/T	T	G	G	G

Nucleotide	1047	1048	1053	1054	1076	1085	1086	1092	1093	1094	1095	1096
HLA-A	A	A	T	G	T	C	A	T	A	A	A	G
HLA-B	G	T	C	G	T	C	A	A	-	-	-	-
HLA-C	G	T	C	A	A	T	C	T	A	A	A	G

Nucleotide	1097	1098	1099	1100	1101
HLA-A	T	G	T	G	A
HLA-B	-	-	-	-	-
HLA-C	C	C	T	G	A

Figure 3. HLA-A, -B, and -C locus specific nucleotides. Nucleotide positions at which either one of HLA-A, -B, or -C alleles can be differentiated are given. '-' denotes the insertion of spaces to enable comparisons of sequences from different loci. HLA-B46 differs from other -B locus alleles at nucleotides indicated '*' as it has an identical α1 helix to HLA-Cw1 due to an intergenic conversion.

A.

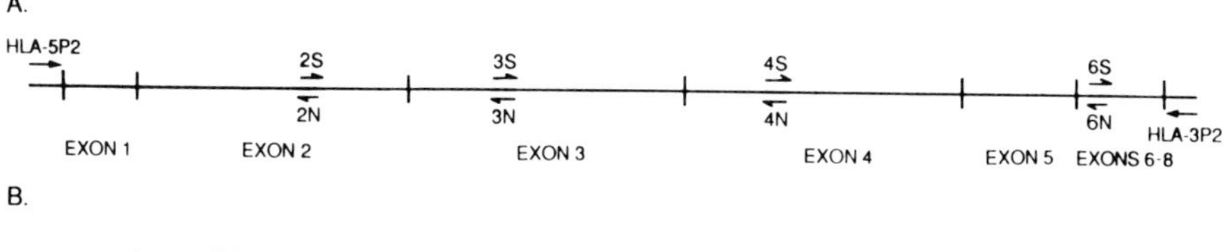

B.

sequencing oligonucleotides		
2S	exon 2 236–255	AGGGGCCGGAGTATTGGGAC
2N	exon 2 236–255	GTCCCAATACTCCGGCCCCT
3S	exon 3 429–450	CGGCAAGGATTACATCGCCCTG
3N	exon 3 429–450	CAGGGCGATGTAATCCTTGCCG
4S	exon 4 703–724	GCGGAGATCACACTGACCTGGC
4N	exon 4 703–724	GCCAGGTCAGTGTGATCTCCGC
6S	exon 6 1023–1043	AGGGAGCTACTCTCAGGCTGC
6N	exon 6 1023–1042	TGCAGCCTGAGAGTAGCTCCCT

Figure 4. A. Schematic of a HLA class I cDNA PCR product showing PCR primers (large arrows), sequencing primers (small arrows), and exon boundaries (vertical lines). Arrowheads are at the 3′ hydroxyl end of each oligonucleotide primer and point in the direction of polymerase extension. B. Sequences of sequencing oligonucleotides. S oligonucleotides are derived from the sense strand and are used with M13 mp19 clones. N oligonucleotides are derived from the anti-sense strand and used with M13 mp18 clones. Adapted from reference 31.

Protocol 10. Annealing reaction

1. For each clone to be screened, label a screw cap microcentrifuge tube.
2. Prepare annealing cocktail of $2(n + 1)$ μl 5 × reaction buffer (200 mM Tris-HCl pH 7.5, 100 mM $MgCl_2$, 250 mM NaCl) and $(n + 1)$ μl primer where n = number of clones to be screened. Mix 7 μl DNA and 3 μl annealing cocktail gently with pipette tip taking care not to introduce any bubbles.
3. Place tubes in a plastic or foam rack and submerge in a beaker filled with 72°C water. Place a second breaker also containing 72°C water on top of the tubes to ensure the tubes are completely submerged.
4. After 2–5 min switch off water-bath and allow temperature to drop to 30°C or less by placing beaker containing tubes on the bench, this takes about 30 min.
5. Pulse spin tubes to concentrate solution from sides of tubes, either use immediately (keep on ice) for sequencing or store at −20°C for up to seven days.

3.3.2 Chain-termination sequencing

A DNA strand is synthesized from the site at which the sequencing primer anneals by a DNA polymerase using the single stranded recombinant M13 template. This synthesis is terminated when a 2′,3′-dideoxynucleoside-5′-triphosphate (ddNTP) is incorporated. This nucleotide will not support further DNA elongation due to a missing 3′-OH group. Each of the four termination mixes contains either ddGTP, ddATP, ddTTP, or ddCTP plus all four dNTPs. Together the complete sequence of a DNA fragment is obtained. We use a 96 well plate for the sequencing reactions (see *Figure 5*). The labelling reaction occurs in wells labelled '*'. First the primer is extended with limited concentrations of dNTPs including [^{35}S]dATP; this step continues to completely incorporate all radiolabelled dATP into DNA chains of various lengths. In the 'GATC' wells, the concentration of dNTPs is increased, synthesis continues, and every so often a ddNTP is incorporated, e.g. ddGTP in well 'G', until all growing chains are terminated by a ddNTP. The reactions are stopped by the addition of stop solution (95% formamide, USB). Each of the four GATC reactions is loaded on a single track in the sequencing gel. A ladder of bands is obtained from each reaction where each band defines a fragment which ends with a ddNTP.

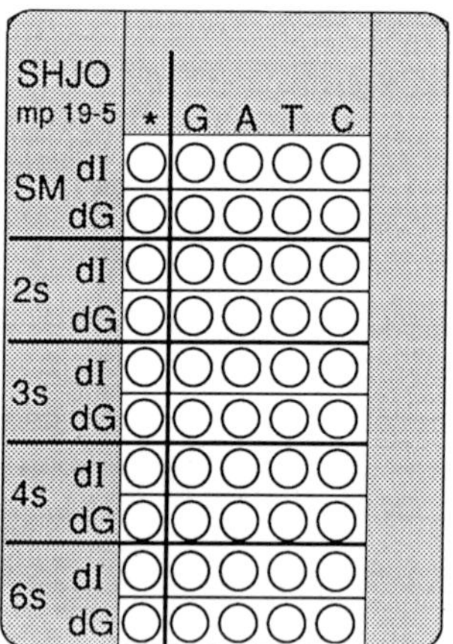

Figure 5. A diagrammatic representation of a 96 well microtitre plate which is used to perform the sequencing reactions. Five wells per sample are required. Unused wells are covered with tape and used to label the wells in use. Here the plate is designed for the sequencing of an mp19 clone, number 5 from the cell line SHJO. The five sequencing primers: SM (universal M13 sequencing primer), 2S, 3S, 4S, and 6S (HLA specific primers, see *Figure 4*) are indicated. Both dITP and dGTP reactions will be performed for each annealing.

Protocol 11. Sequencing reactions

- n = number of reactions (i.e. annealings)
- mix solutions gently with pipette tip and avoid introducing bubbles

1. Prepare and label a 96 well plate (see *Figure 5*).
2. On ice, label three microcentrifuge tubes; 'A cocktail', 'B dilute labelling mix', and 'C dilute enzyme'.
3. Add $0.5n$ μl labelling mix (USB) and $2n$ μl dH_2O to tube B, mix with pipette tip.
4. Add $n + 1$ μl 0.1 M dithiothrietol (DTT) to tube A.
5. Add $2(n + 1)$ μl B to A, mix with pipette tip.
6. Add $0.5(n + 1)$ μl [α-^{35}S]dATP to A and mix.
7. Add $7n/3$ μl TE and $n/3$ μl Sequenase 2.0 (USB) to C, mix. Remove $2(n+ 1)$ μl and add to A, mix.
8. Add 10 μl annealing mix to wells marked '*'.
9. Add 2.5 μl termination mix (USB) to wells 'GAT and C', (e.g. add to G well G etc.).
10. Start timer for a 10 min count.
11. Add 5.5 μl cocktail (tube A) to DNA in each well labelled '*', mix three times with pipette tip.
12. After 6 min, remove 3.5 μl from well labelled '*' and add to the side of well 'G', taking care not to touch termination mix. Repeat for wells 'A, T, and C'. Complete within 10 min.
13. Mix contents of wells GAT and C by centrifugation using buckets specifically made to carry the 96 well plates. Bring speed up to 800 g then switch centrifuge off.
14. Carefully float plates in a 37°C water-bath for 5 min.
15. Remove plates, add 4 μl stop solution to the side of each GAT and C well using an automatic dispensing pipette. Mix as before (step **13**). Store plates at 4°C for up to seven days prior to running the sequencing gels, although in order to obtain the cleanest banding patterns the reactions should be analysed as soon as possible.
16. Denature the DNA by heating the plates in an 80°C oven for 5 min. Place plates on ice and load sequencing gels immediately.

Urea acrylamide gels are mostly used for the sequencing electrophoresis but more recently Hydro-Link 'Long-Ranger' gel solution (At Biochem) has been used. This comes as a 50% liquid concentrate containing a 'chemically modified acrylamide monomer with a novel cross-linker'. By using Long-Ranger, we are able to reduce the time of electrophoresis from three to two hours for the 'short' gel runs, and from seven to three hours 15 minutes for the 'long' gel runs. The resolution with these gels is excellent and more sequence information is obtained in the 'long' gel runs compared to that obtained using acrylamide, thus a greater overlap in the sequences is obtained.

Protocol 12. Preparation and running sequencing gels

1. Wash glass plates with scouring powder, (e.g. 'Comet', 'Vim', 'Ajax'), rinse in dH_2O, and wash again with alcohol.
2. Siliconize one side of one of the two sequencing gel plates. In a fume hood pour 5–10 ml siliconizing agent, e.g. 'SurfaSil' (Pierce, 42801) on the plate, spread evenly, and allow to dry for 5 min. Rinse with alcohol. Keep this side labelled.
3. Arrange gel sandwich according to manufacturer's instructions, or if using a home-made apparatus insert clean spacers along sides in between the two plates. Seal the sides with electrical tape and then seal the bottom. Hold plates together and in place by using one inch Bulldog (binder) clips at the sides.
4. Gel solution for one gel (35 × 41 cm plate and 35 × 39 cm plate). Standard gel: 34.7 g urea, 11.25 ml 40% acrylamide (19:1), 7.5 ml 10 × TBE, 20 ml ddH_2O. Long-Ranger gel: 31.5 g urea, 7.5 ml Long-Ranger, 9 ml 10 × TBE, 20 ml dH_2O.
5. (a) Dissolve urea by warming solution whilst stirring.
   (b) Increase volume to 75 ml dH_2O for both acrylamide and Long-Ranger gels and filter through a Whatman no. 1 filter paper.
   (c) Add 90 μl TEMED and 90 μl 25% APS for acrylamide gels. For Long-Ranger gels add 375 μl 10% APS and 37.5 μl TEMED.
   (d) Pour immediately holding plates vertically on one corner. Using a 60 ml plastic syringe (no needle is required) inject the gel down the side of the glass plate leading to the corner resting on the bench. Lower opposite corner allowing the acrylamide to cross the bottom. As the acrylamide level rises lower the plates (still pouring) to an almost horizontal position, the top of the gel should eventually rest at a slight incline to the bench (place a Bulldog clip underneath the top of the gel). Take care not to introduce bubbles. Steady hands, practice, and a uniform flow should ensure this.
6. Insert the flat side of the combs (opposite side to teeth) and leave to polymerize (about 1 h). The gel can be left at room temperature overnight.

7. Remove combs, rinse top of the gel with dH_2O. Remove tape from the bottom and clips from the side.
8. Cover the top of the gel with 1 × TBE and carefully insert sequencing combs (Bethesda Research Laboratories, Sharkstooth combs). Teeth should only enter the gel 1–2 mm. Clamp combs to larger plate using mini Bulldog clips. These combs stay inserted throughout the gel run. Remove tape from the bottom of the gel.
9. Connect plates to the gel stand using minimal amount of grease and Bulldog clips to hold in place. Add 600 ml TBE to the top tank for acrylamide gels and check there are no leakages before adding 400 ml TBE to the lower tank. For Long-Ranger gels the manufacturer's protocol suggests using 0.6% TBE in both buffer tanks, however more recently we have found better resolution with undiluted TBE, and this modification only increases the time for electrophoresis by about 5 min.
10. Rinse wells with TBE from the top tank, ensuring that there are no bubbles in any of the wells.
11. Run acrylamide gels 55 mA for 30 min before loading gels. For Long-Ranger gels run at 45 mA.
12. Load samples in groups of four, keeping the same order, e.g. GATC. It is important not to pause in the middle of one group. Four groups are probably the maximum to load in one step. Loading quickly and correctly takes practice and it is easier to start by loading two groups at a time. Allow samples to run into the gel in between loading groups. Never load samples when the power is switched on. Loading a 96 well sequencing gel (24 sequencing reactions) can take 30–60 min.
13. When using acrylamide gels add 200 ml 3 M sodium acetate (in TBE) to the lower tank buffer of 'short' gel runs so that the final concentration is 1 M sodium acetate. Run acrylamide gels at 55 W for 3 h for a short run and 7 h for a long run. Run Long-Ranger gels at 45 W for 2 h or whenever the first dye front reaches the bottom of the gel for 'short' gel runs, and for 3 h 15 min for 'long' gel runs.
14. After the gel run carefully dismantle the apparatus. The gel ought to stay attached to the non-siliconized glass plate and this is placed in a large fibre glass tray gel face up. Both acrylamide and Long-Ranger gels are treated identically. Slowly add fixing solution (10% methanol, 10% acetic acid) to the tray gradually covering the gel. Leave for 10 min without any shaking. Carefully remove glass plate and allow excess fix solution to run off the plate. Slowly lower a sheet of dry Whatman 3MM paper on the gel and remove the gel from the glass plate by inverting it.
15. Dry with heat (80 °C) in vacuum slab drier, (e.g. BioRad), and expose to X-ray film in cassettes with intensifying screens for 24 to 48 h at room temperature.

The sequences are examined by hand and clones of interest are identified as described above. DNA is prepared from clones containing the desired sequences by following the maxi-prep procedure in *Protocol 13* in order to obtain sufficient DNA to fully sequence the gene.

Protocol 13. 'Maxi preparation' of ss M13 DNA for full insert sequencing

1. Add 10 μl supernatant from desired clone (from *Protocol 9*) to 10 ml 2 × YT in a 50 ml Falcon tube. Add 100 μl JM109 culture (from *Protocol 9*) and grow 10 to 15 h shaking at 37°C.
2. Decant culture into 13 ml centrifuge tube (Sarstedt, 60.540) and pellet bacteria by centrifugation at 7500 r.p.m. for 10 min (Beckman JA20 rotor).
3. Transfer supernatant to a clean tube. Add 2.5 ml PEG/NaCl solution and follow steps **4–5**, *Protocol 9*, except centrifuge firstly at 12 000 r.p.m. and second at 10 000 r.p.m. (Beckman JA20 rotor).
4. Resuspend phage pellet in 600 μl TE and transfer to a microfuge tube.
5. Add 300 μl phenol and follow step **7**, *Protocol 9*.
6. Remove 575 μl aqueous phase and ether extract (600 μl) as step **9**, *Protocol 9*.
7. Split TE solution into two microcentrifuge tubes and add 15 μl 3 M sodium acetate pH 5.2 and 700 μl 100% ethanol, precipitate DNA as in *Protocol 5*.
8. Resuspend DNA in 85 μl TE. Calculate DNA concentration. Usually between 150 to 500 ng/μl DNA is obtained, if the concentration is greater than 250 ng/μl, dilute with TE.

When sequences containing many dG and dC are not fully denatured, 'compression' or larger spacing between bands can occur. Therefore we run both dITP, a dGTP analogue, which sharpens bands and eliminates compressions, and dGTP reactions in parallel. However for the initial screening sequencing only dGTP reactions are performed. Since both dGTP and dITP reactions are run for full length sequencing, double the quantity of reagents for each annealing reaction.

3.4 Sequencing analysis

Sequencing autoradiograms can be read either by hand, automatically by machine, or with the help of a digitizer linked to a computer. We use a VAX 8550 running VMS version 5.2 and the GCG programme version 7 (36). The nucleotide sequence is read from the bottom of the gel to the top. Read each sequence at least two times to avoid incorporating reading errors.

There are areas in the nucleotide sequence which are more difficult to read than others. These areas, which may contain gaps and compressions are found in regions rich in the nucleotides dGTP and dCTP. Compressions formed by the secondary structure of the DNA can go unnoticed (see *Figure 6*). This is alleviated by running dITP reactions and also by sequencing genes in both directions, i.e. sequencing both mp18 and mp19 clones per allele. However as dITP reactions may accentuate pauses made by the Sequenase enzyme at sites of different secondary structure causing gaps to appear, always run dITP and dGTP reactions in parallel.

Assemble the ten nucleotide sequences obtained for each clone together by joining overlapping sequences. Correct, or verify, regions of non-identity within the overlapping regions by returning to the original autoradiograms and re-reading the area in question by hand. If a correction is necessary, incorporate this into the assembly of overlapping sequences. Once the ten fragments are correctly assembled and overlapping, a consensus sequence is derived. 'Line-up' the consensus sequences obtained for each clone, remembering to reverse the sequences obtained for anti-sense mp18 clones. A table is generated whereby each sequence forms one line, and each nucleotide represents a column. Regions of identity are found within the same columns. As before, correct or verify areas where there are mismatches between the consensus sequences of the clones being compared by checking back to

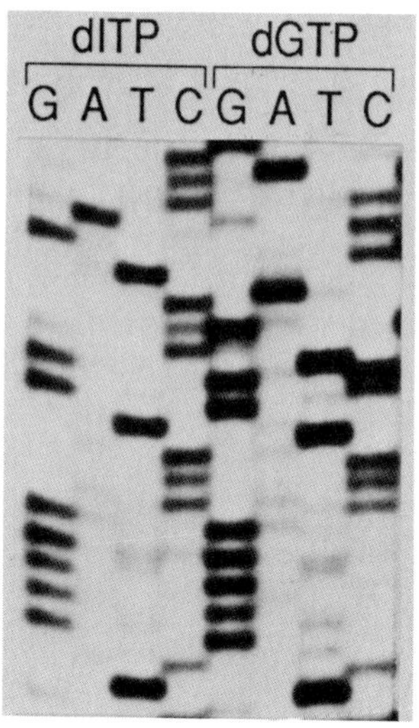

Figure 6. Recognition of a sequencing compression. Both dITP and dGTP sequences are shown for nucleotides 40 through 60 in exon 1 for the A*2402 sequence, from the cell line SHJO. The universal M13 −20 sequencing primer was used.
dITP sequence (from *bottom* to *top*): -TCGGGGGCCCTG GCC CTGACC-
and the dGTP sequence: -TCGGGGGCCCTG CG CTGACC-
In this example the triplet of dCTP nucleotide in the dGTP reactions has been compressed.

autoradiograms. Misincorporations which have occurred during the PCR are identified at this point, i.e. if one of the clones has a different nucleotide at a position where all other clones are identical this would be attributed to a misincorporation by the DNA polymerase during the PCR reaction. It is usually necessary to sequence three to six clones per allele in order to obtain an accurate consensus sequence. Once this sequence is obtained, compare it to other sequences. All published and WHO recognized HLA class I sequences are available through GenBank. No corrections to the consensus sequence should be necessary at this stage. If corrections are necessary, e.g. due to mis-alignments of sequences as a result of a nucleotide insertion or deletion, these are more than likely to be the result of mishandling the data and care must be taken to thoroughly check autoradiograms and sequences making sure not to introduce bias from comparisons with known sequences.

3.4.1 Analysis of nucleotide sequences

Comparison of nucleotide sequences permits identification of alleles most closely related to the new allele. These alleles may not encode the most related serological specificity, e.g. comparison of the nucleotide sequences for five B22 alleles; B*5401, B*5501, B*5502, B*5601, and B*5602 with 29 other B locus sequences (28) demonstrated this group of alleles to be most closely related to each other, as would be expected from the serological data. However the next most related allele to the B22 group is the non-serologically related B*7801 and not a member of the B7 cross reacting group with which B22 specificities are associated by serology. This is not surprising as the serological cross reactions between the B22 group, B7 and B42 is probably due solely to the α1 helices which are identical in amino acid sequence.

It is necessary to compare patterns of nucleotide substitutions within a large group of alleles. If the same pattern of nucleotide substitution is observed in other alleles this suggests that segmental exchange has occurred where a stretch of nucleotides has been replaced by the homologous region from a second allele. Such segmental exchanges are common amongst class I alleles, e.g. the five subtypes so far defined for the HLA-B22 group appear to have evolved from a common ancestor by simple segmental exchange. Donor alleles for these events, except for B*5401, can be found in other HLA-B locus alleles. B*5401 most likely evolved via an intergenic conversion event between B*5502 and C*0101. This is the second example of an intergenic conversion, the first being the formation of B*4601, which is identical to B*1501 (B62) except for the α1 helix which has been replaced with the α1 helix from C*0101 (37).

Some nucleotide substitutions are silent, i.e. they do not cause a change in the amino acid encoded by the codon. The accumulation of silent substitutions is a reflection of the age of the allele. The older an allele the more likely it is to have more silent substitutions, and comparison of the number of silent substitutions within and between species is one method of defining the age of alleles.

3.4.2 Analysis of protein sequences obtained

Nucleotide sequences are translated into protein sequences and again these sequences are lined-up and compared with other protein sequences. Amino acid substitutions which make the 'new' class I allele unique are defined and their probable effect on antigen presentation are analysed. When choosing sequences to compare, the obvious choices are sequences from related alleles such as those which cross react serologically, e.g. B75 was initially compared with the sequences of B62, B35, B72, B46, and B53. Very few amino acid substitutions in class I molecules are unique, i.e. usually another allele can be found with the same substitution but which has other additional non-identical substitutions. However there are exceptions, e.g. HLA-A*2901 which has a glutamine at position 63 not observed in other human class I HLA-A, -B, or -C sequences, all of which possess either a glutamate or an asparagine residue.

Conserved residues include those which are important for the 3D conformation of the class I HC polypeptide backbone and its association with β_2-m, e.g. those involved in the formation of disulphide bonds: cysteine 101 and 164 in the α2 domain, and 203 and 259 in the α3 domain. Asparagine 86 is also invariant as are the adjacent residues glutamine and serine which together form the site acceptor for N-linked glycosylation. Further descriptions of invariant and variant residues are reviewed in reference 1.

The 3D X-ray crystal structure of HLA-A2 determined at 3.5Å resolution and refined at 2.5Å has provided invaluable information on the location and probable function of polymorphic residues in the class I molecule (38). The majority of residues which contribute to class I polymorphism are localized in or around the peptide binding cleft, the sides of which are formed by two α helices which lie on top of a β pleated sheet platform. Residues which affect both allorecognition and T cell recognition have been found at the bottom of the peptide binding cleft supporting the notion that peptide binding is important, for both self and nonself recognition.

Serological and MoAb epitopes can be mapped through the identification of regions unique to a group of cross reacting alleles. For the five members of the HLA-B22 group, which has recently been characterized by sequencing analysis (28), the α1 helix was shown to be identical to HLA-B7 and -B42 which most certainly accounts for their serological cross reactivity. In order to compare differences between related alleles and to understand how these may affect function, alleles can be compared to one another using a modified version of the ribbon diagram of the α1 and α2 domains of the HLA-A2 structure (see *Figure 7A* and *7B*). In *Figure 8* residues which are responsible for variability within the five alleles so far defined for the B22 group are shown.

Six pockets A to F have been identified within the peptide binding site (38). These pockets may serve to interact with side chains from the bound peptide. Residues which contribute to these pockets are described in *Table 1*. Polymorphic residues which lie within the peptide binding cleft can be further

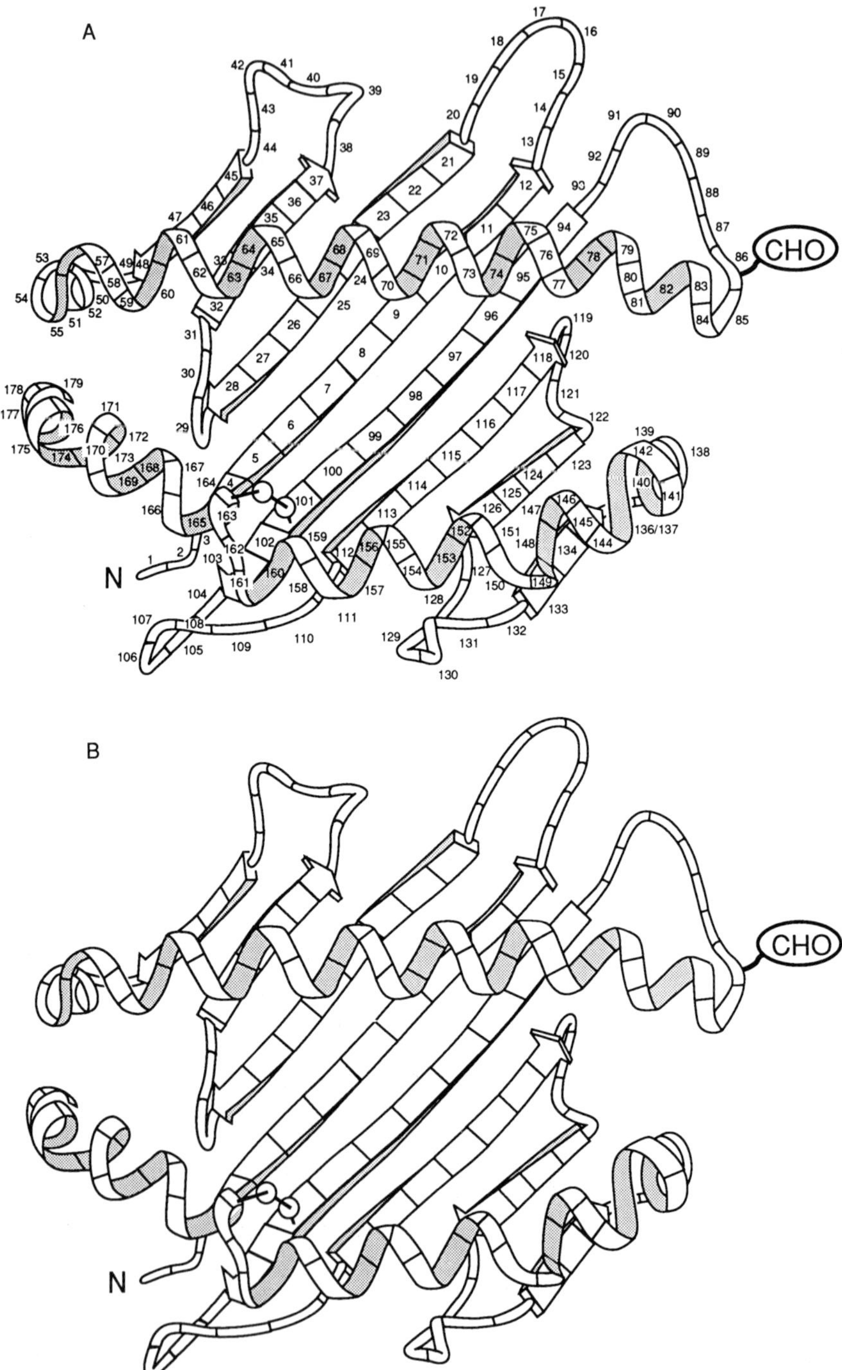
A
CHO
N
1
2
3
4
5
6
7
8
9
10
11
12
13
14
15
16
17
18
19
20
21
22
23
24
25
26
27
28
29
30
31
32
33
34
35
36
37
38
39
40
41
42
43
44
45
46
47
48
49
50
51
52
53
54
55
57
58
59
60
61
62
63
64
65
66
67
68
69
70
71
72
73
74
75
76
77
78
79
80
81
82
83
84
85
86
87
88
89
90
91
92
93
94
95
96
97
98
99
100
101
102
103
104
105
106
107
108
109
110
111
112
113
114
115
116
117
118
119
120
121
122
123
124
125
126
127
128
129
130
131
132
133
134
136/137
138
139
140
141
142
144
145
146
147
148
149
150
151
152
153
154
155
156
157
158
159
160
161
162
163
164
165
166
167
168
169
170
171
172
173
174
175
176
177
178
179
B
CHO
N

Figure 7. Ribbon diagrams depicting the backbone polypeptide structure of the α1 and α2 domains of the class I HLA molecule. Each segment corresponds to an amino acid which are numbered in A, but left blank in B. Copies of B can be used to shade in residues that are critical for serological epitopes, involved in peptide interactions, distinguish particular alleles, or other functions. An example is shown in *Figure 8*. CHO denotes the site of carbohydrate attachment at position 86. Adapted from Figure 2a in reference 1.

characterized as to which pockets they interact with, and the possible effect this may have on the nature of peptides which bind to the class I molecule.

Using *Table 1*, amino acids from the allele under study can be assigned to each position within the six pockets. These assignments can be compared with other alleles. The size of the amino acid side chains and how these may affect peptide binding has to be considered. The presence of charged amino acids may also affect peptide binding, e.g. a characteristic of B22 alleles is the predominance of neutral specificity pockets, suggesting that hydrophobic and nonpolar peptides may interact more favourably with the B22 alleles than with other HLA alleles which have more charged pockets such as HLA-B27.

B22 Variability

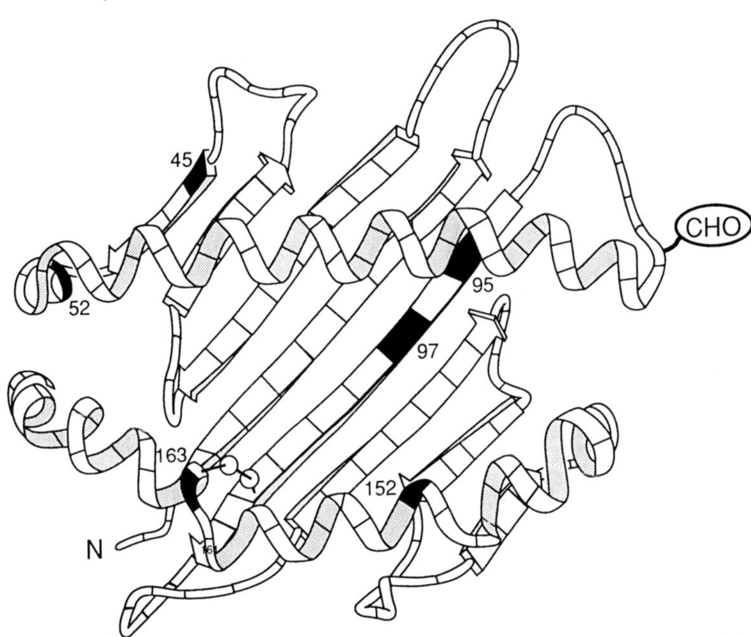

Figure 8. Ribbon diagram depicting all the positions of substitution found for the five alleles of the B22 family: B*5401, B*5501, B*5502, B*5601, and B*5602.

Table 1. Amino acids which contribute to the six specificity 'pockets' located in the peptide binding cleft. Polymorphic residues are indicated in bold

	Peptide binding cleft 'pockets'					
Location	**A**	**B**	**C**	**D**	**E**	**F**
Surface	59, 63, 66, 99, 159, **163**, 167	7, 63, 66 **70, 99**	**70**, 73, **97**	**99, 114, 155, 156** 159	**97, 114,** 147, **152, 156**	**77, 80,** 84, 143, 146, 147
Inside and bottom	5, 7, 171	**9**, 23, **24,** 34, **45, 67**	**9, 74**	113, 160	133	**116**, 123

Although the majority of polymorphic residues are located in the α1 and α2 domains, substitutions do occur in other domains and these should not be overlooked in the analysis. Compared with the α1 and α2 domains the α3 domain is relatively conserved. HLA-A*6801 and -A*6802 both have valine instead of alanine at position 245 in the α3 domain which has been shown to affect CD8 binding and presumably interactions with TCR (39).

4. Class I typing—the future

Routine class II typing methods based on comparisons of nucleotide sequences are increasingly being used in tissue typing laboratories. Some laboratories have reported success with class I typing although the analysis is at an early stage and still dependent on serological data.

Class I DNA typing has to involve methodology which will allow the identification of new alleles, particularly as more ethnic groups come under scrutiny. Using available class I sequences, oligonucleotide probes can be generated to cover the polymorphic α1 and α2 domains where the majority of substitutions affecting function are found. Analysis of class I oligotyping data will not be too different from analysis of serological data. Instead of comparing the reaction patterns of cells with antisera for identification of individual antigenic specificities, patterns of reactivity of class I genes with oligonucleotides will be compared to identify each class I allele. Broad patterns of reactivity will be defined and further characterized into narrower specific allelic assignments, e.g. if the DNA being 'typed' reacted with oligonucleotides defining the α1 helix of B22 alleles it would also react with oligonucleotides defining the α1 helix of HLA-B7 and -B42. If this gene was B*5602, then it would be distinguished from the other B22 alleles by failing to react with an oligonucleotide covering nucleotides 351 through 369, in which region B*5602 differs from the other B22 alleles by six nucleotides, resulting in a two amino acid difference in the peptide sequence.

Ideally oligonucleotide probes would be generated to cover the complete class I gene. The number of different probes required to cover nucleotides encoding the α3, transmembrane, and cytoplasmic domains would be less than that required for the α1 and α2 domains due to the reduced variation in these regions. This approach would maximize the discovery of the many class I alleles which so far have escaped detection.

Acknowledgements

We thank Monica P. Belich, William H. Hildebrand, Alejandro J. Madrigal, and Jacqueline Zemmour for helpful comments and technical contributions.

References

1. Bjorkman, P. J., Saper, M. A., Samraoui, B., Bennett, W. S., Strominger, J. L., and Wiley, D. C. (1987). *Nature,* **329,** 506.
2. Bjorkman, P. J. and Parham, P. (1990). *Ann. Rev. Biochem.,* **59,** 253.
3. Lopez de Castro, J. A. (1989). *Immunol. Today,* **10,** 239.
4. Castano, A. R., Lauzurica, P., Domenech, N., and Lopez de Castro, J. A. (1991). *J. Immunol.,* **146,** 2915.
5. Biddison, W. E., Ward, F. E., Shearer, G. M., and Shaw, S. (1980). *J. Immunol.,* **124,** 548.
6. Opelz, G., Mytilineos, J., Scherer, S., Dunckley, H., Trejaut, J., Chapman, J., Middleton, D., Savage, D., Fischer, O., Bignon, J.-D., Bensa, J.-C., Albert, E., and Noreen, H. (1991). *Lancet,* **338,** 461.
7. Fleischhauer, K., Kernan, N. A., O'Reilly, R. J., Dupont, B., and Yang, S. Y. (1990). *New Engl. J. Med.,* **323,** 1818.
8. Sanfilippo, F., Vaughn, W. K., Light, J. A., and LeFor, W. M. (1985). *Transplantation,* **39,** 151.
9. Kato, K., Trapani, J. A., Allopenna, J., Dupont, B., and Yang, S. Y. (1989). *J. Immunol.,* **143,** 3371.
10. Trapani, J. A., Mizuno, S., Kang, S. H., Yang, S. Y., and Dupont, B. (1989). *Immunogenetics,* **29,** 25.
11. Gilliland, B. C. (1987). In *Harrison's Principles of Internal Medicine,* 11th Edition, p. 1434. McGraw Hill, New York.
12. Hill, A. V. S., Allsop, C. E. M., Kwiatkowski, D., Anstey, N. M., Greenwood, B. M., and McMichael, A. J. (1991). *Lancet,* **337,** 640.
13. Belich, M. P., Madrigal, J. A. Hildebrand, W. H., Zemmour, J., Williams, R. C., Luz, R., Petzl-Erler, M. L., and Parham, P. (1992). *Nature*, **357,** 326.
14. Watkins, D. I., McAdam, S. N., Liu, X., Strang, C. R., Milford, E. L., Levine, C. G., Garber, T. L., Dogon, A. L., Lord, C., Ghim, S. H., Troup, G. M., Hughes, A. L., and Letvin, N. L. (1992). *Nature*, **357,** 329.
15. Neefjes, J. J., Breur-Vriesendorp, B. S., van Seventcr, G. A. Ivanyi, P., and Ploegh, H. L. (1986). *Hum. Immunol.*, **16,** 169.
16. Yang, S. Y. (1989). In *Immunobiology of HLA*, Volume I, *Histocompatibility Testing 1987* (ed. B. Dupont), p. 332. Springer-Verlag, New York.
17. Yang, S. Y., Morishima, Y., Collins, N. H., Alton, T., Pollack, M. S., Yunis, E. J., and Dupont, B. (1984). *Immunogenetics,* **19,** 217.

18. Barnstable, C. J., Bodmer, W. F., Brown, G., Galfre, G., Milstein, C., Williams, A. F., and Ziegler, A. (1978). *Cell,* **14,** 9.
19. Reekers, P., Coates, D., Doxiadis, I., Ellis, S., Klouda, P., Hajek-Rosenmayr, A., van der Horst, A., Little, A.-M., Madrigal, J. A., and Rinke de Wit, T. (1989). In *Immunobiology of HLA, volume I: Histocompatibility Testing 1987* (ed. B. Dupont), p. 353. Springer-Verlag, New York.
20. Guttridge, M. G. and Klouda, P. T. (1989). *Immunogenetics,* **30,** 506.
21. Guttridge, M. G., Gordon, D. E., and Klouda, P. T. (1990). *Tissue Antigens,* **36,** 127.
22. Neefjes, J. J., Doxiadis, I., Stam, N. J., Beckers, C. J., and Ploegh, H. L. (1986). *Immunogenetics,* **23,** 164.
23. Stam, N. J., Spits, H., Ploegh, H. L. (1986). *J. Immunol.,* **137,** 2299.
24. Dobbe, L. M. E., Stam, N. J., Neefjes, J. J., and Giphart, M. J. (1988). *Immunogenetics,* **27,** 203.
25. Yang, S. Y. (1989). In *Immunobiology of HLA, volume I: Histocompatibility Testing 1987* (ed. B. Dupont), p. 43. Springer-Verlag, New York.
26. Hajek-Rosenmayr, A. and Doxiadis, I. (1989). In *Immunobiology of HLA, volume I: Histocompatiblity Testing 1987* (ed. B. Dupont), p. 338. Verlag-Springer, New York.
27. Little, A.-M., Madrigal, J. A., and Parham, P. (1992). *Immunogenetics,* **35,** 41.
28. Hildebrand, W. H., Madrigal, J. A., Little, A.-M., and Parham, P. (1992). *J. Immunol.,* **148,** 1155.
29. Sambrook, J., Fritsch, E. F., and Maniatis, T. (1989). In *Molecular Cloning, A laboratory Manual*, 2nd edition. Cold Spring Harbor Laboratory Press, New York.
30. Bragado, R., Lauzurica, P., López, D., and López de Castro, J. A. (1990). *J. Exp. Med.,* **171,** 1189.
31. Ennis, P. D., Zemmour, J., Salter, R. D., and Parham P. (1990). *Proc. Natl. Acad. Sci. USA.,* **87,** 2833.
32. Zemmour, J., Little, A.-M., Schendel, D. J., and Parham, P. (1992). *J. Immunol.,* **148,**1941.
33. Messing, J. (1983). In *Methods in Enzymology 101 (part C): Recombinant DNA* (ed. R. Wu, L. Grossman, and K. Moldave), p. 20. Academic Press, New York.
34. Hanahan, D. (1985). In *DNA Cloning volume I* (ed. D. M. Glover), p. 109. IRL Press, Oxford.
35. Zemmour, J. and Parham, P. (1991). *Immunobiology,* **182,** 347.
36. Devereux, J., Haeberli, P., and Smithies, O. (1984). *Nucleic Acids Res.,* **12,** 387.
37. Parham, P., Lawlor, D. A., Lomen, C. E., and Ennis, P. D. (1989). *J. Immunol.,* **142,** 3937.
38. Saper, M. A., Bjorkman, P. J., and Wiley, D. C. (1991). *J. Mol. Biol.,* **219,** 277.
39. Salter, R. D., Benjamin, R. J., Wesley, P. K., Buxton, S. E., Garrett, T. P. J., Clayberger, C., Krensky, A. M., Norment, A. M., Littman, D. R., and Parham, P. (1990). *Nature,* **345,** 41.

Protocol 2. Preparing cryostat sections

1. Cut sections of tissue 6 μm thick (we cut at −20°C).
2. Allow to melt on to Multispot slides (C. A. Hendley Ltd.).
3. Leave slides to air-dry for about 30 min.
4. Fix tissue in acetone, at room temperature for 10 min.
5. Sections may be stored at low temperatures for long periods of time before staining, it is important however that they are free of moisture. We bind slides back to back with cling film and store at −70°C.

2.2 Antibodies

2.2.1 Monoclonal antibodies

The MoAb commonly used to determine the distribution of HLA antigens in frozen tissue are given in *Table 1*. Although often donated by colleagues, where the monoclonals are commercially available, this information is provided.

2.2.2 Storage

Antibodies purchased from commercial sources should be stored according to manufacturer's instructions. Pre-diluted antibodies and conjugates should be stored at 4–8°C because freeze-thawing is likely to have a deleterious effect on these proteins. Commercial antibodies usually contain a small amount (0.1%) of sodium azide as a preservative. Monoclonal antibodies in the form of ascites fluid, serum, or tissue culture supernatant should be aliquoted and stored at −20°C or below to prevent cycles of repeated freezing and thawing. Tissue culture supernatant usually contains about 20 μg/ml of protein, anything less than this should receive addition of bulk protein prior to aliquoting

Table 1. Anti-HLA MoAb used for immunocytochemistry

MoAb	Specificity	Source and catalogue number	Reference
W6/32	Class I, monomorphic	Dako M736	(9)
PA2.6	Class I, monomorphic		(10)
NFK1	Class II, monomorphic including DR, DP, and DQ		(11)
CA22	Class II monomorphic		(12)
L227	Class II monomorphic		(12)
L-243	Class II, DR	Becton Dickinson, 7360	(13)
B7.21	Class II, DP	Becton Dickinson, 7730	(14)
Tu 22	Class II, DQ		(15)
Leu-10	Class II, DQ	Becton Dickinson, 7450	(16)

and storage. This is to prevent loss of active antibody by polymerization and absorption on to the container, and can be achieved by addition of 0.1% to 1% bovine albumin.

2.3 Staining methods

After addition of MoAb to tissues there are a number of ways of visualizing and amplifying the signal. Fluorochrome or enzyme linked secondary reagents are used. Fluorochromes are of limited use compared to the enzyme methods because positively stained structures have to be viewed against a background of darkness. In contrast, enzymatic methods have developed rapidly in recent years, they can be highly sensitive, and give elegant pictures of the whole tissue. Immunoenzymatic methods utilize enzyme-substrate reactions to convert colourless chromogens into coloured products. The enzymes used most commonly are horseradish peroxidase, calf intestine alkaline phosphatase, glucose oxidase, and beta-galactosidase.

The site of MoAb fixation is detected by using directly conjugated enzyme – MoAb, or by indirect methods such as addition of secondary or tertiary enzyme linked antibodies (*Figure 1*). The best amplification is achieved using soluble enzyme immune complexes (*Figure 2*) or avidin–biotin methods (*Figure 3*). The soluble enzyme complex method is named after the particular enzyme used. For example, the two most commonly used methods peroxidase anti-peroxidase (PAP) and alkaline phosphatase anti-alkaline phosphatase (APAAP) use peroxidase and alkaline phosphatase complexes respectively. The PAP complex consists of three molecules of peroxidase and two antibodies against peroxidase enzyme (*Figure 2*), and the APAAP complex

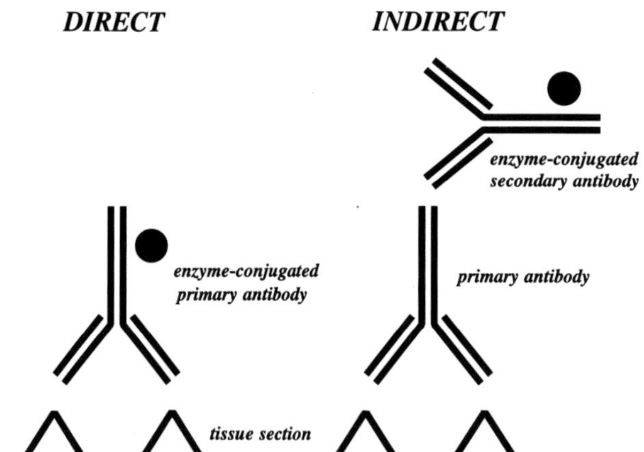

Figure 1. Diagrammatic representation of direct (where primary antibody is conjugated) and indirect (where secondary antibody is conjugated) staining methods. (Modified from 'Handbook of Immunocytochemical Staining Methods', 1989. Ed. Naish, S. J. Published by DAKO Corp. Kind permission of Dakopatts Ltd.)

consists of two molecules of alkaline phosphatase and one antibody against the enzyme.

In this laboratory we routinely use the soluble enzyme immune complex method with alkaline phosphatase (APAAP) or the avidin–biotin complex with horseradish peroxidase (ABC-HRP).

2.3.1 Alkaline phosphatase anti-alkaline phosphatase method

In this method, the primary antibody is unconjugated mouse monoclonal directed against human antigens, secondary antibody being unlabelled rabbit anti-mouse immunoglobulin, and the third layer consisting of alkaline-phosphatase mouse monoclonal anti-alkaline phosphatase complex. The secondary antibody (also called the link antibody) must fulfil two conditions;

(a) be directed against immunoglobulins of the species producing the primary antibody

(b) be added in excess so that one of its Fab sites bind to the primary antibody leaving the other Fab site free to bind to antibody from the enzyme immune complex (see *Figure 2*).

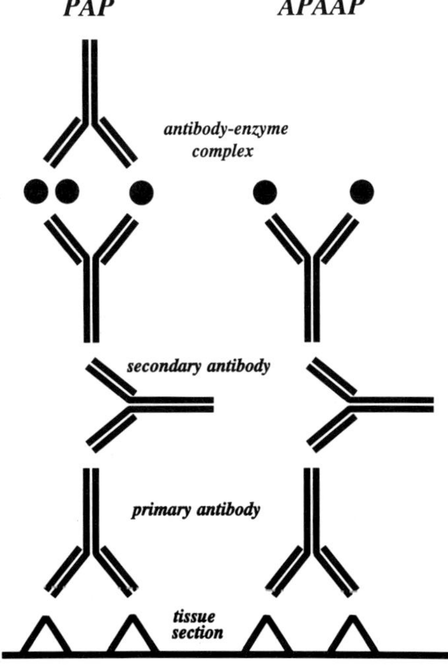

Figure 2. Diagrammatic representation of PAP and APAAP methods. Amplification is achieved using soluble enzyme/antibody complexes. The PAP method uses peroxidase anti-peroxidase complexes and the APAAP method uses alkaline phosphatase anti-alkaline phosphatase complexes. (Modified from 'Handbook of Immunocytochemical Staining Methods', 1989. Ed. Naish, S. J. Published by DAKO Corp. Kind permission of Dakopatts Ltd.)

Protocol 3. Staining with APAAP

Add 6.06 g Tris-HCl and 1.39 g Tris base (Sigma) to 100 ml distilled water, adjust pH to 7.6, and add 100 ml of above to 900 ml saline (8.77 g NaCl).

1. Add 50 μl of chosen mouse MoAb (see *Table 1*) (appropriately diluted in Tris-buffered saline, TBS) to section on slide. Ensure all of the tissue section is adequately covered but avoid touching any tissue with pipette tip.
2. Place slides in humidified slide chamber. Humidification is usually achieved by placing pieces of wet tissue paper in the bottom of the box. Leave slides to incubate for 30–45 min at 22 °C.
3. Rinse slides gently in jar of TBS by agitation for several minutes. Remove excess TBS carefully from the slide with clean tissue paper, wiping around the section and under the slide.
4. Add 50 μl of appropriately diluted rabbit anti-mouse immunoglobulin (Dako Ltd., Z259) to each section.
5. Leave in humidified slide chamber for 30 min.
6. Rinse slides as in step **3**.
7. Add 50 μl of alkaline phosphatase mouse anti-alkaline phosphatase complex (Dako Ltd., DC51) to each section and leave to incubate for 30 min.
8. Rinse slides as in step **3**.
9. Repeat steps **4–8** one more time.

The substrate most commonly used for alkaline phosphatase is naphthol AS-MX phosphate, it can be used in its acid form or as the sodium salt. The chromogens Fast Red TR and Fast Blue BB produce a bright red or bright blue end product respectively. Both are soluble in alcoholic solvents so aqueous mounting media must be used. New Fuchsin also gives a red end product. Unlike the other chromogens it is insoluble in alcohol allowing the specimens to be dehydrated before making permanent mounts.

Protocol 4. Preparation and use of naphthol AS-MX phosphate and Fast Red TR

Prepare 1 litre of 0.1 M Tris-buffer pH 8.2 by using 7.08 g Tris-HCl and 6.68 g Tris base. Dissolve 10 mg of naphthol AS-MX phosphate (Sigma N4875) in 1 ml N,N-dimethylformamide (Sigma D4254). Add 49 ml of 0.1 M Tris-buffer pH 8.2. Add to this 6 mg of levamisole (Sigma L9756). This stock solution has a shelf life of one week at 4°C.

1. Immediately before use weigh 10 mg Fast Red TR salt (Sigma, F1500), add 10 ml of stock solution, and filter.
2. Add one drop of this solution to each section and allow to incubate for 10–15 min.

3. Wash in TBS.
4. Wash in distilled water.
5. Counter stain in Harris's Haematoxylin 1–2 min.
6. Wash in distilled water and mount in an aqueous medium such as Apathy's mounting medium (BDH, 36172).

2.3.2 Avidin–biotin complex horseradish peroxidase method

This method (*Figure 3*) utilizes the high affinity of avidin or streptavidin for biotin. The sequence or reagent application is mouse MoAb, followed by biotinylated secondary antibody, followed by pre-formed avidin–biotin–enzyme complex. Open sites on avidin from the avidin–biotin complex bind to the biotin on the link antibody. The strong affinity of avidin for biotin and the mild biotinylation process makes this method very sensitive.

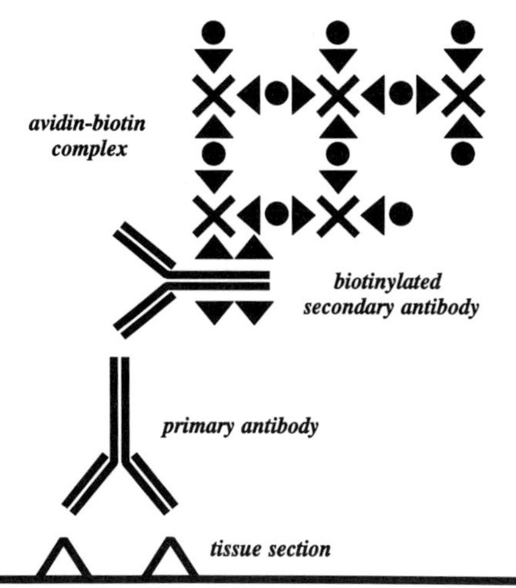

Figure 3. Diagrammatic representation of ABC-HRP method. Amplification is achieved using biotinylated secondary antibody and complexes of avidin, biotin, and peroxidase. (Modified from 'Handbook of Immunocytochemical Staining Methods', 1989. Ed. Naish, S. J. Published by DAKO Corp. Kind permission of Dakopatts Ltd.)

Protocol 5. ABC-HRP staining

1. Add 50 μl of mouse MoAb appropriately diluted in PBS to each tissue section. Ensure tissue is completely covered with antibody and avoid touching section with pipette tip.

Protocol 5. *Continued*

2. Leave to incubate for 30 min in humidified slide chamber.
3. Rinse slides gently in jar of PBS to wash.
4. Dry slides by wiping around section and under the slides with a clean piece of tissue paper.
5. Add 50 μl of appropriately diluted biotinylated goat or rabbit anti-mouse immunoglobulin (Dako Ltd., E354) to each section and leave to incubate for 45 min. This antibody should be diluted in a 1:1 mixture of PBS: pooled normal human serum (obtainable from a Blood Transfusion Centre).
6. Repeat steps **3** and **4**.
7. Add 1–2 drops of pre-formed avidin–biotin–peroxidase complex (from Dako Ltd., K355) to each section. Allow to incubate in humidified slide chamber for 60 min.
8. Wash in PBS.

Note. Slides can be kept in PBS while fresh substrate is prepared.

The substrate for use with peroxidase is hydrogen peroxide and the chromogen most commonly used is 3,3′ diaminobenzidine tetrahydrochloride (DAB, Sigma D5637). The atomic oxygen resulting from the breakdown of H_2O_2 oxidizes DAB to produce a brown coloured end product which is highly insoluble in alcohol. Although DAB tablets are available (Sigma D5905) and are probably safer to use, the author has always found using the powder gives more satisfactory results.

Protocol 6. Preparation and use of DAB

Note. DAB is carcinogenic and should be handled with gloves.

1. Add 15 mg of DAB to 50 ml of PBS.
2. Add 17.5 μl of 30% hydrogen peroxide to above mixture, mix, and pour into Coplin jar.
3. Stand slides in this mixture for 10 min.
4. Remove slides, wash in PBS, counterstain in Harris's Haematoxylin (BDH 34124), dehydrate through a series of alcohols of increasing concentration, and mount in a non-aqueous mountant such as DPX (BDH 36029).

2.3.3 Assessment of non-specific staining

Some cells and tissues have endogenous peroxidase or phosphatase activities. In particular tissues containing large numbers of macrophages or neutrophils (both of which contain endogenous peroxidase) should not be stained with a peroxidase technique. Most tissues contain some phosphatase activity and addition of 6 mg of levamisole to the substrate/chromogen mixture (see *Protocol 4*) should inhibit non-specific phosphatase. All experiments should include control tissue sections which have completed all of the staining procedure, but where primary antibody and primary and secondary antibody have been omitted.

3. Distribution of HLA antigens in normal and diseased tissue

A detailed study of the normal distribution of HLA class I and class II was performed by Daar and colleagues (2), using MoAb PA2.6 and NFK1 against numerous tissues and organs derived from a single cadaveric kidney donor and surgical specimens (see *Tables 2* and *3*). These results serve to demonstrate that HLA class I distribution is not as ubiquitous as previously thought, and that HLA class II is well represented on normal human endothelia and epithelial tissues. These results however give no indication of the variation which can be found in normal tissue and the effects of transplantation.

There have been numerous detailed studies describing the distribution of Mhc antigens before and after solid organ transplantation. There follows a brief description of the normal distribution in heart, lungs, kidney, and liver and the main changes in expression which occur during rejection episodes.

3.1 Heart

Immunocytochemical staining of normal heart taken prior to transplantation demonstrates that all the interstitial structures are HLA class I positive (see *Figure 4a*) and the large majority are also HLA class II positive (see *Figure 4b*). In contrast, the myocardium plasma membrane is negative for Mhc antigens (6). Some studies have reported faint staining of the intercalated discs for class I in normal heart (2, see *Table 2*). The HLA class II positive interstitial structures in normal heart have been identified as being mostly capillary endothelial cells with a small minority being cells of the monocyte/macrophage/dendritic series. In addition venules and arterioles can be distinguished in cardiac biopsies and these are normally class II antigen positive. Larger vessels, sections of which may appear within an endomyocardial biopsy, tend to be negative for class II antigens as does the endocardium.

Using MoAb specific for the sub-determinants of class II antigens (17) it was shown that HLA-DR and -DP are strongly expressed on the endothelial cells of normal heart, but HLA-DQ was barely visible.

During rejection (see *Table 4*) there is a dramatic up-regulation of Mhc class I antigens on the myocardial plasma membrane and intercalating discs (*Figure 4c*). This is almost always accompanied by an infiltrate, most of which

Table 2. Detailed tissue distribution of HLA class I antigens in normal human tissues

Tissue	Staining with anti-HLA-A, -B, -C
Cells of the endocrine system:	
Thyroid (follicular, parafollicular)	+ [b]
Parathyroid (chief, oxyphil.)	+
Pituitary (acidophils, basophils, chromophobes)	+
Pancreatic islets of Langerhans	+
Adrenal	
3 cortical zones	++ [a]
Medulla	++
Gastrointestinal tract:	
Epithelium of	
Tongue	++
Esophagus	++ [c]
Stomach	
Fundus	+
Antrum	+
Duodenum	++
Brunners glands	+/– [d]
Ileum	++
Appendix	++
Colon	++
Rectum	++
Gall bladder	+
Liver	
Sinusoidal lining cells	++
Hepatocytes	+/–
Biliary epithelium	++
Respiratory and cardiovascular systems:	
Epiglottis	
Lingual surface epithelium	++
Mixed glandular tissue	++
Trachea	
Surface epithelium	++
Mixed glandular tissue	++
Tonsillary epithelium	++
Lung (bronchial and alveolar epithelium)	++
Heart	
Myocardium	+
Intercalated discs	++
Endothelium	
Capillaries	++
Larger vessels	++
Nervous system	
Peripheral	++
Central	
Neurones	–
Occasional unidentified cell	++
Dura	++
Urogenital system:	
Kidney glomeruli (endothelium, mesangium)	++
Kidney tubules	++
Epithelium	
Ureter	++
Bladder	++
Prostate	++
Urethra	++
Testis	
Germ cell line, spermatozoa	+
Sertoli and Leydig cells	+
Epidydymis	+
Epithelium	+
Spermatozoa	++
Miscellaneous:	
Breast epithelium	
Ductal	++
Glandular	++
Pancreas	
Exocrine portion	–
Ductal epithelium	++
Parotid	
Acinar epithelium	–
Ductal epithelium	++
Muscle	
Skeletal	± [e]
Smooth	+
Cornea	
Outer squamous epithelium	++
Endothelium, Descemet's membrane, and substantia propria	–
Possible dendritic cells at periphery	++
Langerhans cells, interstitial dendritic cells	++
Lymphatics	++
Fibroblasts	++
Placenta—villous trophoblast	–
Epidermis	+

[a] (++) = strong staining.
[b] (+) = weak staining.
[c] basal or deeper layer(s) positive.
[d] (–) = no staining.
[e] (±) = very weak staining.

Table 3. Detailed tissue distribution of HLA class II antigens in normal human tissues

Tissue	**Staining with anti-HLA Class II**
Cells of the endocrine system:	
Thyroid (follicular, parafollicular	–[a]
Parathyroid (chief, oxyphil)	–
Pituitary (acidophils, basophils, chromophobes)	–
Pancreatic islets of Langerhans	–
Adrenal: 3 cortical zones	–
medulla	–
Gastrointestinal tract:	
Epithelium	
Tongue	+[b,c]
Esophagus	–
Stomach:	
Fundus	–
Antrum	–
Duodenum	++[d,e]
Brunner's glands	–
Ileum	++[e]
Appendix	++[e]
Colon	–
Rectum	–
Gall bladder	ND[f]
Liver	
Sinusoidal lining cells	++
Hepatocytes	–
Biliary epithelium	–/±[g]
Respiratory and cardiovascular systems:	
Epiglottis	
Lingual surface epithelium	++[c]
Mixed glandular tissue	++
Trachea	
Surface epithelium	++
Mixed glandular tissue	++
Tonsillary epithelium	++[c]
Lung (bronchial and alveolar epithelium)	ND
Heart	
Myocardium	–
Intercalated discs	–
Endothelium	
Capillaries	++[h]
Larger vessels	+/–/±
Nervous system	
Peripheral	–
Central	
Neurones	–
Occasional unidentified cell	++
Dura	++
Urogenital system:	
Kidney glomeruli (endothelium, mesangium)	++
Kidney tubules	++/–/±
Epithelium	
Ureter	–
Bladder	–
Prostate	–
Urethra	++
Testis	–
Germ cell line, spermatozoa Sertoli and Leydig cells	–
Epididymis	
Epithelium	++/–
Spermatozoa	–
Miscellaneous:	
Breast epithelium	
ductal, including myoepithelium	++
glandular	++
Pancreas	–
Exocrine portion	–
Ductal epithelium	
Parotid	
Acinar epithelium	–
Ductal epithelium	–
Muscle	
Skeletal	–
Smooth	–
Cornea	
Outer squamous epithelium	–
Endothelium, Descemet's membrane, and substantia propria	–
Possible dendritic cells at periphery	++
Langerhans cells, interstitial dendritic cells	++
Lymphatics	++
Fibroblasts	–
Placenta—villous trophoblast	–
Epidermis	–

[a] (–) = no staining.
[b] (+) = weak staining.
[c] Basal or deeper layer(s) positive.
[d] (++) = strong staining.
[e] Cells deep in crypts of Leiberkühn negative.
[f] ND = not determined.
[g] (±) = very weak staining.
[h] Except in brain and placenta, and possibly testis.

(Reproduced from Daar, Fuggle, Fabre, Ting and Morris, 1984 (6) with kind permission from the authors and publishers, Williams and Wilkins, Baltimore.)

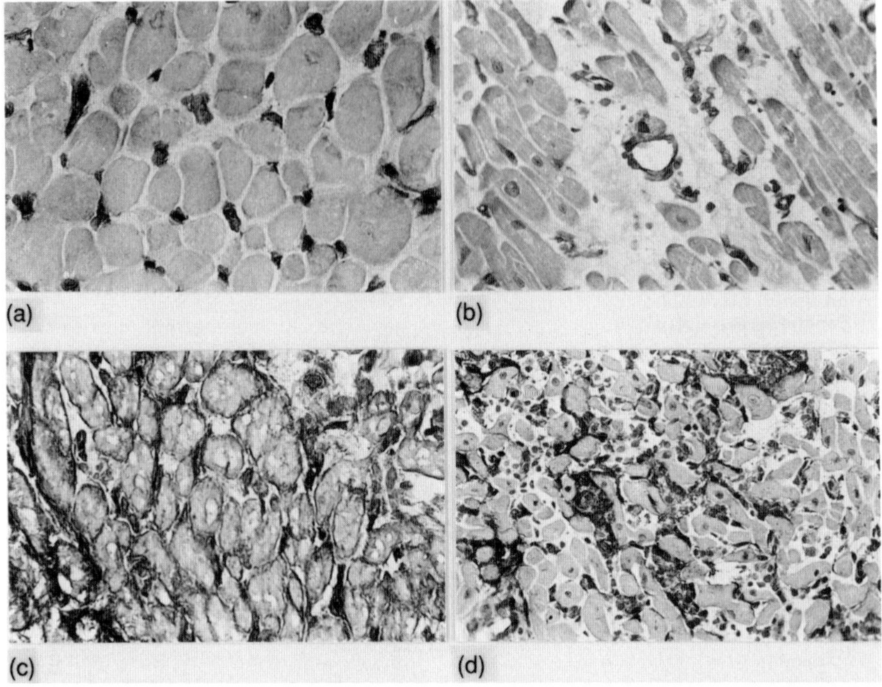

Figure 4. Normal piece of right ventricular biopsy taken from donor prior to transplantation, 6 μm sections stained with MoAb against class 1 (a) or HLA-DR determinants (b). Right ventricular endomyocardial biopsy taken from transplant patient showing histological signs of rejection, 6 μm section stained with MoAb against class I (c) or HLA-DR determinants (d). All sections were stained using the immunoperoxidase ABC method, and counterstained with Harris's Haematoxylin.

is also HLA class II positive (*Figure 4d*). The close apposition of the normally HLA class II positive endothelial cells and HLA class II positive infiltrating cells makes it difficult to determine whether there is up-regulation of class II antigens on the capillary endothelial cells. There is stronger expression of class II antigens on venules, arterioles, and the endocardium. The myocardium remains negative for HLA class II. The above studies used MoAb against common determinants of HLA class II or the -DR determinant. Using MoAb against HLA-DQ determinants, which are normally scarcely expressed on endothelial cells, clear up-regulation is observed associated with rejection episodes.

The correlation between induction of Mhc class I on the myocardium and clinical rejection is not absolute. One hundred percent of first rejection episodes (7/7 biopsies) and 79% (11/14) of subsequent rejection episodes coincided with class I expression (17). Steinhoff (18) reported that 57 out of 78 rejection episodes were characterized by induction of class I on the normally negative myocardium.

Table 4. Distribution of HLA class I[a] and class II[b] in normal heart and endomyocardial biopsies showing rejection

	Normal		**Rejection**	
	Class I	**Class II**	**Class I**	**Class II**
Myocardial plasma membrane	−	−	++	−
Intercalating discs	+−	−	++	−
Capillaries	+	+	+	++?
Venules	+	+−	+	+
Arterioles	+	+−	+	+
Endocardium	+	−	+	+
Coronary endothelium	+	+	+	+

[a] Using MoAb W6/32
[b] Using MoAb L243

− negative staining
+− weak positive staining
+ medium positive staining
++ strong positive staining

MoAb against polymorphic determinants have been used immunocytochemically to follow the fate of class I or class II antigens of donor or recipient origin, after transplantation. MoAb 17.3.3S, specific for HLA-DR7, shows strong staining of the interstitial structures of heart from patients who are -DR7 positive (*Figure 5a*), but no binding to sections of heart from -DR7 negative individuals (*Figure 5b*). Studies in the heart (19) and kidney have shown, surprisingly, persistence of donor class II determinants on parenchymal cells (shown to be endothelial) for long periods after transplantation.

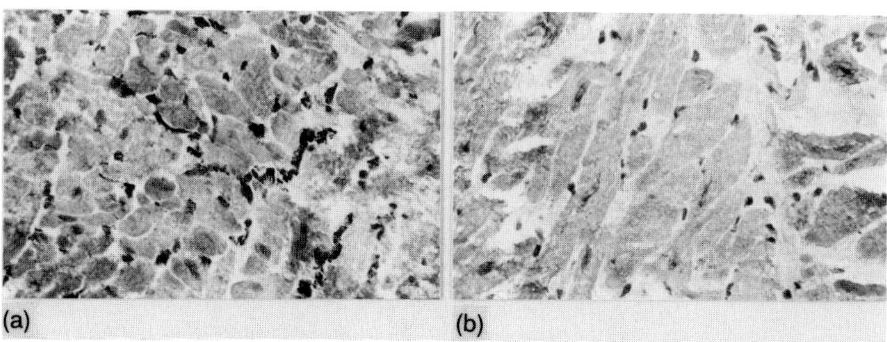

Figure 5. Binding of MoAb against polymorphic determinant HLA-DR7 (MoAb 17.3.3S) to cardiac biopsy from donor heart before transplantation (a), and to cardiac biopsy from the recipient heart explanted at the time of transplantation (b). The donor was typed as being HLA-DR7 positive and the recipient was -DR7 negative. HLA-DR7 positive structures are found in the donor (a) but not in the recipient (b).

3.2 Lungs

Lungs contain large numbers of alveolar macrophages which contain endogenous peroxidase, therefore it is not possible to use a peroxidase technique to visualize antibody staining. Here we routinely use alkaline phosphatase soluble complexes (the APAAP) for immunocytochemical studies of the lungs.

Human lungs 'normally' show a strong and extensive distribution of Mhc antigens. Class II antigens being present on all alveolar macrophages, and tracheal epithelium but being variably expressed on endothelial cells, tracheal, and bronchiolar epithelium (*Table 5*). As in the heart, HLA-DR and -DP antigens are more heavily expressed than HLA-DQ in normal lung. All the above structures were invariably class I positive as well. The strong expression of these antigens in 'normal lungs' makes changes following transplantation or during diseased states difficult to discern.

A study of nine lungs removed from heart-lung recipients because of development of obliterative bronchiolitis reported enhanced and consistent expression of all HLA class II determinants (-DR, -DP, and -DQ) on endothelial cells, and also on bronchiolar epithelium and type I and type II alveolar pneumocytes (see *Table 5*). Using serial dilutions of MoAb directed against HLA-DR, -DP, and -DQ determinants, Yousem *et al.* (20) also observed increased expression of HLA-DR and -DP on all cells within transbronchial biopsies of patients after lung transplantation.

It is much less easy to know whether the immunocytochemical changes observed in the lungs following transplantation are due to rejection, than it is after heart or renal transplantation. The majority of lung transplant patients

Table 5. Distribution of HLA class I[a] and class II[b] in normal lungs and lungs showing signs of rejection

	Normal		**Rejection**	
	Class I	**Class II**	**Class I**	**Class II**
Vascular endothelium	+	15/20*	+	9/9*
Bronchiolar epithelium	+−	+−	+	+
Alveolar epithelium	+	+	+	+

[a] Using MoAb W6/32
[b] Using MoAb L243

* Vascular endothelium was positive in 15/20 specimens of normal lung and 9/9 specimens of transplanted lung.

− negative staining
+− weak positive staining
\+ medium positive staining
++ strong positive staining

experience bacterial and cytomegalovirus infections, which directly affect the transplanted organ. A differential diagnosis of rejection and infection is difficult to achieve in the transplanted lungs. The changes described above almost certainly reflect a variety of immunological and infectious stimuli.

3.3 Liver

In normal liver Mhc class I antigens are invariably expressed on the bile duct epithelium, endothelial cells, cells lining sinusoids (including Kupffer cells), and interstitial leukocytes. In contrast hepatocytes have been described as negative or very weakly stained (see *Table 6*). Mhc class II has been found to be absent or weakly expressed by bile duct epithelium, and to be present on all or most of the endothelial cells, and absent on hepatocytes.

Table 6. Distribution of HLA class I and class II in normal liver and liver showing signs of rejection

	Normal		Rejection	
	Class I	**Class II**	**Class I**	**Class II**
Bile duct epithelium	+	−	+	++
Endothelium	+	+	+	+
Sinusoidal cells (Kupfer cells)	+	++	+	++
Hepatocytes	−	−	++	+−

− negative staining
+− weak positive staining
+ medium positive staining
++ strong positive staining

The major changes during rejections are strong induction of class I antigens on the hepatocytes and class II antigens on biliary epithelial cells. Moreover, using MoAb against β_2-microglobulin which can be used on paraffin embedded tissue, induction of β_2-microglobulin on hepatocytes has been reported (21). Induction of class II antigens on hepatocytes has been reported in severe rejection.

3.4 Kidney

Mhc class I antigens are detected on all structures in the normal human kidney (scc *Tablc 7*). The glomeruli, capillary endothelial cells, and endothelium of the larger vessels are strongly stained whereas the renal tubules express a much lower level of class I antigen. Using MoAb against the HLA-DR determinant of class II, positive staining is consistently found on glomerular endothelium, mesangium, and intertubular structures. Expression of HLA class II in the proximal renal tubules is variable, a single study of 46 normal

Table 7. Distribution of HLA class I and class II in normal kidneys and kidneys showing signs of rejection

	Normal		Rejection	
	Class I	**Class II**	**Class I**	**Class II**
Glomeruli	+	+	+	+
Capillary endothelium	+	++	+	+
Larger vessel endothelium	+	−	+	+
Tubular epithelium	+−	+−	++	++

− negative staining
+− weak positive staining
+ medium positive staining
++ strong positive staining

kidneys (11) found class II antigens in the proximal tubules in 77% of kidneys. The distal tubules are always negative. Variation in expression of class II antigen between individuals and between studies is not surprising, in view of its inducibility by different cytokines.

During rejection episodes, the most dramatic finding is of up-regulation of HLA class II on cytoplasmic and membrane components of all the renal tubules (see *Table 7*). There is also induction of class II antigens on the normally negative endothelial cells of the large vessels. Analysis of the induced molecules with polymorphic and locus specific antibodies (22) showed it was of donor origin and had HLA-DR, -DQ, and -DP components. Mhc class I is also induced during rejection episodes there being increased intensity on the renal epithelial cells.

The question arises whether induction of Mhc antigens is correlated with clinical rejection episodes. As in the heart, in general there is a correlation but it is not absolute. Thus induced HLA class II is not always detected in biopsies during allograft rejection, it can also be found during stable graft function. Although it has been suggested that induced class II antigens would be helpful in distinguishing between renal dysfunction caused by rejection and cyclosporine nephrotoxicity, the lack of complete correlation between class II antigen expression and rejection means class II is not reliable as a differential marker.

3.5 Alterations in Mhc expression in non-transplant disease

Altered expression of Mhc antigens has been described in inflammatory states, autoimmune disorders, viral diseases, and malignancies.

There is induction of class I antigen on the myocardium in active myocarditis and on the normally negative plasma membrane of skeletal muscle in a

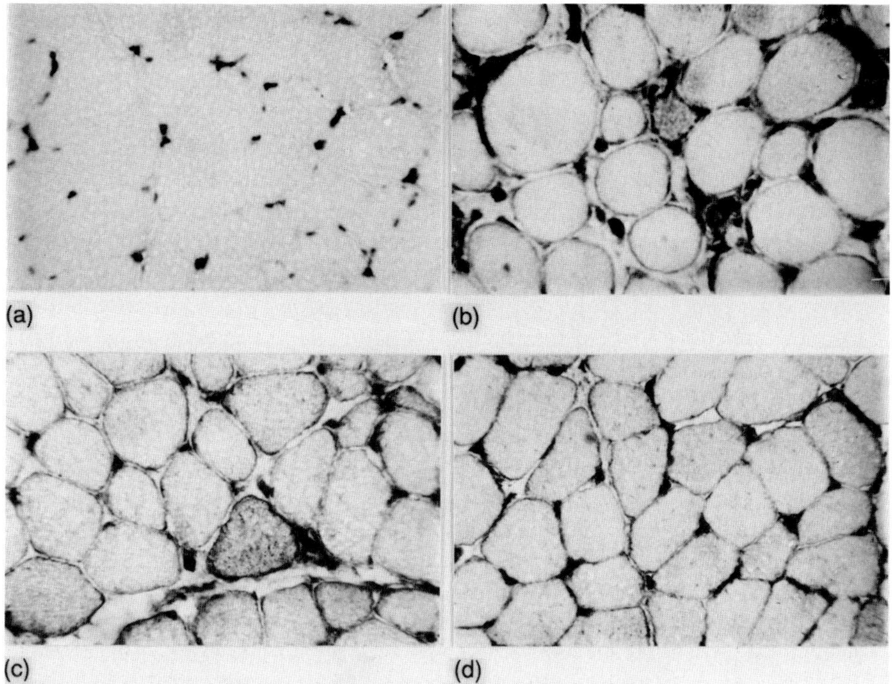

Figure 6. Binding of W6/32 against class I determinants to 6 μm cryostat sections of normal skeletal muscle biopsy (a), and muscle biopsies from patients with Duchenne Muscular Dystrophy (b), Becker Muscular Dystrophy (c), and Juvenile Dermatomyositis (d). All sections were developed with the immunoperoxidase ABC technique. (Photographs reprinted with kind permission of Dr Rhoda McDouall and Dr Michael Dunn.)

number of inflammatory disorders of skeletal muscle (4, *Figure 6*). It is perhaps not surprising to find induction of class I antigen in the presence of an active infiltrate, in this situation it is likely that induction was caused by local production of cytokines. A more interesting question is whether class I antigen once induced, can persist once the infiltrate has disappeared, or whether chronic viral infection can result in persistent aberrant expression of Mhc antigens.

Aberrant expression of Mhc antigens has been implicated in the pathogenesis of autoimmune disorders. It is thought that expression of class I antigen on the normally negative beta cells of the pancreas precedes destruction of these cells leading to Type I Diabetes. Aberrant expression of HLA-DR on thyroid epithelial cells has been associated with development of thyroiditis. The reason for the aberrant expression in the first instance, whether induced by an inflammatory infiltrate or whether the result of direct viral infection is not known.

There is no doubt that viruses can modulate Mhc antigen expression of infected cells *in vitro*. The viruses can act directly on Mhc transcription factors, viral proteins may block transport of Mhc antigens to the cell surface, or signalling via viral oncogenes may lead to effects on transcription of Mhc genes. It is interesting to note that aberrant expression of autologous Mhc antigens can be **directly** damaging to cells. Thus, induction of class I antigens on murine pancreatic beta cells, by means of genetic manipulation of the embryo (23), resulted in failed insulin production by those cells. Although the effects of inflammatory infiltrates is to cause up-regulation of Mhc antigens, the effect of viral infections is often to cause down-regulation of Mhc antigens. Thus many different cell lines transformed *in vitro* and many cultured tumour lines have lost expression of Mhc antigens. It has been suggested that transformation by viral oncogenes that down-regulate Mhc expression results in tumour cells which evade the immune response. Thus transfection of the gross leukaemia-virus infected AKR leukaemia cell line (normally negative for Mhc antigen expression) with class I or class II genes reduce tumourigenicity (24). However, the clinical picture of Mhc antigen expression on tumour cells is one of great diversity.

Finally, an interesting example of clear Mhc induction in the absence of an infiltrate and the absence of any viral involvement is that of Duchenne Muscular Dystrophy. This is an X-linked gene defect. The plasma membrane of skeletal muscle shows strong expression of HLA class I antigens (4, *Figure 6b*), The cause of this aberrant expression, and any possible consequences remain unknown.

4. Clinical applications

(a) Abnormal expression of Mhc antigens (*de novo* expression or up-regulation) is certainly caused by local release of cytokines, but conceivably could also be caused by persistent viruses or some other mechanism which might interfere with a cell's protein synthesis apparatus. Abnormal Mhc expression can thus be cited as evidence of a local cellular immune response, and/or it can be implicated in the pathogenesis of an autoimmune disease (as described above).

(b) Expression of HLA class I in the heart and HLA class II in kidneys can aid or confirm the histological diagnosis of rejection following transplantation.

(c) Mhc antigens are an integral part of the process of rejection. Cells bearing Mhc antigens initially stimulate the recipients immune system, and the same or different cells are the target of the anti-allograft response. These studies provide essential information, necessary to design new immunosuppressive strategies.

(d) The role of Mhc antigens in controlling tumourigenicity needs to be further explored. It is possible that information about the level of Mhc expression could be of prognostic value for certain tumours.

Acknowledgements

I would like to thank Dr Mike Dunn for producing the figures. The work described in this chapter originating from the author's laboratory has been funded by the British Heart Foundation.

References

1. Fleming, K. A., McMichael, A., Morton, J. A., Woods, J., and McGee, J. (1981). *J. Clin. Path.,* **34,** 779.
2. Daar, A. S., Fuggle, S. V., Fabre, J. W., Ting, A., and Morris, P. J. (1984). *Transplantation,* **38,** 287.
3. Natali, P. G., Bigotti, A., Nicotra, M. R., Viora, M., Manfredi, D., and Ferrone, S. (1984). *Cancer Res.,* **44,** 4679.
4. Appleyard, S. T., Dunn, M. J., Rose, M. L., and Dubowitz, V. (1985). *Lancet,* **1,** 361.
5. Rose, M. L., Coles, M. I., Griffin, R. J., Pomerance, A., and Yacoub, M. H. (1986). *Transplantation,* **41,** 776.
6. Daar, A. S., Fuggle, S. V., Fabre, J. W., Ting, A., and Morris, P. J. (1984). *Transplantation,* **38,** 292.
7. Halloran, P. F., Wadgymar, A., and Autenreid, P. (1986). *Transplantation,* **4,** 413.
8. Forsum, U., Claesson, K., Hjelin, E., Karlsson-Parra, A., Scheynius, A., and Tjernlund, U. (1985). *Scand. J. Immunol.,* **21,** 389.
9. Barnstable, C. J., Bodmer, W. F., Brown, G., Galfre, G., Milstein, C., Williams, A. F., and Zieglar, A. (1987). *Cell,* **14,** 9.
10. Brodsky, F. M., Parham, P., Barnstaple, C. J., Crumpton, M. J., and Bodmer, W. F. (1979). *Immunol. Rev.,* **47,** 3.
11. Fuggle, S. V., Errasti, P., Daar, A. S., Fabre, J. W., Ting, A., and Morris, P. J. (1983). *Transplantation,* **35,** 385.
12. Charron, D. J. and McDevitt, H. O. (1979). *Proc. Natl Acad. Sci. USA,* **76,** 6567.
13. Robbins, P. A., Evans, E. L., Ding, A. H., Warner, N. L., and Brodsky, F. M. (1987). *Hum. Immunol.,* **18,** 301.
14. Watson, A. J., De Mars, R., Trowbridge, I. S., and Bach, F. H. (1983). *Nature,* **3,** 358.
15. Ziegler, A., Heinig, J., Muller, C., Gotz, H., Thinnes, F. P., Uchanska-Ziegler, B., and Wernet, P. (1986). *Immunobiology,* **171,** 77.
16. Chen, Y.-X., Evans, P. L., Pollack, M. S., Lanier, L. L., Phillips, J. H., Rousso, C., Warner, N. L., and Brodsky, F. M. (1984). *Hum. Immunol.,* **10,** 221.
17. Suitters, A. J., Rose, M. L., Higgins, A., and Yacoub, M. H. (1987). *Clin. Exp. Immunol.,* **69,** 575.
18. Steinhoff, G., Wonigeit, K., Schafers, H. J., and Haverich, A. (1989). *J. Heart Transplant.,* **8,** 360.
19. Rose, M. L., Navarette, C., Yacoub, M. H., and Festenstein, H. (1989). *Hum. Immunol.,* **23,** 179.
20. Yousem, S. A., Curley, J. M., Dauber, J., Paradis, I., Rabinowich, H., Zeevi, A., Duquesnoy, R., Dowling, R., Zenati, M., Hardesty, R., and Griffiths, B. (1990). *Transplantation,* **49,** 991.

21. Nagafuchi, Y., Thomas, H. C., Hobbs, K. E. F., and Scheuer, P. F. (1985). *Lancet,* **1,** 552.
22. Fuggle, S. V., McWhinnie, D. L., and Morris, P. J. (1987). *Transplantation,* **44,** 214.
23. Allison, J., Campbell, I. L., Morahan, G., Mandel, T. E., Harrison, L. C., and Miller, J. F. A. P. (1988). *Nature,* **333,** 529.
24. Festenstein, H. and Schmidt, W. (1981). *Immunol. Rev.,* **60,** 85.

9

Paternity testing

ERNETTE DU TOIT

1. Introduction

The primary objective of paternity testing is to determine whether or not a man who is falsely accused of paternity can be excluded from being the biological father of a given child. Examples of disputed parentage and methods for resolving them can be traced back to biblical days. In the Old Testament in I Kings 3 verse 16–28 Solomon made the choice of maternity. Although in Solomon's case maternity was at issue, more commonly the question is one of paternity.

The role of modern science in problems of disputed parentage began with the discovery of the ABO blood groups by Karl Landsteiner in 1900. The recognition that the inheritance of the A, B, and O blood groups followed the genetic rules observed by Gregor Mendel became the cornerstone of all that followed. Until the 1950s the only blood group systems used in paternity studies were ABO, Rh, and MN. Together they provided an exclusion of parenthood in about 50% of disputed cases.

In 1938 the question of the alternative result from paternity testing, where the putative father is not excluded, was first looked at by Essen-Möller (1), who suggested using a likelihood ratio to estimate how probable it was that the non-excluded tested man was the biological father. The use of a statistical estimation of the probability of paternity also referred to as an inclusion probability has now become common all over the world.

During the past decade there has been remarkable progress in this field and evidence based on blood groups is now generally accepted in court. A number of additional tests have been introduced, including the analysis of HLA polymorphisms, red cell enzyme and plasma protein polymorphisms, and most recently DNA testing.

The American Association of Blood Banks recommends that the range of tests used has at least a 95% probability of excluding a falsely accused man (2). However, at the 14th Congress of the International Society for Forensic Haemogenetics held in September 1991, in Mainz, Germany, the general consensus was that, with the advent of DNA typing, the probability of excluding a falsely accused man should be much closer to 99%. It is possible to exclude more than 99% of the falsely accused men in all three major population groups, South African Caucasoid, Cape Coloured (a group of mixed ancestry), and South African Blacks. Each laboratory which performs paternity testing should calculate the probability of excluding a falsely accused man based on the tests it performs. The systems vary markedly in their usefulness in excluding a falsely accused man and they can be ranked and selected accordingly, depending on the population groups tested. Furthermore, the alleged fathers should have the right to ask what his chances are of being excluded by a laboratory, if falsely accused. When a man is not excluded and the probability of paternity is calculated, compared to inclusion of a random man of the same race, we suggest that a probability figure of 99% or more should be reached wherever possible. Where this figure is not reached with the conventional tests we offer additional tests including DNA testing. The accreditation of laboratories to perform certain tests, as is practised in the USA, ensures that standards are set and maintained and only laboratories that have been inspected and found to meet these standards are allowed to perform paternity testing.

This chapter outlines procedures to be followed when collecting samples and reporting test results, and some of the basic techniques and statistical analysis of the data.

2. Collection of samples

This includes several important steps:

- sample collection
- consent
- identification of each individual

The procedures are set out in *Protocol 1*. Photographs are used in court where the proof of identity of the person bled is an issue. Since the tests in the laboratory are performed by different technologists in separate sections a chain-of-custody form accompanies the blood samples.

Protocol 1. Sample collection, consent, and identification

1. Book all cases in advance. Blood samples are taken by registered nurses who have been specially trained to take blood from babies. A medical doctor is available, in case any problem arises, particularly with infants.
2. Take a photograph of all the parties involved in the disputed paternity case with the names and individual case number clearly displayed.
3. In this laboratory blood is taken only from infants aged more than six months.

4. Obtain proper identification, (e.g. driver's licence) and consent for blood to be taken; this is essential due to the medico-legal implications. The mother and putative father sign a form stating:
   - name and address
   - date of birth
   - race (essential for probability calculations if the putative father is not excluded)
   - consent to take a sample of blood (mother as legal guardian signs for herself and child)
5. The mother and the putative father are each required to sign a form certifying that the blood in the tubes is their blood and was taken in each other's presence.
6. The laboratory dispatches parcels containing tubes for collection of blood samples from individuals living in remote areas. Instructions on the procedures to be followed for collection, storage, and mailing of blood, as well as consent and identification forms are included. These cases are booked in advance by telephone.

---

3. Techniques to determine the relevant genetic systems

The genetic systems included in the repertoire and order of tests offered may vary from one institution to another, depending on their particular expertise. For example, in this laboratory HLA-A and -B serological typing is performed first as it is known to give a very high probability of exclusion (95%). Subsequently testing for the red cell antigens and then for the red cell enzyme and plasma protein systems (*Table 1*) is performed.

All the individuals (usually three) belonging to one case should be tested by a single laboratory under identical conditions since only in this way will it be possible to compare any 'extra' reactions that may help to establish the final conclusion. To avoid errors certain tests are performed by two qualified technologists separately; for all tests the results are read and recorded independently by two technologists. All results are subjected to a final careful review in a meeting of senior representatives from each of the individual laboratories involved.

3.1 Serological testing for HLA-A and -B antigens

The HLA-A and -B antigens are determined using well characterized antisera. Each antigen is detected with at least three different antisera in the standard NIH microlymphocytotoxicity test described in Chapter 3. Although tissue typing tests are not performed in duplicate due to scarcity of reagents

Table 1. Methods used to detect the alleles for plasma proteins and red cell enzymes

	Starch	Agarose	Polyacrylamide
HP (haptoglobin)	1,2,2-1M 2-1 Ca	–	–
TF (transferrin)	–	C,D,B	–
BF (properdin factor B)	–	F,S,F1,FN, S07,SN	–
ACP1 (acid phosphatase)	A,B,C,R	–	–
ADA (adenosine deaminase)	1,2	–	–
AK (adenylate kinase)	1,2	–	–
ESD (esterase-D)	1,2	–	–
G6PD (glucose-6-phosphate dehydrogenase)	A,B	–	–
PGD (phosphogluconate dehydrogenase)	A,C,R	–	–
PGM1 (phosphoglucomutase)	1,2	–	1A,1B,2A,2B
CA2 (carbonic anhydrase)	1,2	–	–
GLO (glyoxylase)	1,2	–	–

and the high cost, the tests are read and recorded independently by two technologists, and the results are reviewed by a third technologist. Where necessary the family of the mother or alleged father are typed to clarify the inheritance of the genes.

3.2 Red cell antigens

The blood group systems used includes ABO, MNS, Rh (D, C, E, c, and e), Duffy (Fy^a and Fy^b), and Kell (K and k). Blood groups should be determined independently by two qualified technologists each of whom uses a different 'set' of antisera. Standard methods such as those recommended by Race and Sanger (3) are used. General methods are outlined in *Protocol 2*. Antisera should always be used strictly according to the manufacturer's directions. The test cells must always be washed once in isotonic (0.9%) saline before a 3–5% suspension is made also in isotonic saline. It should be noted that ABO blood grouping of adults is always done in two parts, i.e. grouping of red cells with known antisera (*Protocol 2A*), and determining the antibodies present in the patient's serum by testing with known cells (*Protocol 2B*). Individuals grouped as blood group A may be sub-divided into A_1 or subgroups of A by means of anti-A_1 (*Protocol 2C*). When determining the Rh D antigen a reagent capable of detecting the D^u antigen should be used (*Protocol 2D*). The antisera source for M and N typing may be lectin, rabbit, or human (*Protocol 2F*). In those cases where it is necessary, additional antisera, (e.g. anti-M^g, anti-C^w, anti-Ce) should also be used. Whenever an anti-human globulin (AHG) test is performed it is necessary to do a direct AHG test on the patient's cells by washing the cells three times, adding two drops of AHG, centrifuging and examining for agglutination (*Protocol 2G*).

This is to rule out the possibility of false positive reactions due to the fact that the cells are already sensitized before the addition of antisera. Most laboratories now use monoclonal antibodies (MoAb); the methods are similar regardless of antibody source. For paternity testing we always use commercial antisera (MoAb or alloantisera).

Protocol 2. Red cell antigen testing

Antisera are available from Biotest AG, DiaMed Ag, CNTS, Organon GmbH, and many other companies. Methods can be performed as directed on the package insert, or as indicated below. Appropriate positive and negative controls must be included in each test. Use known heterozygous cells as positive controls. Cells known to be negative from previous tests can be used as negative controls.

A. *ABO (grouping of cells)*

1. Place one drop appropriate antisera (anti-A, anti-B, and anti-AB) into correspondingly labelled tubes.
2. Add to each one drop 3–5% suspension of cells to be tested and mix.
3. Incubate at room temperature for 30 min.
4. Gently dislodge cell button and examine for agglutination.

B. *ABO (grouping of serum)*

1. Place one drop of serum from specimen to be tested into appropriately labelled tubes (A_1, A_2, B, O neg).
2. Add one drop of 3–5% suspension of known A_1, A_2, B, and O negative cells to the correspondingly labelled tubes and mix.
3. Proceed as for *Protocol 2A* steps **3** and **4**.

Note. Since no antibodies are produced during the first months of life, grouping of serum is not reliable on babies under six months old.

C. *Subgroups of A*

These may be determined by the use of anti-A_1 reagent.

D. *Rh D antigen*

1. Prepare the cells as per directions, either in saline or whole blood.
2. Place one drop of anti-D reagent into tube.
3. Add one drop of cell suspension and mix.
4. Incubate at 37 °C for 30 min.
5. Resuspend cell button and examine for agglutination.

Protocol 2. *Continued*

6. If there is no agglutination proceed to test for D^u as set out in *Protocol 2G* steps **4–7**.

E. *Rh C, c, E, e*

Methods vary greatly depending on the antiserum used; follow the directions from the package insert.

F. *M and N typing*

1. As for *Protocol 2E*.

G. *S, s, K, k, Fy^a, Fy^b*

1. Place one drop of antisera in correspondingly labelled tubes.
2. Add one drop of 3–5% cell suspension to each.
3. Incubate at 37°C for 30 min or as recommended.
4. Wash cells three times with isotonic saline.
5. Add two drops AHG and mix well.
6. Centrifuge at 200 *g* for 1 min.
7. Resuspend cells gently and examine for agglutination.

3.3 Red cell enzyme and plasma protein systems

Alleles of these proteins are characterized according to isoelectric point by isoelectric focusing (see Chapter 7), and according to molecular weight. Depending on the protein tested, starch gels, agarose gels, or polyacrylamide gels are used. The sample is applied to the gel, and proteins are separated by subjecting the system to electrophoresis. After separation, the alleles are visualized by protein staining protocols. Specific protocols are used for each genetic marker; a schedule of the methods used in this laboratory are outlined in *Table 1*. The details of the standard methods for determining red cell enzyme phenotypes are described by Harris and Hopkinson (4). The plasma protein phenotyping of transferrin and haptoglobin are carried out according to the methods described by Giblett (5). Properdin factor B typing are tested according to Alper *et al.* (6). These tests are not performed in duplicate, but the results are read and recorded by two independent observers.

Although this methodology is well established and relatively cheap, many systems are not very polymorphic and hence not very informative. It is anticipated that routine use of the less polymorphic systems will diminish in the future.

3.4 DNA typing

The International Society for Forensic Haemogenetics at present attaches great importance to the establishment of definite conditions in the field of DNA testing. These include:

- definition of DNA probe (mapping, stability, purity)
- definition of the basic genetics of the DNA probe (family studies)
- implementation of parallel-studies
- investigation of the mutation rate for each allele

The DNA commission of the Society for Forensic Haemogenetics report of 1991 (7) also states that paternity testing with conventional techniques is a well established procedure for producing evidence in court cases, and can continue to be used either alone or in combination with determination of DNA polymorphisms. They also stated that, provided that a DNA system has been suitably and adequately scrutinized, there is no reason why DNA should not be used alone. Before any laboratory decides to provide DNA typing for paternity testing it is however recommended that data is exchanged with another laboratory recognized in the field, so that comparability of data is verified.

Genetic polymorphism in DNA sequence can result from nucleotide base substitutions or from insertions or deletions of nucleotide(s). The latter can be recognized on the basis of differences in size (kilobases) of DNA fragments. The method of detection is restriction fragment length polymorphism (RFLP) analysis (see Chapter 5) in which allelic polymorphic DNA fragments are generated by digestion of high molecular weight DNA with a restriction enzyme. The resulting fragments are separated according to size by agarose gel electrophoresis. After transfer of the DNA to a membrane (Southern blotting), the presence of a specific allelic DNA fragment may be detected by hybridization with a radiolabelled probe. DNA fragments are then visualized by autoradiography. Among the most informative DNA polymorphisms are insertion/deletion polymorphisms containing variable number of tandem repeat (VNTR) sequences, which can be detected with probes which recognize a core sequence of the repeat unit. There are two main groups of systems for the detection of VNTR polymorphism:

- single-locus systems (8)
- multi-locus systems (9, 10)

The same protocol is followed for both single and multi-locus probes. Methodology varies depending on the probe used. Many of the commercially available probes are patented. Patented 'kits' are available for several stages of the protocol to perform DNA analysis in the routine laboratory (Amersham International, Promega Corporation). When kits are used, the methodology is performed as indicated on the package insert.

3.4.1 Single-locus systems

These systems are more commonly used as the results are easier to interpret. It is recommended that single-locus systems with low mutation rates are used. To be excluded the allele of the putative father and the non-maternal allele of the child should clearly differ in size. For ease of comparison, the samples from the trio of mother, child, putative father are run in adjacent lanes on the agarose gel. Allele frequency should be established for each population group and there should be compliance to Hardy–Weinberg equilibrium expectations. When these criteria have been met, calculated probabilities of paternity can be used. In this laboratory we use the following system.

DNA fragments are generated by digestion with the restriction enzyme *Pvu*II, which has been found to be suitable for use in combination with a range of DNA single-locus probes. We use probes which detect the following allelic VNTR: alpha globin 3′ hyper-variable repeats (HVR) (11), mucin HVR (12), HaRas HVR (13), YNH24 (14), CMM101 (15), and TBQ7 (16) (Promega Corporation). These allelic systems are highly polymorphic and the probes are commercially available.

3.4.2 Multi-locus systems

In these systems, polymorphisms at multiple loci are detected simultaneously resulting in a pattern of visualized migration bands, the 'DNA-fingerprinting' technique (9). The original 'Jeffreys multi-locus probes' are subject to a patent and are commercially available (Cellmark Diagnostics). A drawback of this system is that the results are complicated and therefore hard to interpret. Results obtained with this method are only considered to be reliable if the following criteria are met.

Mutation rates of the detected alleles must be known, and it should be realized that rates of 10^{-2} and higher can occur. Determination of band sizes and/or band patterns should be scored in an objective manner, i.e. an automated reader system should be used. Questions concerning manner of inheritance, allelism, and linkage disequilibrium need to be answered and biostatistical calculations should only be used with caution. Until more international experience has been obtained with this method, it is advisable to give only a verbal opinion on exclusion or non-exclusion to the courts.

4. Analysis of data

Genetic systems used in paternity testing must meet the following criteria:

- have a clear cut hereditary mechanism
- availability and simplicity of test procedures
- give clear reproducible results
- detectable from birth

Listed below are certain assumptions that have to be made in the determination of the genotypes of each individual in the trio, i.e. mother/alleged father/child to arrive at a conclusion.

(a) The mother is in fact the biological mother and one half of the genetic information in the child is maternal in origin and one half paternal.
(b) Only one man can be the biological father of the child and a sperm containing one half the genetic information present in this individual fertilized the ovum containing one half the genetic information of the biological mother.
(c) Mutations do not affect blood group genes (rate less than one in a million).
(d) The marker studied is established as a product of genes which obey Mendel's Laws of inheritance so that:
   i. a child cannot have a genetic marker that is absent in both parents
   ii. a child must inherit one of a pair of markers from each parent
   iii. a child cannot have a pair of identical genetic markers unless both parents have the marker
   iv. a child must have a genetic marker if it is present as an identical pair in one parent
(e) The test method is reliable and reproducible and measures an inherited characteristic.
(f) The genetic system is polymorphic and the gene frequencies have been established by examination of an appropriate random sample of the population.
(g) The different genetic systems used in calculating inclusion and exclusion probabilities segregate independently. For example, the inheritance of the ABO genes is independent of the HLA genes.
(h) No clerical errors occurred in the labelling of tubes and recording of the results.
(i) The reagents used were suitable and specific as indicated by the results of control specimens.
(j) All tests have been performed correctly.

4.1 HLA in paternity testing

The tremendous HLA polymorphism (over 60 HLA-A and -B alleles (see Chapter 11) and the low frequencies of the phenotypes make it extremely useful in solving problems of disputed parentage. The pattern of inheritance of a combination of the antigens from each parent in the form of a haplotype occurs because the genes coding for the HLA-A and -B antigens are closely linked on chromosome 6. The haplotype frequencies for different populations are used in the calculation of the probability of paternity. Although crossing-over (reciprocal exchange of material between chromosomes during meiosis which is responsible for genetic recombination) does occur, it is a relatively rare event (frequency of about one per cent) between the HLA-A and -B loci. It is imperative that reliable haplotype frequencies have been recorded for the population to which the putative father belongs. It is important that the investigator is familiar with the problems encountered in defining antigens in the particular population. For example, HLA-A43 has only been seen in Southern African populations and most laboratories in other parts of the world do not have the antisera which recognize it. A man may thus be falsely accused of being the biological father due to the inability of a laboratory to identify a rare antigen.

HLA typing alone offers an exclusion probability of about 95% (*Table 2*). However, it should be noted that the potential problem of laboratory error is magnified when only a single genetic system is tested, and we therefore always test multiple genetic systems. The addition of the blood group, plasma protein, and red cell enzyme genetic systems, brings the probability of excluding a falsely accused man to between 99.4% and 99.8% depending on the population group tested.

4.2 Statistical analysis

4.2.1 Exclusions

The probability (or power) of exclusion (PE) of non-fathers is a measure of the efficacy of individual systems in determining paternity and is dependent on the number of alleles in the system, the gene frequencies for each allele, and whether the phenotype expressed by the alleles can be detected (17). The observed PE for each genetic system or locus is established by determining the incidence with which each genetic system provided an exclusion. This has to be determined over a sufficiently large number of cases within the same racial group. The observed rate of exclusion for each system obtained should closely approximate the expected (theoretical) PE calculated from the gene frequencies. The exclusion probability formulas depend on the number of alleles identified (17).

As is shown in *Table 2*, the PE for the genetic systems varied from one population group to another. For example, the Rh system alone excluded 43.9% of the Caucasoid non-fathers but only 14.0% of the Black non-fathers. This is a reflection of the variations in gene frequencies between different populations groups. The G6PD allotypes are X-linked and therefore only used in female children. This system is however extremely useful in Blacks where the percentage of putative fathers excluded by G6PD alone was 27.1%; whereas only 6.9% of Cape Coloured men were excluded and no South African Caucasoid men. This is the result of the higher frequency of the A allele seen among South African Blacks compared to the other two groups.

Table 2. The proportion (%) of non-fathers excluded[a] by individual systems and groups of systems, and the cumulative probability of exclusion for the combined systems in the three populations

	South African Caucasoid [b]n = 57	Cape Coloured n = 463	South African Black n = 265
Red cell blood group			
ABO	12.3	13.0	22.3
MNS	31.6	26.0	26.8
Rh	43.9	23.0	14.0
Kell	1.8	1.9	0
Duffy	14.0	15.6	14.6
Cumulative	71.5	59.0	58.2
Plasma proteins[c]			
HP	13.5	19.2	17.7
TF	0	2.5	2.0
BF	7.7	20.9	29.9
Cumulative	20.1	37.7	43.4
Red cell enzymes[c]			
ACP1	9.8	19.5	16.7
ADA	2.0	3.6	0
AK	2.0	2.9	1.2
ESD	2.4	8.1	3.4
G6PD (females)	0	6.9	27.1
PGD	0	4.0	6.2
PGM1	7.8	17.7	17.8
CA2	0	2.6	5.0
GLO	9.1	20.3	18.1
Cumulative	29.1	60.4	65.1
HLA (A and B)	96.4	97.8	93.5
All 18 systems	99.4	99.8	99.5

[a] The observed rate of exclusion for each system is expressed as a percentage of the total number of excluded men when using all 18 systems.
[b] n = The total number of men excluded.
[c] See *Table 1* for the full names of the plasma protein and red cell enzyme systems.

In order to evaluate the efficiency of using multiple marker systems, one must consider the cumulative power of exclusion (CPE) for all the systems tested (17). The cumulative probability that at least one of several tests performed will exclude a falsely accused man can be calculated by employing the formula:

$$CPE = 1 - (1 - PE_1)(1 - PE_2) \ldots (1 - PE_n)$$

where PE_1, PE_2, and PE_n are the probabilities of exclusion for individual systems (17).

We calculated the cumulative chances of excluding a man from paternity in the three populations tested using all 18 systems. It is interesting to note that there is very little difference in the CPEs for the three population groups. The CPE was 99.4% for Caucasoids, 99.8% for Cape Coloureds, and 99.5% for Blacks (*Tables 3, 4, 5*).

Table 3. Probability of exclusion 'top 10' systems in South African Caucasoids

Systems	Observed exclusion (%) (PE)	Cumulative exclusion (%) (CPE)
1 HLA-A and -B	96.43	96.43
2 Rh	43.86	98.00
3 MNS	31.58	98.63
4 Duffy	14.04	98.82
5 HP	13.46	98.98
6 ABO	12.28	99.11
7 ACP1	9.80	99.19
8 GLO	9.09	99.27
9 PGM1	7.84	99.32
10 BF	7.69	99.38
all 18 systems		99.42

Table 4. Probability of exclusion 'top 10' systems in Cape Coloureds

Systems	Observed exclusion (%) (PE)	Cumulative exclusion (%) (CPE)
1 HLA-A and -B	97.78	97.78
2 MNS	25.92	98.36
3 Rh	23.11	98.74
4 BF	20.94	99.00
5 GLO	20.28	99.20
6 ACP1	19.46	99.36
7 HP	19.24	99.48
8 PGM1	17.67	99.57
9 Duffy	15.56	99.64
10 ABO	12.96	99.69
all 18 systems		99.76

Table 5. Probability of exclusion 'top 10' systems in South African Blacks

Systems	Observed exclusion (%) (PE)	Cumulative exclusion (%) (CPE)
1 HLA-A and -B	93.46	93.46
2 BF	29.89	95.41
3 MNS	26.79	97.55
4 ABO	22.26	98.10
5 GLO	18.10	98.44
6 PGM1	17.76	98.72
7 HP	17.65	98.94
8 ACP1	16.67	99.12
9 Duffy	14.58	99.25
10 Rh	13.96	99.35
all 18 systems		99.46

4.2.2 Probability or likelihood of paternity

Blood typing results are usually considerd conclusive if the putative father is excluded. Failure to exclude an alleged father however, does not mean that he has been proved to be the biological father. Absolute proof of paternity cannot be established by any known blood tests. However, when a man is not excluded the likelihood (or probability) of paternity can be calculated and the results considered as additional evidence, which is weighed with other evidence presented (1, 18).

When no exclusion is obtained a comparison is made of the chance (X) the putative father has of being able to fertilize an ovum with a sperm carrying all the known paternal genes, (bearing in mind that on many occasions the genetic contribution from the father could be one of several alleles) with the chance (Y) of a random man fertilizing an ovum with a sperm carrying the same genes. The value of Y is the same as the gene frequency in the population and is determined by testing a sample of the population. These gene frequencies vary in different racial groups. The reliability of the values depends on having a large enough sample. A much larger sample is especially needed to obtain an accurate estimate of the gene frequencies in systems in which there are numerous alleles most of which have a low frequency.

The ratio of X/Y is known as the likelihood ratio. This value indicates the chance that a man with the phenotype observed in the alleged father could transmit the paternal gene compared to a man in the random population. A summary of the results in several systems can be calculated by multiplying the X/Y values for each system used. This figure is called the paternity index (PI), which is the genetic odds in favour of paternity (19).

Another way of presenting the results is to calculate a likelihood of paternity or probability of paternity (W) (1, 18). In order to do this Essen-Moller in 1938 devised a formula expressing the probability of paternity (W) numerically (1).

$$W = \frac{1}{1 + Y/X}$$

This combines the genetic information with the non-genetic information (the assumption that the non-excluded tested man had the opportunity to father the child). The calculation of likelihood of paternity (W) is based on Bayes' Theorem which is the assumption of a prior possibility of 0.5 (non-genetic information), if (in a two-hypothesis case) there is no further information available on the case in dispute, or if one wants to take a neutral basis as one's starting point. As the testing laboratory does not have knowledge of the non-genetic factors most experts use a neutral value of 0.5 prior probability in their calculation. The first suggestion for verbal assessment of different W ranges also came from Essen-Moller. These were later modified by Hummel (18) as shown in *Table 6*.

In *Table 7* an example of two different theoretical child-mother-alleged father trios is given and the PI and W calculated. This emphasizes the importance of having the appropriate population gene frequencies for the calculation. In example 1, a classic South African Caucasoid haplotype was used in the alleged father, while in example 2, a classic Black haplotype was used. In each case the PI and W were calculated using haplotype frequencies for the different populations. It can be seen in example 1 that for South African Caucasoids, where the HLA-A1, B8 haplotype is very common, the alleged father has a low probability of paternity, but if the Black haplotype frequency is used, the probability is high, indicating the alleged father is very likely to be the true father. Conversely, in example 2, if the frequency for the HLA-A30, B42 haplotype in South African Caucasoids is used, the probability of paternity is very high, whereas using the Black haplotype frequency produces a low PI. The results clearly illustrate the importance of using the appropriate population frequencies based on reliable data.

Table 6. Verbal predicates for the different likelihoods of paternity

W (%)	Likelihood of paternity	PI
99.8–99.9	practically proved	>399 to 1
99.0–99.7	extremely likely	> 95 to 1
95.0–98.9	very likely	> 19 to 1
90.0–94.9	likely	> 9 to 1
80.0–89.9	certain hint	> 4 to 1
less than 80	not useful	< 4 to 1

Table 7. Two examples of the calculations of the PI and *W* (%) values in the three South African populations using the HLA haplotype frequencies of these populations

	Example 1		**Example 2**	
Alleged father	A1,3	B7,8	A30,28	B42,70
Child	**A1**,X	**B8**,Y	**A30**,X	**B42**,Y
Mother	A24,X	B27,Y	A23,X	B45,Y
Population	**PI**	***W* (%)**	**PI**	***W* (%)**
South African Caucasoid	7.18	87.78	547.56	99.82
Cape Coloured	21.80	95.61	27.17	96.45
South African Black	46.89	97.91	7.56	88.18

Bias due to insufficient knowledge of the system and inadequate gene frequencies may have an important effect upon the inclusionary estimates of the particular system.

5. Reporting

The interpretation of the results in a case of disputed parentage is the challenge of that person assuming responsibility for the final report. The conclusions are based on a careful review of all the data followed by the application of the laws and principles of genetics, bearing in mind all the pitfalls in interpretation, such as the presence of a 'silent' allele. For example, only one antigen of a particular system may be detected and the individual wrongly assumed to be homozygous, where segregation studies may show it to be as a result of an as yet undetected or non-expressed ('silent') allele.

Although blood tests are not obligatory in all countries, in practice every time paternity is denied on application by any party involved in the proceedings, the court may give a direction for a blood group investigation. The court may direct, not order, because blood samples may not be taken from an individual without first obtaining his or her consent (otherwise it is an assault against the person). In many cases, where the mother and the putative father both consent to blood tests, a formal direction is not made by the court and the results are presented to the lawyers acting on behalf of the parties. However, it may be the knowledge that a negative inference can be drawn from the fact that there was unwillingness to undergo blood tests, that motivates many of these informal arrangements.

Acknowledgements

I wish to thank Dr P. Creemers and Dr J. Rousseau for helpful comments regarding the manuscript. Mrs L. Halliday is thanked for assistance in preparing the manuscript.

References

1. Essen-Moller, E. (1938). *Mitt. Anthropol. Ges. Wein,* **68,** 9.
2. American Association of Blood Banks (1984). *Standards for parentage testing laboratories.* American Association of Blood Banks Press, Arlington, VA.
3. Race, R. R. and Sanger, R. (ed.) (1975). *Blood Groups in Man.* 6th edn. Blackwell Scientific Publications, Oxford.
4. Harris, H. and Hopkinson, D. A. (ed.) (1976). *Handbook of Enzyme Electrophoresis in Human Genetics.* North-Holland, Amsterdam.
5. Giblett, E. R. (ed.) (1969). *Genetic Markers in Human Blood.* Blackwell Scientific Publications, Oxford.
6. Alper, C. A., Boenish, T., and Watson, L. (1972). *J. Exp. Med.,* **135,** 68.
7. Brinkmann, B., Butler, R., Lincoln, P., Mayr, W. R., and Rossi, U. (1991). *Forensic Science International,* **52 (2),** 125.
8. Allen, R. W., Wallhermfechtel, M., and Miller, W. V. (1990). *Transfusion,* **30,** 552.
9. Jeffreys, A. J., Wilson, V., and Thein, S. L. (1985). *Nature,* **316,** 76.
10. Jeffreys, A. J., Wilson, V., and Thein, S. L. (1985). *Nature,* **314,** 67.
11. Higgs, D. R., Goodburn, S. E. Y., Wainscoat, J. S., Clegg, J. B., and Weatherall, D. J. (1981). *Nucleic Acids Res.,* **9,** 4213.
12. Gendler, S., Taylor-Papadimitriou, J., Duhig, T., Rothbard, J., and Burchell, J. (1988). *J. Biol. Chem.,* **263,** 12820.
13. Capon, D. J., Chen, Y. E. L., Levinson, A. D., Seeburg, P. H., and Goeddel, D. V. (1983). *Nature,* **302,** 33.
14. Nakamura, Y., Gillilan, S., O'Connell, P., Leppert, M., Lathrop, G. M., Lalouel, J.-M., and White, R. (1987). *Nucleic Acids Res.,* **15 (23),** 10073.
15. Nakamura, Y., Culver, M., Gill, J., O'Connell, P., Leppert, M., Lathrop, G., Lalouel, J.-M., and White, R. (1988). *Nucleic Acids Res.,* **16 (1),** 381.
16. Nakamura, Y., Carlson, M., Krapcho, K., Kanamori, M., and White, R. (1988). *Am. J. Hum. Genet.,* **43,** 854.
17. Dykes, D. D. (1982). In *Probability of Inclusion in Paternity Testing* (ed. H. Silver), pp. 15–26. American Association of Blood Banks Press, Hartford, Conn.
18. Hummel, K. (1984). *Forensic Sci. Int.,* **25,** 1.
19. Gurtler, H. (1956). *Acta Med. Leg. Soc. Liege,* **9,** 83.

10

Mapping techniques used to isolate novel major histocompatibility complex genes

R. DUNCAN CAMPBELL, JOHN TROWSDALE, and IAN DUNHAM

1. Introduction

The HLA class I and class II regions each encode highly polymorphic families of cell surface glycoproteins involved in immune regulation. The class I region contains at least 17, and possibly as many as 40, highly related genes which include those encoding the classical transplantation antigens (HLA-A, -B, and -C). The class II region (HLA-D) is arranged into subregions HLA-DP, -DQ, and -DR, each containing at least one A and B pair of genes encoding the α and β polypeptide chains of the class II molecule. A number of other class II related sequences (including HLA-DO, -DN, and -DM) have also been defined within the HLA-D region. The class I and class II regions are separated by the class III region which also contains genes with immune related functions such as the complement genes C2, C4, and Bf, the tumour necrosis factor (TNF) A and B genes, and the major heat shock protein HSP70 genes. In order to characterize the HLA-A to -DP region in detail, the two techniques of pulsed-field gel electrophoresis (PFGE) and chromosome walking are especially useful. Used together, they are powerful techniques for elucidating the overall organization of a genomic region, as has been amply shown for the human and murine Mhcs.

Pulsed-field gel electrophoresis (1–3) allows the separation of fragments up to 10 000 kb in size, and can be used to produce long-range restriction maps based on the hybridization of probes to common fragments on Southern blots (4–7). The infrequently cutting enzymes used in PFGE are also useful for identifying HTF (*Hpa*II tiny fragment)-islands. These are CpG rich, short unmethylated stretches of DNA which are often found at the transcriptional start sites of genes (8). Comparison with the distribution of the same enzyme sites in cloned DNA can be used to define which are cleaved at the chromosomal level. Once the position of a potential island structure has been found, flanking probes may then be used to search for transcripts by northern blot hybridization.

Chromosome walking using cosmid (9, 10) and/or yeast artificial chromosome (YAC) vectors (11, 12) to isolate large stretches of genomic DNA allows the construction of a molecular map (13–16). The cloned DNA can be analysed for the distribution of unique sequences, followed by the analysis of single copy probes to define which nucleotide sequences are conserved between species. This represents one approach for the identification of potential genes (17). In addition the cosmid (16) or YAC (18) clones can be used to directly screen cDNA libraries in order to isolate cDNAs corresponding to novel genes, whose location in the cosmid or YAC inserts can be ascertained by using the cDNAs as probes in Southern blot analysis of the cloned DNA.

2. Pulsed-field gel electrophoresis: a method for long-range mapping

Conventional agarose gel electrophoresis has an effective upper limit of separation by molecular sieving of about 50 kb. Efforts to improve resolution above this size by the use of dilute agarose gels (down to 0.035% agarose) have been hampered by poor resolution and the fragility of the gels. Schwartz and Cantor (1) realized that if the DNA was subjected to alternating, approximately perpendicular electric fields, a novel form of separation was achieved. Thus, they were able to separate yeast chromosomal DNAs up to an estimated size of 2000 kb and the technique of PFGE was born. Carle and Olson (2) modified the system to produce orthogonal field alternation gel electrophoresis (OFAGE) with a more convenient electrode configuration, and were also able to separate yeast chromosomes. A number of further modifications to the original idea have been made, in particular a variety of electrode geometries, homogeneous fields, field switching regimes, and angles between fields have been explored to produce linear DNA trajectories. These include a system where the two fields at 180° are alternated in either duration or strength (19), systems where multiple electrode arrays are individually voltage clamped or autonomously controlled by a computer to produce homogeneous fields (20–22), a vertical system with an orthogonal design (23), a vertical system based on the simultaneous application of fixed and cyclically alternating polarity fields at a right angle (24), systems in which the gel is rotated between equal and opposite angles in a homogeneous field (25, 26), or a system where the electrodes are rotated rather than the gel (27).

Parameters which affect the separation in PFGE include temperature, voltage gradient, reorientation angle, and the interval between field switching (3). The most useful variable is switching interval. Briefly, as the switching interval is increased, DNA molecules of increasing size are separated from the bulk of slowly moving unresolved DNA. Maximum resolution between

Table 1. Commonly used rarely cutting restriction endonucleases and their predicted average fragment size in mammalian DNA

Enzyme	Recognition site	Average fragment size (kb)
*Bss*HII	GCGCGC	100
*Cla*I	ATCGAT	100
*Eag*I	CGGCCG	100
*Fse*I	GGCCGGCC	500
*Fsp*I	TGCGCA	100
*Ksp*I	CCGCGG	100
*Mlu*I	ACGCGT	300
*Nae*I	GCCGGC	100
*Nar*I	GGCGCC	100
*Not*I	GCGGCCGC	1000
*Nru*I	TCGCGA	300
*Pvu*I	CGATCG	300
*Rsr*II	CGGWCCG	200
*Sal*I	GTCGAC	100
*Sfi*I	GGCCN5GGCC	200
*Sma*I	CCCGGG	100
*Xho*I	CTCGAG	100

W – A or T
N – any nucleotide

molecules in a given size range occurs closest to the switching interval required to separate the DNA from the unresolved region.

In order to use PFGE for mapping of mammalian DNA, which is at least two orders of magnitude larger than, say, the yeast genome, methods are required to reproducibly fragment the DNA into sizes that can be analysed. A number of restriction endonucleases are known that cleave mammalian DNA rarely enough to produce fragments in the 50–1000 kb range (see *Table 1*). The enzymes may simply have a long, and therefore infrequently occurring, recognition sequence. An example of this is the enzyme *Sfi*I. However, the most useful enzymes for PFGE mapping are those that have either six or eight base pair recognition sites which contain one or more CpG dinucleotides. CpG is rare in the mammalian genome (9) and 70–90% of CpG is methylated. Since many of these enzymes are sensitive to cytosine methylation in their recognition sequences (28), cleavable sites are infrequent. Additionally, unsheared genomic DNA is required for digestion and this can be conveniently prepared in agarose blocks (see *Protocol 1*).

Protocol 1. Isolation of mammalian chromosomal DNA from cultured cells in agarose blocks

Care is essential to avoid any nuclease contamination.

- wear disposable gloves during all manipulations
- autoclave all tubes and dispensing tips

Protocol 1. *Continued*

- where possible autoclave all solutions
- use glass double distilled or double deionized autoclaved water to prepare all solutions which cannot be autoclaved, and then filter-sterilize through a 0.22 μm sterile filter unit

1. Wash tissue culture cells in PBS, count using a haemocytometer, wash in PBS, and resuspend in PBS at 2×10^7 cells/ml. Prepare peripheral blood mononuclear cells as in Chapter 3.
2. Make 2% low gelling temperature (LGT) agarose (FMC Seaplaque—Flowgen, 50101) in PBS and hold at 40°C. If necessary the agarose can be purified by prior treatment with DEAE-cellulose DE52 (Whatman, 4057 050) to remove negatively charged inhibitors of restriction enzymes. Incubate 50 ml of 2% LGT agarose twice with 5 ml of packed DEAE-cellulose (thoroughly washed with PBS) at 50°C for 30 min. Remove the resin by three successive centrifugations at 3000 *g* for 10 min each, holding the agarose at 50°C between each centrifugation to prevent gelling.
3. Stick 25 × 66 mm tape on to one surface of the clean perspex plug mould containing slots of dimensions 9 × 7 × 2 mm, as shown in *Figure 1*, and place on ice to cool. The dimensions of the perspex plug mould can be varied dependent on the dimensions of the comb being used to form the wells of the PFGE gel.
4. Warm cells to 37°C then mix gently, but thoroughly, with an equal volume of the 2% LGT agarose.
5. Dispense cell–agarose mixture immediately into the perspex plug mould on ice and allow blocks to set for 10 min. Avoid air bubbles.
6. Remove backing tape and knock blocks into a tube (Falcon) containing 10 ml of 0.5 M EDTA/10 mM Tris pH 9.5/1% Sarkosyl (NDS), 1 mg/ml with proteinase K (Boehringer, 745 723) using a bent Pasteur pipette.
7. Incubate blocks overnight at 50°C. Decant off solution, add 10 ml fresh NDS-proteinase K solution, and incubate for a further 48 h at 50°C.
8. Wash blocks twice for 30 min each in 10 ml NDS and store in NDS at 4°C. The plugs are stable for several years.

Restriction enzyme digestion of high molecular weight DNA with infrequently cutting enzymes (see *Protocol 2*), separation of the restriction fragments by PFGE, Southern blotting, and hybridization with appropriate probes gives construction of a physical map of a region in an analogous way to conventional restriction mapping.

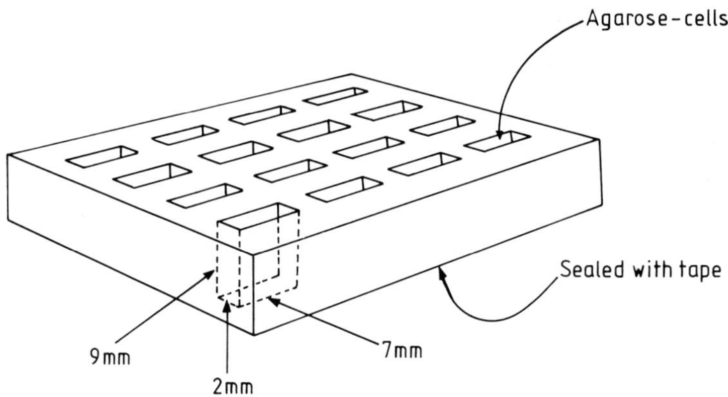

Figure 1. A perspex mould used for preparing DNA embedded in agarose. The dimensions of the wells in the mould can be varied dependent on the dimensions of the comb being used to form the wells in the PFGE gel.

Protocol 2. Restriction enzyme digestion of DNA in agarose blocks and PFGE analysis

(a) Follow the same precautions described in *Protocol 1* in order to avoid nuclease contamination and subsequent DNA degradation.

(b) Make up restriction enzyme buffers as 10 × stock solutions without bovine serum albumin (BSA) (Gibco-BRL, 540-5661UA), spermidine (Sigma, S2501), and dithiothreitol (DTT) (Sigma, D9779) [or β-mercaptoethanol (Sigma, M3148)], autoclave, and store at −20°C.

1. Slice one-third of an agarose block (3 × 7 × 2 mm) containing 5–10 μg of DNA from the blocks prepared as described in *Protocol 1* for each digest. This can be varied such that the dimensions of the slice match the sample loading well of the gel. The best results will be achieved using slices which are long and narrow.
2. Wash blocks three times in 10 ml of 10 mM Tris/0.1 mM EDTA pH 8.0/0.1 mM PMSF (Sigma, P7626) for 30 min at 4°C. If all the blocks are from the same cell line they can be washed together. Otherwise wash separately.
3. Equilibrate in 0.5 ml restriction enzyme buffer without BSA, spermidine, and DTT (β-mercaptoethanol) on ice. The restriction enzyme buffer used is dependent on the enzyme and is that recommended by the supplier.
4. Replace the buffer with 100 μl fresh restriction enzyme buffer containing 500 μg/ml BSA, and the appropriate concentration of DTT (or β-mercaptoethanol). Include 5 mM spermidine if the buffer contains ≥ 50 mM

Protocol 2. *Continued*

NaCl. Add 10–50 U of enzyme and incubate at the appropriate temperature for up to 3 h.

5. For double digestions, carry out the first enzyme digestion as above. Next equilibrate blocks in 0.5 ml of the second restriction enzyme buffer without BSA, spermidine, and DTT (or β-mercaptoethanol). Continue digestion with the second enzyme as in step **4**.
6. After digestion place the blocks at 4°C for at least 10 min, slice in half (1.5 × 7 × 2), and load directly into PFGE gel slots slightly larger than the blocks. Seal the blocks in the gel slots with 0.5% LGT agarose. Store blocks (plus unused half blocks) in NDS at 4°C for up to six months. Low molecular weight DNA may diffuse out of the block on prolonged storage.
7. Conditions for PFGE are dependent on the apparatus being used and the separation range required, and the reader should refer to the manuals supplied by the manufacturers of the PFGE apparatuses. Generally, electrophoresis using 1.5% agarose gels containing 0.5 × TAE (20 mM Tris-acetate/1 mM EDTA pH 8.5) is carried out in 0.5 × TAE at a constant temperature and with a fixed voltage between the electrodes. For most applications the only experimental variables that need to be adjusted are the pulse time and run time. However, the voltage gradient across the electrodes and the gel concentration can also be varied.
8. After electrophoresis stain PFGE gel in 500 ml of 0.5 × TAE containing 200 μl 0.5% ethidium bromide for 30 min. Destain for up to 1 h in water. Photograph gel with a ruler beside it using a UV transilluminator (306 nm), and Polaroid camera or video imaging system, to obtain permanent record of run.
9. Wash PFGE gel twice in 0.25 M HCl for 20 min each at room temperature to depurinate the DNA. Blot DNA on to nylon or nitrocellulose membranes (Amersham, Hybond-N and Hybond-E) following protocol suggested by supplier.
10. Probe blots by hybridization (see Chapter 5).

In order to estimate the size of the restriction fragments detected in Southern blot analysis it is necessary to include appropriate standards in the PFGE run (see *Table 2* and *Figure 2*). At the lower size range these are usually a *Hin*dIII digest of lambda DNA (0.5–23 kb) together with concatemers of lambda DNA which yield a series of markers which vary in size by one molecule, (i.e. 48.5–~1000 kb). At the higher size range both lambda concatemers and yeast chromosomes would be used as size markers, while chromosomes from *Schizosaccharomyces pombe* can act as size markers at the limits of PFGE resolution (Promega, G3011, G3021, D1951; FMC BioProducts, 50401, 50411, 50421; Pharmacia, 27-4530-01, 27-4520-01; New England BioLabs,

Table 2. Standards commonly used in PFGE analysis (all sizes in kb)

***Hind*III digest of λ-DNA**	**λ-concatemers**	***S. cerevisiae* chromosomes* (strain YPH80)**	***S. pombe* chromosomes**
23.7	824.5	1900	5700
9.5	776	1640	4600
6.6	727.5	1100	3500
4.3	679	1080	
2.3	630.5	950	
2.0	582	910	
0.6	533.5	820	
	485	790	
	436.5	750	
	388	680	
	339.5	600	
	291	550	
	242.5	440	
	194	360	
	145.5	280	
	97	220	
	48.5		

* can differ between different strains.

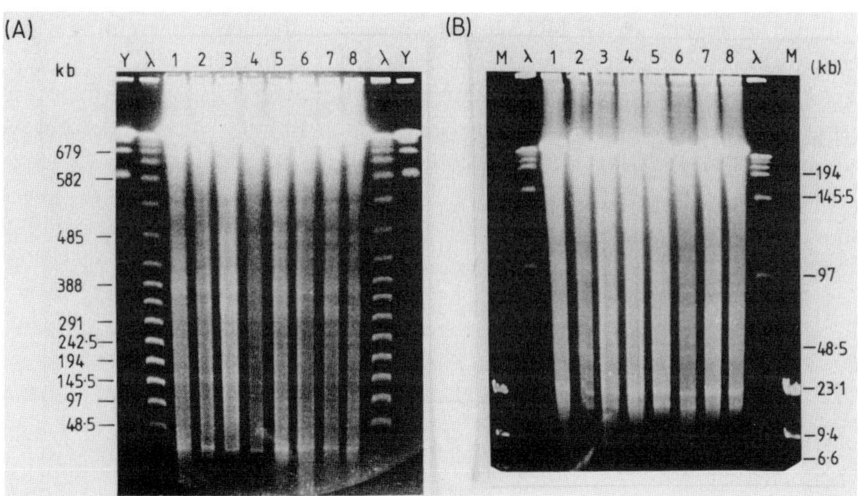

Figure 2. Examples of PFGE separations using different pulse times. (A) Ethidium bromide stained gel showing separation of eight cell line DNAs digested with *Mlu*I by PFGE using a 30 sec switching interval. The markers are yeast chromosomes (Y) and λ concatemers (λ). (B) Ethidium bromide stained gel showing separation of eight cell line DNAs digested with *Bss*HII by PFGE using a 7.5 sec switching interval. The markers are λ concatemers (λ) and λ DNA digested with *Hin*dIII (M). Standard PFGE running conditions for both gels were 22 cm diameter × 0.5 cm thick 1.5% agarose (Sigma Type I) gel in 0.5 × TAE, run at a constant 150 V and 18°C for 30 hr.

340, 345). Estimation of restriction fragment size is then based upon comparing the mobility of the fragment detected against the mobility of the standards on the same gel.

In order to be able to compare accurately the sizes of large DNA fragments, it is necessary to have a PFGE gel system that has straight lanes which enable a large number of samples to be electrophoresed on the same gel. A number of PFGE designs have been described which give straight lanes (19–27). The system used extensively in the authors' laboratories include the crossed field gel electrophoresis system described by Southern *et al.* (25) (Tribotics, Waltzer II, 2600 series). *Figure 2* illustrates a typical result using the 'Waltzer' PFGE box showing that both a large number of samples can be loaded and the lanes are straight. *Figure 3* shows a plot of mobility against molecular weight for the lambda concatemers separated on the PFGE gel in *Figure 2*. The plot illustrates the characteristic pattern of separation obtained with PFGE gels. There is a region of good resolution where mobility is linearly related to molecular weight which in this case is up to the ninth lambda concatemer. Increasing the length of the switching interval extends this region of linear separation to higher molecular weight, but decreases the resolution. There is then a short region of maximum resolution where separation of the markers is increased (for this gel between the ninth and thirteenth step of the ladder). Finally at high molecular weight there is a region where little or no separation of DNA is obtained. The molecular weight at which resolution is lost again is dependent upon the switching interval.

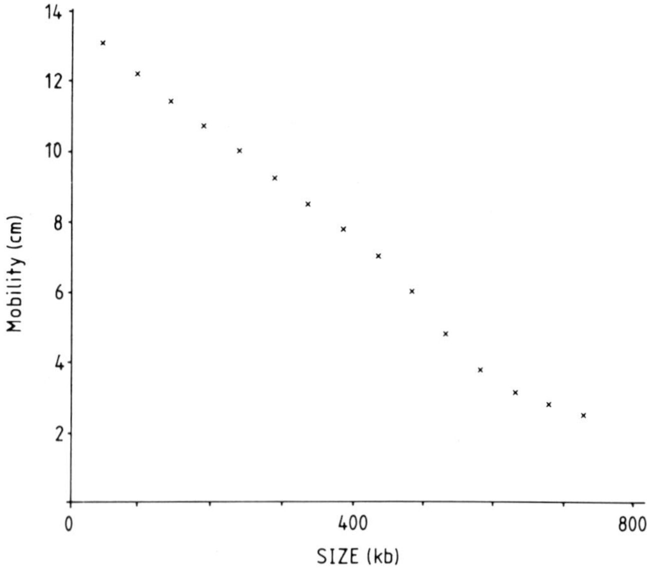

Figure 3. Plot of mobility against molecular weight for the λ concatemers separated on the PFGE gel shown in *Figure 2(A)*. Crosses mark the point for each concatemer.

Further information on the preparation and processing of samples and standards, gel processing, DNA transfer, and hybridization can be found in Anand and Southern (3), Sealey and Southern (29).

2.1 Application of PFGE to mapping of the Mhc

In order to construct a restriction map of the Mhc in man, genomic DNA from HLA homozygous cell lines, (e.g. Ice 5, HLA-A2, B7, DR2; BfS, C2C, C4A3, C4BQO) (5, 6), or a monosomy 6 mutant cell line (7), has been used to minimize mapping problems caused by possible haplotype specific RFLPs. The linkage of probes on one DNA fragment was established by observation of hybridization of both probes to the same DNA restriction fragment when the same PFGE filter was used. The possibility that the observed hybridization to a common fragment was due to comigration of two different fragments was reduced by the use of data from a range of restriction enzyme digests mapped relative to one another by double digestion, and by separating the same DNA samples at different switching intervals. Construction of a physical map of the Mhc utilized this single and double digestion data, the known genetic data, and information of the positions of probes and restriction sites from cosmid cloning.

The Mhc spans about 4000 kb of chromosomal DNA (see *Figure 4*). The class II genes at the centromeric end are contained within 750 kb, while the class I genes at the telomeric end are contained within 2000 kb of DNA. The most telomeric class II gene HLA-DRA, and the most centromeric class I gene HLA-B are separated by 1100 kb of chromosomal DNA which can be termed the class III region. These mapping experiments also established the orientation and position of the complement and TNF gene clusters with respect to the class I and class II regions. The HLA-DRA to CYP21B distance is 400 kb, the C2 to TNF A distance is estimated at 350 kb, while the HLA-B locus lies 220 kb telomeric of the TNF genes. PFGE has also been used to estimate gene copy number carried by different Mhc haplotypes (30–33), to ascertain the extent of restriction fragment length variation in genomic DNA prepared from blood samples of an unselected population (34), and to carry out comparative mapping of the Mhc in different racial groups (35).

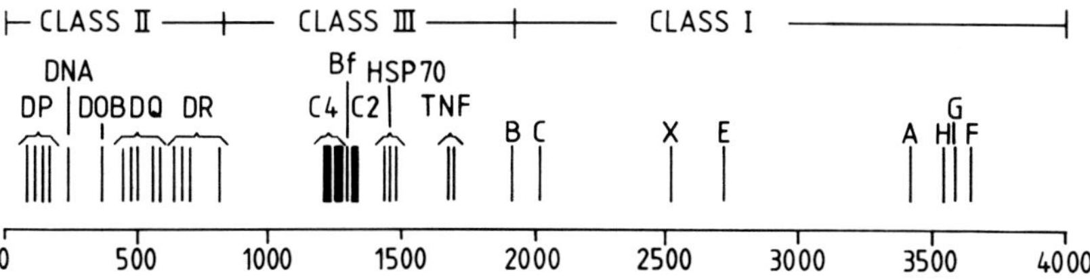

Figure 4. Physical map of the human Mhc. Variations in DNA content have been reported in the DRB region, at the C4 loci, and between HLA-E and HLA-A.

2.1.1 The class II region

The first molecular map of the class II region was determined for the HLA-DR4 haplotype (4). This established the relative order of the subregions and a complete linkage map of the genes including HLA-DOB and -DNA. The HLA-DRA gene was separated from the HLA-DPB2 gene on this haplotype by 900 kb of chromosome DNA. This analysis also indicated that the HLA-DR to -DQ distance was similar to the HLA-DQ to -DP distance. This is of particular interest as no recombination between the HLA-DQ and -DR genes has yet been reported and alleles of these genes are in very strong linkage disequilibrium. In contrast, a high recombination frequency of 1–3% occurs between HLA-DP genes and HLA-DQ and -DR genes. Furthermore, there is weak linkage disequilibrium between HLA-DP and -DR alleles. As the distance between HLA-DR and -DQ is similar to the distance separating HLA-DQ and -DP, this suggests that a recombination 'hot spot' exists between HLA-DP, and HLA-DQ and -DR.

Comparative analysis of restriction site mapping data in the HLA class II region has suggested that the chromosomal DNA organization in the HLA-DR3, 5, and 6 haplotypes is the same (see Chapter 11). However, in cell lines possessing the HLA-DR2 haplotype evidence has been obtained for the presence of an extra 20–30 kb of chromosomal DNA in the HLA-DRB region when contrasted to the HLA-DR3, 5, and 6 haplotypes (36). Similarly the HLA-DR4 haplotype appears to have a large additional segment of chromosomal DNA (120 kb), irrespective of subtype, compared to the HLA-DR3, 5, and 6 haplotypes in the region containing the HLA-DRB genes (36, 37). In addition, a number of groups have studied the HLA-DR7 and -DR9 haplotypes (38, 39) which are related to the HLA-DR4 haplotype by virtue of sharing the -DR53 specificity, and have obtained results consistent with 100–150 kb more chromosomal DNA present in the region containing the HLA-DRB and -DQA genes. This has subsequently been re-estimated by Kendall *et al.* (37) at 120 kb. Therefore, there appears to be consistency in the observations of the amount of chromosomal DNA present in cell lines possessing the same haplotype. However, there are a number of variations in the positions of cleavable restriction sites between even the related cell lines which might be due to methylation or chromosomal DNA sequence differences.

The nature of the alterations in chromosomal DNA organization observed between haplotypes is open to question. The HLA-DR3, 5, and 6 haplotypes have previously been shown to be related in the HLA-DRB region since they share common RFLPs, whereas the HLA-DR2 and -DR4 haplotypes have distinct DRB restriction fragment patterns (see Chapter 5). It is therefore possible that there could be size variation between the haplotypes because of differences in the HLA-DRB gene organization, or because of the presence of different numbers of duplicated DRB genes. A further possibility for the differences is that there are variations in the amount of uncharted DNA present in the HLA class II region between haplotypes. The observation of these differences may be of significance for the maintenance of linkage disequilibrium within the HLA-DR and -DQ subregions, and for the lack of recombination events between HLA-DR and -DQ. It seems reasonable to propose that these organizations have been maintained because the differences in size of the HLA class II region favour recombination between related haplotypes rather than between haplotypes with different structures. This may go some way to preserving ancestral haplotypes in the HLA-DQ and -DR subregions.

2.1.2 The class III region

PFGE analysis with a range of different enzymes has established that the class III region spans 1100 kb of chromosomal DNA (5–7). Large DNA fragment RFLPs for the enzymes *Bss*HII, *Mlu*I, and *Sac*II (or its isoschizomer *Ksp*I) have been observed at the complement loci (30–32). The size of the observed fragment with C4 or cytochrome 21-hydroxylase (CYP21) probes for these enzymes is directly related to the number and length of C4 genes present in the DNA samples analysed (30). For example, as shown in *Figure 5* individual B has *Bss*HII fragments of 115 kb and 105 kb and these correspond to

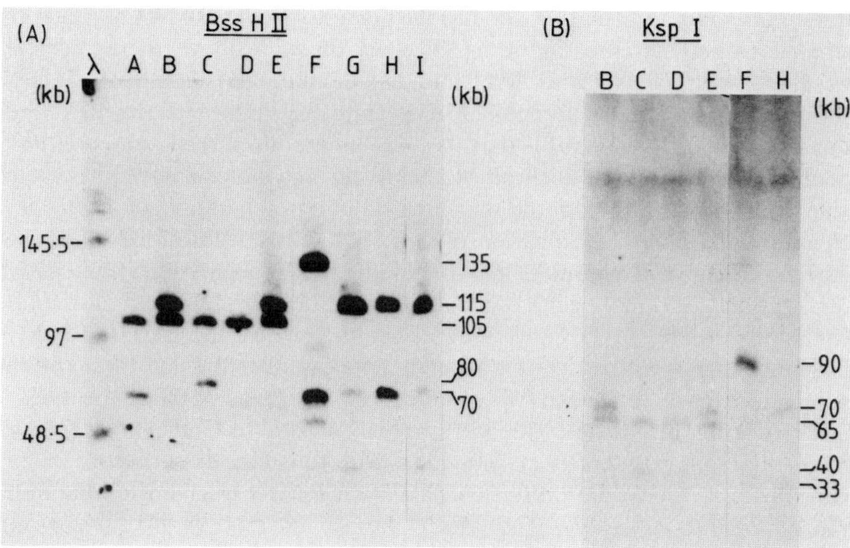

Figure 5. C4 genotyping by PFGE. (A) Autoradiograph of a PFGE separation of *Bss*HII digested DNA from nine unrelated individuals blotted on to Hybond N and hybridized with a C4 cDNA probe. (B) Autoradiograph of a PFGE separation of *Ksp*I digested DNA from six of the unrelated individuals shown in panel (A) blotted on to Hybond N and hybridized with a C4 cDNA probe. The sizes of the fragments detected are shown on the right of each panel. Both gels were run in 0.5 × TAE at a constant 150 V and 18°C for 30 h using a 7.5 sec switching interval.

Table 3. Analysis of C4 genotype using PFGE

C4 genotype	Fragment size (kb) observed with	
	*Bss*HII	*Ksp*I
1 short C4 gene	75	33
1 long C4 gene	85	40
1 long and 1 short C4 gene	105	65
2 long C4 genes	115	70
1 long and 2 short C4 genes	135	90
2 long and 1 short C4 genes	140*	95*
4 long C4 genes	180	130*

* predicted size

haplotypes with two long C4 genes, and one long and one short C4 gene, respectively, while individual F has *Bss*HII fragments of 135 kb and 70 kb and these correspond to haplotypes with three C4 genes (one long and two short), and one short C4 gene, respectively (see *Table 3*). *Ksp*I can also be used to derive the same information and has the advantage over *Bss*HII of generating smaller fragments (see *Figure 5*). The relationship between fragment size observed with *Bss*HII and *Ksp*I and the number and length of C4 genes present is summarized in *Table 3*. The major advantage of the RFLPs detected by *Bss*HII and *Ksp*I is that they can be used to directly observe the C4 gene organization on both chromosomes using high molecular weight chromosomal DNA isolated from the whole blood of an individual (see *Protocol 1*). In combination with the previously described polymorphism for *Taq*I, these RFLPs can give a complete picture of the C4 gene organization, and by implication the CYP21 gene organization, for an individual without the need for family studies or DNA cloning. Since the size of the *Bss*HII (or *Ksp*I) fragment observed is altered with gene copy number present on a chromosome, it is easy to identify the number of C4 genes without the need to interpret band intensities. This was amply illustrated by Collier *et al.* (31) who have been able to identify an individual with four C4 genes present on each chromosome due to the presence of a 180 kb *Bss*HII fragment on the PFGE blots.

Comparative analysis of a number of different haplotypes has revealed that there are no gross differences in the DNA organization between haplotypes in the class III region, apart from the known differences in C4 and CYP21 gene number (40). The resolution of the PFGE technique in the fragment size range studied was sufficient to be able to accurately detect fragment size differences of 2 kb and over. This is in contrast to Tokunaga *et al.* (38) who have suggested that there may be 'deletions' of DNA (10–20 kb) within the class III region in HLA-B8 -DR3 and -B18 -DR3 haplotypes, contained within the *Pvu*I fragment that hybridized to the CYP21 probe. Similarly they have predicted a 40 kb deletion telomeric to the TNF A and B genes, but these have yet to be confirmed by molecular cloning studies.

2.1.3 The class I region

PFGE analysis of the class I region has revealed that the relative order of the genes is roughly consistent with that obtained by recombination analysis. The HLA-B and HLA-C genes are 130 kb apart, while the HLA-A gene lies 1400 kb telomeric to HLA-C. A number of non-classical class I genes have now been defined within the class I region. The HLA-E gene lies 700 kb from HLA-C, while the HLA-G, HLA-F, and HLA-H genes all map telomeric to HLA-A (5–7, 41–44). By recombination analysis HLA-G and HLA-F appear to lie 8 cM telomeric to HLA-A, yet physical mapping indicates that the two genes are within 250 kb of HLA-A. The large discrepancy between the genetic distance and the physical distance between these genes suggests that there is a recombination hot spot telomeric of HLA-A. PFGE analysis using HLA-A, HLA-B, and -C locus specific probes has shown that the DNA content of the class I region in different haplotypes is relatively invariant (44). This study reported that, despite the high nucleotide polymorphism in this region, no restriction fragment length polymorphisms were observed between the different haplotypes using the rare cutting endonuclease *Sfi*I, and only two haplotypes showed variant bands with *Mlu*I. However, variation in the size of *Not*I fragments detected using an HLA-A probe have been reported (34), and haplotype differences of up to 90 kb have been observed between HLA-E and HLA-A (42).

3. Characterization of the Mhc genomic structure

Overlapping cosmid and YAC clones have now been isolated which cover the entire Mhc. These have allowed highly detailed maps of the Mhc to be constructed and have aided in the location of novel genes within the Mhc. The cosmid and YAC clones were isolated either by using the available HLA class I, class II, and class III region probes, or by techniques of chromosome walking.

One approach which can be used to locate genes is to isolate cDNA clones by hybridizing whole cosmid genomic inserts (16), or YAC clones (18), on to a cDNA library (see *Protocol 3*) using a pre-annealing procedure to prevent cross-hybridization by repetitive DNA elements (see *Protocols 4* and *5*).

Protocol 3. Plating of cDNA library and preparation of replica filters for screening

1. Spread 3 ml of L-broth containing approximately 1.25×10^5 recombinant bacteria evenly and carefully on to 20×20 cm nitrocellulose filters (Hybond-C extra, Amersham, RPN 2020E) on L-agar plates containing

Protocol 3. *Continued*

20 μg/ml ampicillin and 20 μg/ml tetracycline when using MC1061/p3 cells. Normally four master plates are prepared containing in total 5×10^5 colonies. Incubate the plates at 37°C for 16–20 h until the colonies are no larger than 0.5 mm in diameter.

2. Make two copies of each of the master plates on to fresh 20 × 20 cm Hybond-C extra filters, pre-wetted on L-broth/ampicillin/tetracycline plates, by placing the copy carefully on top of the master filter which is put colony side up on a sheet of Whatman 3MM paper. Cover with a second sheet of Whatman 3MM paper and press the sandwich between two glass plates. Remove the top plate and key the filters with indelible ink, before they are peeled apart and placed on to fresh plates. Allow the master to recover for 2 h between each round of replication. Take the second copy off the master plate as described above, then store the master plates at 4°C to provide a copy from which the positive colonies are taken after screening.
3. Incubate the copies at 37°C until the colonies are 0.5 mm in size.
4. Lyse the colonies *in situ* by placing the copies, colony side up, sequentially on 3MM Whatman papers soaked in (a) 10% sodium dodecyl sulphate (SDS), (b) 0.5 M NaOH, (c) 1 M Tris-HCl pH 7.4. Incubate for 3 min at each step.
5. Wash the copies in 2 × SSC (20 × saline sodium citrate (SSC) is 3 M NaCl/0.3 M Na_3 citrate) for 15 min at room temperature with gentle rocking.
6. Fix the DNA by baking at 80°C for 2 h.

Protocol 4. Preparation and labelling of cosmid genomic insert

Avoid unacceptable background on the library filters by ensuring that the genomic DNA insert is free of vector sequences

- TE: 0.1 mM EDTA 10 mM Tris-HCl pH 7.4
- glycerol dye mix: 30% v/v glycerol, 5 mM EDTA, 0.1% w/v bromophenol blue, 0.1% w/v xylene cyanol
- 10 × TBE: 0.9 M Tris, 0.9 M boric acid, 25 mM EDTA

1. Digest 10 μg of cosmid DNA with an appropriate restriction enzyme that will excise the genomic DNA insert from the cosmid vector. Most cosmid vectors contain unique sites, (e.g. *Sal*I in pDVCOS) that flank the cloning site (*Bam*HI in pDVCOS). However, if sites exist within the insert then, providing that they are not too frequent, the resulting fragments can be mixed back together after gel electrophoresis to be used as probe. This

protocol is for those cosmid vectors that contain *Sal*I sites flanking the cloning site. Set up the following digest:

cosmid DNA	10 (1 μg/μl)
10 × buffer	10
100 mM spermidine	5
dH_2O	70
*Sal*I (10 U/μl)	5

Incubate at 37 °C for 2 h. Check a small aliquot on a 1% agarose mini-gel to ensure complete digestion.

2. Phenol extract the sample by adding 100 μl phenol (saturated with 1 mM EDTA/10 mM Tris-HCl pH 7.4). Vortex briefly, then microcentrifuge for 4 min. Remove aqueous layer to a fresh tube. Re-extract the phenol layer with 50 μl TE. Vortex, microcentrifuge for 4 min, then combine aqueous layers.

3. Add 150 μl of chloroform/isopropanol (24:1 v/v), vortex briefly, then continue as for step **2**.

4. Add 20 μl 3 M Na acetate pH 7.0 and mix, followed by 550 μl cold absolute ethanol. Mix then precipitate DNA at −70 °C for 30 min.

5. Recover precipitate by microcentrifugation for 5 min, remove supernatant, and wash precipitate with 500 μl cold 70% ethanol. Microcentrifuge for 3 min, remove ethanol, and dry precipitate *in vacuo*. Resuspend in 20 μl TE.

6. Prepare a 1% LGT agarose (SeaPlaque) mini-gel (approximately 10 × 10 × 0.5 cm) with wells of 9 × 4 × 1 mm. Add 3 μl glycerol dye mix to the sample, mix, load into the well, and electrophorese using 1 × TBE at 10 mA and 4 °C until the vector band is well separated from the insert band(s), usually overnight.

7. Cut out agarose slice containing band over UV transilluminator and transfer to microfuge tube. Add 300 μl dH_2O and incubate at 70 °C to melt agarose. Mix thoroughly.

8. Incubate 10 μl of mix (containing 20–50 ng DNA) plus 20 μl of dH_2O in a boiling water-bath for 5 min, prior to labelling using the Amersham MultiPrime Kit (RPN1601) and 50 μCi [α-^{32}P]dCTP.

9. Separate labelled probe from unincorporated radioactivity by gel filtration on 2 ml Sephadex G50 column using 0.2 M NaCl, 1 mM EDTA, 10 mM Tris-HCl pH 7.4 as elution buffer. Typically specific activities of 5×10^8 c.p.m./μg of DNA are obtained.

Protocol 5. Screening of cDNA library filters using cosmid genomic inserts

- hybridization buffer: 50% deionized formamide, 1 M NaCl, 50 mM Tris-HCl pH 7.4, 0.2% Ficoll 400 (Pharmacia, 17-0400-01), 0.2% BSA (Sigma,

Protocol 5. *Continued*

A-7638), 0.2% polyvinylpyrrolidone (Sigma, PVP-360), 0.1% sodium pyrophosphate, 10% dextran sulphate (Pharmacia, 17-0340-01), 200 μg/ml sonicated salmon sperm DNA, 0.5% SDS

- 10 × Hogness buffer: 40% v/v glycerol, 36 mM K_2HPO_4, 13 mM KH_2PO_4, 20 mM Na_3 citrate, 10 mM $MgSO_4$

1. Pre-wash filters in 500 ml of 1 M NaCl, 1 mM EDTA, 50 mM Tris-HCl pH 8.0 with 0.5% SDS at 42°C for 1 h.
2. Pre-hybridize filters in pairs (but not the replicas from one plate together) at 42°C for at least 6 h in either heat sealed bags or in bottles, (e.g. Hybaid Oven) in 15 ml of hybridization buffer, which has been boiled for 5 min and allowed to cool before being added to the filters.
3. Boil the labelled probe in 40 ml fresh hybridization buffer containing 200 μg/ml sonicated human placental DNA for 5 min, then allow repetitive sequences to pre-anneal at 42°C for 90 min.
4. Remove pre-hybridization buffer from filters and replace with 10 ml of hybridization mix. Hybridize for 24 to 48 h.
5. Wash filters in 2 × SSC, 0.1% SDS for 10 min (three times) at room temperature, followed by 0.2 × SSC, 0.1% SDS for 30 min (two times) at 65°C. Cover filters in Saran Wrap and expose to hypersensitized X-ray film (Kodak X-Omat S) between two intensifying screens at −70°C for 4 to 72 h.
6. Match autoradiographs against the filters, and transfer the keying marks to the X-ray film. Duplicates are compared to establish whether potential positives are detected on both filters.
7. To remove positive recombinants align master filter with autoradiographs on a light box, using a layer of ethanol washed cling film to separate them. Scrape an area of about 2–3 mm in diameter from the positive area of the master filter using a sterile tooth-pick, or the cut end of an adapted Gilson pipette tip of similar cross-section. Place the colonies into 500 μl L-broth and recover for 1–2 h at room temperature.
8. To re-screen positive recombinants dilute between 10^2 and 10^6 times in L-broth the stock prepared by scraping the positive area from the master filter. From each dilution plate 200 μl on to 82 mm diameter Hybond-C (extra) filters on 9 cm L-broth/ampicillin/tetracycline plates, and incubate at 37°C overnight. The remaining stock is stored at −70°C in L-broth containing 1 × Hogness buffer.
9. Make two copies of the re-screen filters following steps **2** to **6** in *Protocol 3*.
10. Wash, pre-hybridize, and hybridize re-screen filters following steps **1** to **7** above. In this case up to 30 filters can be hybridized together either in a

glass dish or in a bottle, (e.g. Hybaid Oven). An appropriate volume of hybridization buffer is used to just cover the filters.

11. Repeat re-screening until single positives are obtained.

12. Grow single positive colonies in 10 ml L-broth/ampicillin/tetracycline, and prepare plasmid DNA according to standard 'mini-prep' protocols.

We have screened a number of cDNA libraries in the plasmid vector CDM8 which were generously given to us by David Simmonds, Institute of Molecular Medicine, Oxford. These have been constructed using mRNA isolated from various cell types, though in general most of the cDNAs corresponding to the novel MHC linked genes have been isolated from a cDNA library constructed using mRNA from the cell line U937. The methods for preparation of the double stranded cDNA, cloning into the *Bst*XI site of the vector, and transformation into competent MC1061/p3 cells is described in reference 47. *Protocol 3* assumes that the recombinant MC1061/p3 cells are ready for plating. This approach allows the possibility of identifying the maximum number of coding regions, including non HTF-island associated genes, with the minimum number of probings of the cDNA library. Once isolated the cDNA inserts can be mapped back on to the relevant cosmid or YAC insert to define the location of the corresponding gene. This strategy has been successfully used to locate seven novel genes in a 160 kb segment of DNA extending centromeric of the C4 genes (16).

A number of methods have been described that allow the detection of coding sequences within cloned DNA. Unique sequence probes from cosmid clones can be used to probe northern blots. Alternatively, ^{32}P-labelled cDNA synthesized from total mRNA has been hybridized to cosmid clones to identify putative coding sequences (45, 46), but this method appeared to favour identification of low-copy repeated elements. In certain cases it may be possible to genetically map cloned transcripts to a region of interest and then identify the precise position by hybridization to cloned genomic DNA or PFGE blots.

A more generally applicable strategy has been described by Monaco *et al.* (17). In this method it is assumed that coding sequences are likely to be evolutionarily conserved and therefore single copy probes that contain exons should hybridize at high stringency to DNA from other animal species on Southern blots ('zoo blots'). This approach has been successfully used to locate the coding region of the Duchenne Muscular Dystrophy gene (17), to map the novel RD gene into the complement gene cluster in the mouse and human Mhcs, and to locate the HSP70 genes in the HLA class III region.

When a large region of cloned DNA is to be screened for coding sequences, it is useful to be able to rapidly identify the positions of candidate sequences. Bird (8) has proposed that the positions of many genes could be identified by making use of the presence of CpG-rich islands at their 5′ ends. In bulk vertebrate chromosomal DNA the frequency of the dinucleotide CpG is suppressed to 0.2–0.25 of the frequency expected from the base composition, and 60–90% of the CpG is methylated at the 5 position on the cytosine ring. However, a discrete fraction of the genome (1%), which is extensively cleaved with the methylation sensitive enzyme *Hpa*II (recognition sequence CCGG), contains 15% of the total number of CpG dinucleotides and is unmethylated. The sequences from this *Hpa*II tiny fragment (HTF) fraction belong to islands of chromosomal DNA, 500 bp to at least 2 kb long, with a C + G content of >50% and no CpG suppression so that the dinucleotide occurs at the frequency expected from the base composition. The majority of island sequences are 'unique' by the criterion of DNA reassociation. Studies have indicated the presence of a large number of clustered sites for infrequently cutting enzymes within the class II and class III regions, indicative of the presence of a large number of novel genes. The sites for these enzymes are not randomly distributed, but are clustered at CpG-rich islands which show no CpG suppression and are unmethylated.

4. Novel genes in the class II region

In the class II region a number of HTF-islands have been defined between the HLA-DOB and -DNA genes which have been found to be associated with transcripts. *Figure 6* summarizes our current knowledge of the location of novel genes in the class II region.

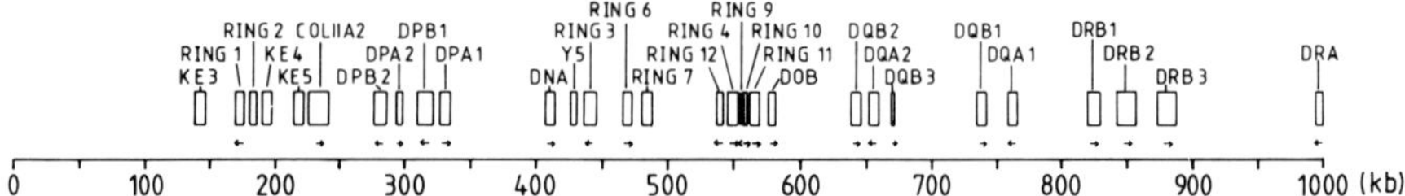

Figure 6. Molecular map of the class II region of the human Mhc. The map is a compilation of physical mapping and cloning data from a number of laboratories.

An HTF-island 25 kb centromeric of the DOB gene defines a novel gene transporter asociated with antigen processing 1 (TAP 1, formerly RING4) which belongs to the 'ABC' superfamily of transporters. A second gene which encodes another 'ABC' transporter-like protein (TAP 2, formerly RING11) is located close by. It is likely that the TAP 1 and TAP 2 products form a heterodimer which is involved in the transport of peptides from the cytoplasm across the membrane where they meet class I molecules. There is evidence that mutations in this region of the Mhc result in defects in peptide presentation by class I molecules (47, 48).

Lying centromeric of each transporter gene are the novel genes LMP 2 (formerly RING12) and LMP 7 (formerly RING10). The products of these genes share homology with a group of low molecular weight proteins which make up a large protein complex called the proteasome (also known as macropain or multi-catalytic protease). The proteasome is a multi-catalytic protease and it has been proposed that the Mhc encoded members of the complex are involved in producing peptides to feed into the transporter in the endoplasmic reticulum (49). Novel genes RING 6 and RING 7 and their products appear to have a class II structure, but their sequences are as homologous to class I as they are to class II.

An HTF-island 30 kb telomeric of the HLA-DNA gene is associated with a gene (RING 3 or Y4) whose sequence is related to that of a Drosophila gene, female sterile homeotic (*fsh*), as well as to some other proteins which may be involved in cell division. Also within the class II region are the genes Y5 and RING 9 which have yet to be characterized.

A cluster of novel genes have been identified immediately centromeric of the HLA-DP subregion (50). These include a collagen gene (COL11A2), the human equivalents of previously described mouse genes (KE3 and KE5), plus three other genes (RING 1, 2, and 5). RING 5 is equivalent to KE4, a mouse gene centromeric of H-2K. The RING 1 gene contains an arrangement of cysteine and histidine residues in a conserved motif, found in a group of proteins including the human V(D)J recombination activating gene RAG-1. This motif is similar to those of zinc finger protein sequences. It may define a new family of metal-dependent DNA binding proteins.

The genes centromeric of HLA-DP are not yet associated functionally with the Mhc or the immune system. Mice have class I H-2K genes in this position. The mouse arrangement is due to a simple insertion of about 60 kb of DNA containing the K genes, as the two genes flanking H-2K in the mouse, KE3 and RING 1, are adjacent in man and in the same order as in mouse in relation to the Mhc (51).

5. Novel genes in the class III region

In addition to the complement genes C2, Factor B, C4A, and C4B which are involved in the humoral immune response, other genes within the class III region that have functions associated with the immune system include the tumour necrosis factors (TNF) A and B genes. The TNF A and B genes are about 1.2 kb apart and lie about 220 kb centromeric of the HLA-B gene. However, recently it has become apparent that these genes form only the tip of the iceberg with respect to the total number of genes in the class III region. *Figure 7* summarizes our current knowledge of the location of novel genes in the class III region.

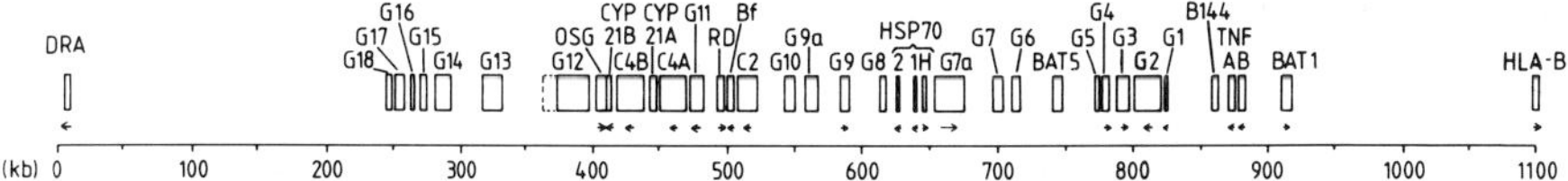

Figure 7. Molecular map of the class III region of the human Mhc. The map is a compilation of physical mapping and cloning data from a number of laboratories.

An example where *zoo blot* analysis has led to the identification of genes in the class III region is the case of the HSP70 genes. A 0.8 kb DNA fragment lying 92 kb telomeric of the C2 gene was found to cross-hybridize strongly to the DNA from a number of different species including chicken and shark. DNA sequencing of the probe revealed that it was derived from a gene encoding the major 70 000 Da heat shock protein HSP70. The probe was found to be duplicated and it has now been shown that there are two intronless genes, HSP70-1 and HSP70-2, which lie 12 kb apart, and which encode an identical protein product of 641 amino acids. A third intronless gene HSP70-Hom, which is located 4 kb telomeric of the HSP70-1 gene, encodes a more basic protein of 641 amino acids that shares 90% identity with HSP70-1.

PFGE mapping in the class III region has established the presence of a large number of HTF-islands (13, 16). Two are located within the complement gene cluster and define the 5′ ends of the novel genes G11 and RD, while 12 are located between the C2 and TNF A genes. Genomic DNA fragments derived from or close to these HTF-islands detected transcripts in northern blot analysis, and these together with cosmid genomic inserts have been used as hybridization probes to isolate the corresponding cDNA clones (13, 14, 16). The region between the C4 and HLA-DRA genes has also been shown to contain HTF-islands, where seven have been mapped within a 160 kb stretch of genomic DNA (16). Seven novel genes were identified by the isolation of cDNA clones using cosmid genomic inserts as hybridization probes, and northern blot analysis, at least five of which appear to be associated with HTF-islands. Thus far 36 genes have been located in a 680 kb stretch of genomic DNA which makes the class III region of the Mhc the most densely packed segment of the human genome so far characterized.

Sequence analysis of cDNA clones corresponding to some of the novel genes has begun to reveal their protein products. For example the single copy G7a gene encodes a 1265 amino acid protein of molecular weight 140 457 Da that shares 49% identity with yeast valyl-tRNA synthetase, indicating that this gene probably encodes human valyl-tRNA synthetase. The G1 gene encodes a 93 amino acid polypeptide which shares 35% identity with the intracellular Ca^{2+} binding protein calmodulin. This similarity is based around two putative Ca^{2+} binding sites in the G1 protein and suggests that G1 is a novel type of Ca^{2+} binding protein. The G13 protein has also been characterized and appears to encode a 77 000 Da protein that shows homology with CREB (cyclic AMP response element binding) protein suggesting that G13 could be a novel type of transcription factor. A novel gene has also been located immediately adjacent to the CYP21 gene. In this case the gene is encoded on the opposite strand to that encoding CYP21 and overlaps with the 3′ end of the CYP21 gene. A 2.7 kb incomplete OSG cDNA clone appears to specify a protein characterized by the presence of repeating fibronectin type III repeats that show significant homology to tenascin (51) a large glycoprotein found in extracellular matrices. The OSG is part of the unit of duplication at the C4 loci and part of the sequence is contained between the CYP21P and C4B genes. Whether this segment of DNA is expressed as mRNA or protein remains to be established. The G11 gene which is located immediately upstream of the C4A gene encodes a 28 000 Da protein, while the RD gene encodes a 42 000 Da intracellular protein with a central core of 52 amino acids consisting of a dipeptide repeat made up of a basic residue (Arg or Lys) next to an acidic one (Asp or Glu). The function of the RD and G11 proteins is unknown. Two of the other novel genes in the class III region, BAT2(G2) and BAT3(G3) encode large proline-rich proteins of molecular weights 228 000 Da and 110 000 Da, respectively. These do not appear to be members of any known family of proteins.

6. Implication of novel genes in HLA disease associations

Given that a large number of novel non-HLA genes are being found in the Mhc it is appropriate to consider their possible involvement with diseases as indicated by association with HLA polymorphisms. There are no data on functions of most of the new genes and in these cases little can be deduced from the translated cDNA sequences. However, there is much interest in the genes (TAP) thought to be involved with antigen processing. The finding of a cluster of genes in the class II region which may affect presentation of peptides through class I molecules may be the reason behind the marked linkage disequilibrium observed in the Mhc.

Since the TAP 1 and TAP 2 transporter genes are polymorphic it is possible that different variants affect the presentation of peptides through HLA class I. Faustman and her colleagues (53) claimed to demonstrate an association of low levels of cell surface HLA class I with diabetes in mice and on diabetic human lymphocytes which they speculate may be due to deficiencies in the TAP 1/TAP 2 transporter functions. Polymorphisms of the transporter genes in groups of diabetic and coeliac disease patients also showed a skewed distribution of some transporter alleles in diabetics (54). These studies need to be confirmed and extended to other HLA associated diseases. There are two variants of each of the LMP genes in the mouse and there is evidence for a similar degree of polymorphism in man.

The identification of HSP70 genes in the class III region is of particular interest since members of the HSP70 protein family are known to be involved in the immune response. The HSP70 proteins are major antigens in many bacterial and parasitic infections and both antibodies and T cells directed against autologous stress proteins have been identified in patients suffering from a number of autoimmune diseases. In addition there are a number of potential roles for the HSP70 proteins in antigen processing and presentation. In the mouse H-2 complex experimental autoimmune orchitis appears to map to a small segment of the S region that contains the mouse Mhc linked HSP70 genes. Polymorphisms in expression of TNF have been implicated in susceptibility to IDDM. A gene conferring susceptibility to IgA deficiency and Common Variable Immunodeficiency has been mapped to the class III region, possibly lying between the C2 and C4B genes (55). Two of the novel genes G11 and RD are located between these genes and either one could be involved in these diseases. However, it is clear that in this and the other cases of Mhc associated diseases further work needs to be done in order to define the contribution that the novel genes play in disease susceptibility.

Acknowledgements

We would like to sincerely thank Elaine Kendall for permission to use her data concerning the genotyping of C4 using *Ksp*I. We would also like to thank Caroline Milner for constructive criticism of the manuscript.

References

1. Schwartz, D. C. and Cantor, C. R. (1984). *Cell,* **37,** 67.
2. Carle, G. F. and Olson, M. V. (1984). *Nucleic Acids Res.,* **12,** 5647.
3. Anand, R. and Southern, E. M. (1990). In *Gel Electrophoresis of Nucleic Acids* (ed. D. Rickwood and B. D. Hames), pp. 101–23. IRL Press, Oxford.
4. Hardy, D. A., Bell, J. I., Long, E. O., Liindsten, T., and McDevitt, H. O. (1986). *Nature,* **323,** 453.
5. Dunham, I., Sargent, C. A., Trowsdale, J., and Campbell, R. D. (1987). *Proc. Natl Acad. Sci. USA,* **84,** 7237.
6. Carroll, M. C., Katzman, P., Alicot, E. M., Koller, B. H., Geraghty, D., Orr, H. T., Strominger, J. L., and Spies, T. (1987). *Proc. Natl Acad. Sci. USA,* **84,** 8535.
7. Ragoussis, J., Bloemer, K., Pohla, H., Messer, G., Weiss, E. H., and Ziegler, A. (1989). *Genomics,* **4,** 301.
8. Bird, A. P. (1987). *Trends Genet.,* **3,** 342.
9. Little, P. F. R. (1992). In *DNA Cloning: A Practical Approach* (ed. D. M. Glover), Vol III, pp. 19–42. IRL Press, Oxford.
10. DiLella, A. G. and Woo, S. L. C. (1987). In *Guide to Molecular Cloning Techniques,* (ed. S. L. Berger, and A. R. Kimmel), pp. 447–51. Academic Press, San Diego.
11. Greene, E. D. and Olson, M. V. (1990). *Science,* **250,** 94.
12. Schlessinger, D. (1990). *Trends Genet.,* **6,** 248.
13. Sargent, C. A., Dunham, I., and Campbell, R. D. (1989). *EMBO J.,* **8,** 2305.
14. Spies, T., Bresnahan, M., and Strominger, J. L. (1989). *Proc. Natl Acad. Sci. USA,* **86,** 8955.
15. Ragoussis, J., Monaco, A., Mockridge, I., Kendall, E., Campbell, R. D., and Trowsdale, J. (1991). *Proc. Natl Acad. Sci. USA,* **88,** 3753.
16. Kendall, E., Sargent, C. A., and Campbell, R. D. (1990). *Nucleic Acids Res.,* **18,** 7251.

17. Monaco, A. P., Neve, R. L., Colletti-Feener, C., Bertelson, C. T., Kurnit, D. M., and Kunkel, L. M. (1986). *Nature,* **323,** 646.
18. Elvin, P., Slynn, G., Black, D., Graham, A., Butler, R., Riley, J., Anand, R., and Markham, A. F. (1990). *Nucleic Acids Res.,* **18,** 3913.
19. Carle, G. F., Frank, M., and Olson, M. V. (1986). *Science,* **234,** 65.
20. Chu, G., Vollrath, D., and Davis, R. W. (1986). *Science,* **234,** 1582.
21. Clark, S. M., Lai, E., Birren, B. W., and Hood, L. (1988). *Science,* **241,** 1203.
22. Bancroft, I. and Wolk, C. P. (1988). *Nucleic Acids Res.,* **16,** 7405.
23. Gardiner, K., Laas, W., and Patterson, D. (1986). *Somat. Cell Mol. Genet.,* **12,** 185.
24. Kolble, K. and Sim, R. B. (1991). *Anal. Biochem.,* **192,** 32.
25. Southern, E. M., Anand, R., Brown, W. R. A., and Fletcher, D. S. (1987). *Nucleic Acids Res.,* **15,** 5925.
26. Serwer, P. (1987). *Electrophoresis,* **8,** 301.
27. Ziegler, A., Geiger, K.-H., Ragoussis, J., and Szalay, G. (1987). *J. Clin. Chem. Clin. Biochem.,* **25,** 578.
28. Nelson, M. and McClelland, M. (1991). *Nucleic Acids Res.,* **19** (Suppl), 2045.
29. Sealey, P. G. and Southern, E. M. (1990). In *Gel Electrophoresis of Nucleic Acids: A Practical Approach,* (ed. D. Rickwood and B. D. Hames), pp. 51–99. IRL Press, Oxford.
30. Dunham, I., Sargent, C. A., Dawkins, R. L., and Campbell, R. D. (1989). *J. Exp. Med.,* **169,** 1803.
31. Collier, S., Sinnott, P. J., Dyer, P. A., Price, D. A., Harris, R., and Strachan, T. (1989). *EMBO J.,* **8,** 1393.
32. Partanen, J., Kere, J., Wessberg, S., and Koskimies, S. (1989). *Genomics,* **5,** 345.
33. Zhang, W. J., Degli-Esposti, M. A., Cobain, T. J., Cameron, P. U., Christiansen, F. T., and Dawkins, R. L. (1990). *J. Exp. Med.,* **171,** 2101.
34. Lawrance, S. K. and Smith, C. L. (1990). *Genomics,* **8,** 394.
35. Tokunaga, K., Saveracker, G., Kay, P. H., Christiansen, F. T., Anaud, R., and Dawkins, R. L. (1988). *J. Exp. Med.,* **168,** 933.
36. Dunham, I., Sargent, C. A., Dawkins, R. L., and Campbell, R. D. (1989). *Genomics,* **5,** 787.
37. Kendall, E., Todd, J. A., and Campbell, R. D. (1991). *Immunogenetics,* **34,** 349.
38. Tokunaga, K., Kay, P. H., Christiansen, F. T., Saueracker, G., and Dawkins, R. L. (1989). *Hum. Immunol.,* **26,** 99.
39. Inoko, H., Tsuji, K., Groves, V., and Trowsdale, J. (1989). In *Immunobiology of HLA,* (ed. B. Dupont) Vol. II, pp. 83–6. Springer-Verlag, New York.
40. Dunham, I., Sargent, C. A., Kendall, E., and Campbell, R. D. (1990). *Immunogenetics,* **32,** 175.
41. Shukla, H., Gillespie, G. A., Srivastava, R., Collins, F., and Chorney, M. J. (1991). *Genomics,* **10,** 905.
42. Kahloun, A. E., Vernet, C., Jouanolle, A. M., Boretto, J., Mauvieux, V., Le Gall, J.-Y., David, V., and Pontarotti, P. (1992). *Immunogenetics,* **35,** 183.
43. Chimini, G., Boretto, J., Marguet, D., Lanau, F., Lanquin, G., and Pontarotti, P. (1990). *Immunogenetics,* **32,** 419.
44. Pontarotti, P., Chimini, G., Nguyen, C., Boretto, J., and Jordan, B. R. (1988). *Nucleic Acids Res.,* **16,** 6767.

45. Trowsdale, J., Kelly, A., Lee, J., Carson, S., Austin, P., and Travers, P. (1984). *Cell,* **38,** 241.
46. Guillemot, F., Billault, A., Pourquie, O., Behar, G., Chausse, A. M., Zoorob, R., Kreibich, G., and Auffray, C. (1988). *EMBO J.,* **7,** 2775.
47. Spies, T. and DeMars, R. (1990). *Nature,* **351,** 323.
48. Kelly, A., Powis, S. H., Kerr, L. A., Mockridge, I., Elliott, T., Bastin, J., Uchanska-Ziegler, B., Ziegler, A., Trowsdale, J., and Townsend, A. (1992). *Nature,* **355,** 641.
49. Parham, P. (1990). *Nature,* **348,** 674.
50. Hanson, I. M., Poustka, A., and Trowsdale, J. (1991). *Genomics,* **10,** 417.
51. Hanson, I. M. and Trowsdale, J. (1991). *Immunogenetics,* **34,** 5.
52. Matsumoto, K. I., Arai, M., Ishihara, N., Ando, A., Inoko, H., and Ikemura, T. (1992). *Genomics,* **12,** 485.
53. Faustman, D., Li, X., Lin, Y., Fu, G., Eisenbarth, J., Abruch, J., and Guo, J. (1991). *Science,* **254,** 1756.
54. Colonna, M., Ferrara, G. B., Strominger, J. L., and Spies, J. L. (1992). *Proc. Natl Acad. Sci. USA,* **89,** 3932.
55. Volanakis, J. E., Zhu, Z.-B., Schaffer, F. M., Macon, K. J., Barger, B. O., Go, R., Campbell, R. D., Schroeder, H. W., and Cooper, M. D. (1992). *J. Clin. Invest.*, **89,** 1914.

11

Data tables

Table 1. Genes in the HLA region (WHO 1991)

Name	**Molecular characteristics**
HLA-A	Class I α-chain
HLA-B	Class I α-chain
HLA-C	Class I α-chain
HLA-E	associated with class I 6.2-kb *Hin*dIII fragment
HLA-F	associated with class I 5.4-kb *Hin*dIII fragment
HLA-G	associated with class I 6.0-kb *Hin*dIII fragment
HLA-H	Class I pseudogene associated with 5.4-kb *Hin*dIII fragment
HLA-J	Class I pseudogene associated with 5.9-kb *Hin*dIII fragment
HLA-DRA	DR α-chain
HLA-DRB1	DR β-chain determining specificities DR1, DR2, DR3, DR4, DR5 etc.
HLA-DRB2	pseudogene with DR β-like sequences
HLA-DRB3	DR β3-chain determining DR52 and Dw24, Dw25, Dw26 specificities
HLA-DRB4	DR β4-chain determining DR53
HLA-DRB5	DR β5-chain determining DR51
HLA-DRB6	DRB pseudogene found on DR1, DR2, and DR10 haplotypes
HLA-DRB7	DRB pseudogene found on DR4, DR7, and DR9 haplotypes
HLA-DRB8	DRB pseudogene found on DR4, DR7, and DR9 haplotypes
HLA-DRB9	DRB pseudogene, isolated fragment
HLA-DQA1	DQ α-chain as expressed
HLA-DQB1	DQ β-chain as expressed
HLA-DQA2	DQ α-chain-related sequence, not known to be expressed
HLA-DQB2	DQ β-chain-related sequence, not known to be expressed
HLA-DQB3	DQ β-chain-related sequence, not known to be expressed
HLA-DOB	DO β-chain
HLA-DMA	DM α-chain
HLA-DMB	DM β-chain
HLA-DNA	DN α-chain
HLA-DPA1	DP α-chain as expressed
HLA-DPB1	DP β-chain as expressed
HLA-DPA2	DP α-chain-related pseudogene
HLA-DPB2	DP β-chain-related pseudogene
TAP 1	ABC transporter
TAP 2	ABC transporter
LMP 2	Proteasome-related sequence
LMP 7	Proteasome-related sequence

Table 2. Complete listing of recognized HLA specificities (WHO 1991)

A	B	C	D	DR	DQ	DP
A1	B5	Cw1	Dw1	DR1	DQ1	DPw1
A2	B7	Cw2	Dw2	DR103	DQ2	DPw2
A203	B703	Cw3	Dw3	DR2	DQ3	DPw3
A210	B8	Cw4	Dw4	DR3	DQ4	DPw4
A3	B12	Cw5	Dw5	DR4	DQ5(1)	DPw5
A9	B13	Cw6	Dw6	DR5	DQ6(1)	DPw6
A10	B14	Cw7	Dw7	DR6	DQ7(3)	
A11	B15	Cw8	Dw8	DR7	DQ8(3)	
A19	B16	Cw9(w3)	Dw9	DR8	DQ9(3)	
A23(9)	B17	Cw10(w3)	Dw10	DR9		
A24(9)	B18		Dw11(w7)	DR10		
A2403	B21		Dw12	DR11(5)		
A25(10)	B22		Dw13	DR12(5)		
A26(10)	B27		Dw14	DR13(6)		
A28	B35		Dw15	DR14(6)		
A29(19)	B37		Dw16	DR1403		
A30(19)	B38(16)		Dw17(w7)	DR1404		
A31(19)	B39(16)		Dw18(w6)	DR15(2)		
A32(19)	B3901		Dw19(w6)	DR16(2)		
A33(19)	B3902		Dw20	DR17(3)		
A34(10)	B40		Dw21	DR18(3)		
A36	B4005		Dw22			
A43	B41		Dw23	DR51		
A66(10)	B42					
A68(28)	B44(12)		Dw24	DR52		
A69(28)	B45(12)		Dw25			
A74(19)	B46		Dw26	DR53		
	B47					
	B48					
	B49(21)					

B50(21)
B51(5)
B5102
B5103
B52(5)
B53
B54(22)
B55(22)
B56(22)
B57(17)
B58(17)
B59
B60(40)
B61(40)
B62(15)
B63(15)
B64(14)
B65(14)
B67
B70
B71(70)
B72(70)
B73
B75(15)
B76(15)
B77(15)
B7801

Bw4
Bw6

Table 3. Designations of HLA-A, -B, -C, and -E alleles (WHO 1991)

HLA-A alleles	HLA-A specificity	HLA-B alleles	HLA-B specificity	HLA-Cw alleles	HLA-Cw specificity	HLA-E alleles
A*0101	A1	B*0701	B7	Cw*0101	Cw1	E*0101
A*0201	A2	B*0702	B7	Cw*0102	Cw1	E*0102
A*0202	A2	B*0703	B703	Cw*0201	Cw2	E*0103
A*0203	A203	B*0801	B8	Cw*02021	Cw2	E*0104
A*0204	A2	B*1301	B13	Cw*02022	Cw2	
A*0205	A2	B*1302	B13	Cw*0301	Cw3	
A*0206	A2	B*1401	B14	Cw*0302	Cw3	
A*0207	A2	B*1402	B65(14)	Cw*0401	Cw4	
A*0208	A2	B*1501	B62(15)	Cw*0501	Cw5	
A*0209	A2	B*1502	B75(15)	Cw*0601	Cw6	
A*0210	A210	B*1503	B72(70)	Cw*0701	Cw7	
A*0211	A2	B*1504	B62(15)	Cw*0702	Cw7	
A*0212	A2	B*1801	B18	Cw*0801	Cw8	
A*0301	A3	B*2701	B27	Cw*0802	Cw8	
A*0302	A3	B*2702	B27	Cw*1201	–	
A*1101	A11	B*2703	B27	Cw*1202	–	
A*1102	A11	B*2704	B27	Cw*1301	–	
A*2301	A23(9)	B*2705	B27	Cw*1401	–	
A*2401	A24(9)	B*2706	B27			
A*2402	A24(9)	B*2707	B27			
A*2403	A2403	B*3501	B35			
A*2501	A25(10)	B*3502	B35			
A*2601	A26(10)	B*3503	B35			
A*2901	A29(19)	B*3504	B35			
A*2902	A29(19)	B*3505	B35			
A*3001	A30(19)	B*3506	B35			
A*3002	A30(19)	B*3701	B37			
A*31011	A31(19)	B*3801	B38(16)			

A*31012	A31(19)	B*3901	B3901
A*3201	A32(19)	B*3902	B3902
A*3301	A33(19)	B*4001	B60(40)
A*3401	A34(10)	B*4002	B40
A*3402	A34(10)	B*4003	B40
A*3601	A36	B*4004	B40
A*4301	A43	B*4005	B4005
A*6601	A66(10)	B*4101	B41
A*6602	A66(10)	B*4201	B42
A*6801	A68(28)	B*4401	B44(12)
A*6802	A68(28)	B*4402	B44(12)
A*6901	A69(28)	B*4403	B44(12)
A*7401	A74(19)	B*4501	B45(12)
		B*4601	B46
		B*4701	B47
		B*4801	B48
		B*4901	B49(21)
		B*5001	B50(21)
		B*5101	B51(5)
		B*5102	B5102
		B*5103	B5103
		B*5201	B52(5)
		B*5301	B53
		B*5401	B54(22)
		B*5501	B55(22)
		B*5502	B55(22)
		B*5601	B56(22)
		B*5602	B56(22)
		B*5701	B57(17)
		B*5702	B57(17)
		B*5801	B58(17)
		B*7801	B7801
		B*7901	–

Table 4. Designations of HLA-DR, -DQ, and DP alleles (WHO 1991)

HLA-DR alleles	HLA-DR serological specificities	HLA-D-associated (T-cell-defined) specificities	HLA-DQ alleles	HLA-DQ serological specificities	HLA-D-associated (T-cell-defined) specificities	HLA-DP alleles	Associated HLA-DP specificities
DRA*0101	–	–	DQA1*0101	–	Dw1,w9	DPA1*0101	–
			DQA1*0102	–	Dw2,w21,w19	DPA1*0102	–
DRA*0102	–	–	DQA1*0103	–	Dw13,w12,w8,Dw'FS'	DPA1*0103	
			DQA1*0104	–		DPA1*0201	–
DRB1*0101	DR1	Dw1	DQA1*0201	–	Dw7,w11	DPA1*02021	–
DRB1*0102	DR1	Dw20	DQA1*03011	–	Dw4,w10,w13,w14,w15	DPA1*02022	–
DRB1*0103	DR103	Dw'BON'	DQA1*03012	–	Dw23	DPA1*0301	–
DRB1*1501	DR15(2)	Dw2	DQA1*0302	–	Dw23	DPA1*0401	–
DRB1*1502	DR15(2)	Dw12	DQA1*0401	–	Dw8,Dw'RSH'		
DRB1*1503	DR15(2)	–	DQA1*0501[a]	–	Dw3,w5,w22	DPB1*0101	DPw1
DRB1*1601	DR16(2)	Dw21	DQA1*05011	–	Dw3	DPB1*0201	DPw2
DRB1*1602	DR16(2)	Dw22	DQA1*05012	–	Dw5	DPB1*02011	DPw2
DRB1*0301	DR17(3)	Dw3	DQA1*05013	–	Dw22	DPB1*02012	DPw2
DRB1*0302	DR18(3)	Dw'RSH'	DQA1*0601	–	Dw8	DPB1*0202	DPw2
DRB1*0303	DR18(3)	–				DPB1*0301	DPw3
DRB1*0401	DR4	Dw4	DQB1*0501	DQ5(1)	Dw1	DPB1*0401	DPw4
DRB1*0402	DR4	Dw10	DQB1*0502	DQ5(1)	Dw21	DPB1*0402	DPw4
DRB1*0403	DR4	Dw13	DQB1*05031	DQ5(1)	Dw9	DPB1*0501	DPw5
DRB1*0404	DR4	Dw14	DQB1*05032	DQ5(1)	Dw9	DPB1*0601	DPw6
DRB1*0405	DR4	Dw15	DQB1*0504			DPB1*0801	–
DRB1*0406	DR4	Dw'KT2'	DQB1*0601	DQ6(1)	Dw12,w8	DPB1*0901	–
DRB1*0407	DR4	Dw13	DQB1*0602	DQ6(1)	Dw2	DPB1*1001	–
DRB1*0408	DR4	Dw14	DQB1*0603	DQ6(1)	Dw18,Dw'FS'	DPB1*1101	–
DRB1*0409	DR4	–	DQB1*0604	DQ6(1)	Dw19	DPB1*1301	–
DRB1*0410	DR4	–	DQB1*0605	DQ6(1)	Dw19	DPB1*1401	–

DRB1*0411	DR4	–	DQB1*0606			DPB1*1501	–
DRB1*0412	DR4	–	DQB1*0201	DQ2	Dw3,w7	DPB1*1601	–
DRB1*11011	DR11(5)	Dw5	DQB1*0301	DQ7(3)	Dw4,w5,w8,w13	DPB1*1701	–
DRB1*11012	DR11(5)	Dw5				DPB1*1801	–
DRB1*1102	DR11(5)	Dw'JVM'	DQB1*0302	DQ8(3)	Dw4,w10,w13,w14	DPB1*1901	–
DRB1*1103	DR11(5)	–	DQB1*03031	DQ9(3)	Dw23	DPB1*2001	–
DRB1*11041	DR11(5)	Dw'FS'	DQB1*03032	DQ9(3)	Dw23,w11	DPB1*2101	
DRB1*11042	DR11(5)	–	DQB1*0304	DQ7(3)	–		
DRB1*1105	DR11(5)	–	DQB1*0401	DQ4	Dw15	DPB1*2201	–
DRB1*1201	DR12(5)	Dw'DB6'	DQB1*0402	DQ4	Dw8,Dw'RSH'	DPB1*2301	–
DRB1*1202	DR12(5)	–				DPB1*2401	–
DRB1*1301	DR13(6)	Dw18				DPB1*2501	–
DRB1*1302	DR13(6)	Dw19				DPB1*2601	–
DRB1*1303	DR13(6)	Dw'HAG'				DpB1*2701	–
						DPB1*2801	–
DRB1*1304	DR13(6)	–				DPB1*2901	–
DRB1*1305	DR13(6)	–					
DRB1*1306	DR13(6)	–				DPB1*3001	–
DRB1*1401	DR14(6)	Dw9				DPB1*3101	–
DRB1*1402	DR14(6)	Dw16				DPB1*3201	–
DRB1*1403	DR1403	–				DPB1*3301	–
DRB1*1404	DR1404	–				DPB1*3401	–
DRB1*1405	DR14(6)	–				DPB1*3501	–
DRB1*1406	DR14(6)	–				DPB1*3601	–
DRB1*1407	DR14(6)						
DRB1*1408	DR14(6)						
DRB1*1409	DR14(6)	–					
DRB1*1410	–	–					
DRB1*0701	DR7	Dw17					
DRB1*0702	DR7	Dw'DB1'					
DRB1*0801	DR8	Dw8.1					

Table 4. (*continued*)

HLA-DR alleles	HLA-DR serological specificities	HLA-D-associated (T-cell-defined) specificities	HLA-DQ alleles	HLA-DQ serological specificities	HLA-D-associated (T-cell-defined) specificities	HLA-DP alleles	Associated HLA-DP specificities
DRB1*08021	DR8	Dw8.2					
DRB1*08022	DR8	Dw8.2					
DRB1*08031	DR8	Dw8.3					
DRB1*08032	DR8	Dw8.3					
DRB1*0804	DR8	–					
DRB1*0805	DR8	–					
DRB1*09011	DR9	Dw23					
DRB1*09012	DR9	Dw23					
DRB1*1001	DR10	–					
DRB3*0101	DR52	Dw24					
DRB3*0201	DR52	Dw25					
DRB3*0202	DR52	Dw25					
DRB3*0301	DR52	Dw26					
DRB4*0101	DR53	Dw4,Dw10,Dw13,Dw14 Dw15,Dw17,Dw23					
DRB5*0101	DR51	Dw2					
DRB5*0102	DR51	Dw12					
DRB5*0201	DR51	Dw21					
DRB5*0202	DR51	Dw22					

Table 5. HLA-Bw4 and -Bw6 associated specificities (WHO 1991)

The following specificities are generally agreed inclusions of HLA epitopes Bw4 and Bw6.

Bw4:	B5, B5102, B5103, B13, B17, B27, B37, B38(16), B44(12), B47, B49(21), B51(5), B52(5), B53, B57(17), B58(17), B59, B63(15), B77(15)
and	A9, A23(9), A24(9), A2403, A25(10), A32(19)
Bw6:	B7, B703, B8, B14, B18, B22, B35, B39(16), B3901, B3902, B40, B4005, B41, B42, B45(12), B46, B48, B50(21), B54(22), B55(22), B56(22), B60(40), B61(40), B62(15), B64(14), B65(14), B67, B70, B71(70), B72(70), B73, B75(15), B76(15), B7801

Table 6. HLA loci: HLA-A, -B, -C, -DR, and -DQ allele frequencies in different ethnic groups. Refer to Table 9 for the names of ethnic groups. (Reproduced with permission from *HLA 1991*, © Oxford University Press and the Committee of the 11th International Histocompatibility Workshop and Conference, 1992)

HLA-A

Ethnic	10	10	10	10	10	10	10	10	10	10	10	20	20	30	30	30	30	30	30	30	30	30	30	30	30	30
code	100	102	200	201	202	204	206	207	210	400	600	100	202	101	102	103	104	105	106	109	110	111	113	114	115	117
N =	59	99	103	103	65	90	61	70	105	447	116	99	51	208	141	98	60	64	117	101	96	150	321	295	231	551
A1	3.6	5.1	2.4	1.0	9.1	1.7	4.4	5.3	9.9	5.3	5.6	2.0	0.0	14.3	10.3	16.8	12.7	13.0	15.0	17.3	17.4	13.0	13.7	18.0	9.2	14.0
A2	13.6	19.5	14.9	18.4	9.5	17.1	13.8	20.6	16.5	16.7	12.8	15.7	0.0	26.4	15.6	28.1	24.6	20.0	24.1	39.6	29.1	30.3	21.3	27.8	26.1	24.4
A3	5.9	6.1	6.2	15.5	3.2	3.3	7.0	1.5	5.7	8.9	8.0	6.6	0.0	10.3	17.8	15.3	7.0	20.2	15.3	12.4	12.3	13.6	13.3	14.1	8.6	12.9
A11	4.2	0.0	0.0	0.0	0.8	0.0	0.0	0.0	5.3	2.3	3.4	4.5	12.1	6.0	5.5	5.1	7.5	5.0	7.3	5.0	5.5	5.6	6.2	4.6	8.8	6.1
A23	10.6	18.3	9.7	12.6	14.7	9.2	19.3	9.7	2.9	8.1	14.4	6.1	0.0	2.9	4.0	2.6	5.0	2.3	2.6	1.0	3.1	2.0	2.6	2.7	3.0	1.5
A24	5.4	1.0	4.1	0.5	3.1	0.6	1.6	3.8	6.2	4.7	3.5	31.8	65.7	10.1	15.3	9.7	10.0	6.3	6.2	5.4	11.3	8.3	8.9	7.5	11.3	9.7
A25	1.7	0.0	0.0	0.0	0.8	0.6	0.0	0.0	0.0	0.6	1.3	0.5	0.0	1.2	0.4	1.5	0.0	2.3	1.4	0.5	3.6	2.3	2.3	2.4	1.2	2.4
A26	3.4	4.4	5.0	0.5	4.0	4.4	1.6	2.1	7.0	1.6	2.6	1.0	0.0	3.4	5.2	2.0	2.0	4.2	5.6	2.0	5.5	4.0	4.4	4.7	6.9	4.3
A28	11.9	6.8	11.1	10.7	6.4	7.8	10.1	12.6	9.2	10.9	9.0	3.0	0.0	3.0	4.7	4.1	0.8	4.7	4.5	4.0	2.1	5.7	5.6	5.1	4.1	3.6
A29	3.4	5.1	5.9	0.0	6.4	5.6	3.4	2.3	4.3	3.7	3.2	2.5	0.0	1.0	1.8	4.6	11.3	3.9	4.3	5.9	1.6	3.3	6.4	2.2	3.0	3.9
A30	9.3	14.3	13.6	20.4	13.1	31.6	13.5	11.8	11.3	9.5	10.9	1.0	0.0	1.9	0.7	3.1	11.3	1.6	2.6	3.0	1.6	1.3	2.3	2.4	1.9	3.9
A31	0.8	1.5	3.9	0.5	3.1	0.0	1.7	0.7	1.4	1.7	5.4	1.5	0.0	3.5	3.3	2.0	1.8	3.9	2.6	1.0	2.1	2.0	3.7	2.2	1.5	3.9
A32	1.7	2.0	2.0	6.8	0.8	1.7	2.6	1.4	1.4	1.0	0.9	3.0	0.0	8.7	4.6	2.0	0.8	4.7	4.7	2.0	1.0	5.0	4.2	3.2	9.0	5.6
A33	8.0	10.0	1.5	0.0	0.0	2.2	5.9	15.2	6.0	8.1	6.0	1.0	0.0	2.2	2.8	1.5	0.8	1.6	1.7	0.5	1.6	2.6	3.1	3.1	1.7	2.1
A34	4.5	2.5	6.9	0.0	0.0	5.0	3.3	2.1	4.8	5.1	3.1	19.2	19.1	0.7	0.0	0.5	0.0	0.9	0.0	0.5	0.0	0.3	0.8	0.2	0.4	0.1
A36	4.5	0.0	1.0	0.0	0.0	4.1	0.8	0.0	0.0	2.6	0.9	0.5	0.0	0.2	1.5	0.5	0.0	0.0	0.0	0.0	0.5	0.0	0.0	0.0	0.2	0.1
A43	0.0	0.0	2.4	11.7	17.4	1.2	0.0	0.0	2.4	0.1	0.0	0.0	0.0	0.0	0.0	0.0	0.0	0.0	0.0	0.0	0.0	0.0	0.0	0.0	0.2	0.0
AX	1.8	0.0	0.0	1.5	0.8	1.1	1.6	6.4	3.8	2.5	0.9	0.0	0.0	0.0	0.4	0.5	0.0	1.6	0.0	0.0	0.0	0.3	0.3	0.0	0.0	0.2
ABL	5.7	3.4	9.4	0.0	7.0	2.9	9.2	4.5	1.9	6.7	8.1	0.0	3.0	4.2	6.2	0.0	4.3	4.0	2.3	0.0	1.6	0.2	0.9	0.0	0.6	1.3

Ethnic	30	30	30	30	30	30	30	30	30	30	30	30	30	30	30	30	30	30	30	30	30	40	40	40	40	40
code	119	120	121	122	124	125	127	128	129	130	131	141	400	401	406	413	601	701	702	801	902	101	104	201	202	204
N =	50	125	91	97	264	95	83	97	98	98	61	75	99	50	73	50	63	246	197	114	351	1023	261	145	138	69
A1	10.0	10.3	12.0	5.7	11.3	19.1	10.9	7.2	17.8	15.8	13.9	28.1	11.1	20.9	16.5	12.0	20.3	16.9	18.6	5.3	9.9	0.7	2.5	4.7	0.4	1.4
A2	34.0	26.6	28.3	32.2	27.6	38.7	30.2	25.4	26.4	23.5	29.0	11.8	12.1	12.0	7.2	18.5	25.3	28.3	26.7	23.2	26.0	24.4	29.3	36.7	33.7	42.8
A3	16.5	12.7	15.1	3.4	9.0	12.0	16.5	15.3	11.7	12.6	15.2	5.7	7.9	19.9	6.8	6.0	7.8	12.2	11.3	7.0	6.9	0.6	1.3	2.4	1.1	0.7
A11	3.4	5.8	8.2	9.7	7.0	3.9	8.4	9.7	4.1	9.0	1.6	25.6	15.9	8.8	8.6	5.0	1.6	5.5	4.1	4.8	4.7	10.4	9.4	20.3	31.9	29.7
A23	1.0	2.0	2.7	2.6	2.5	1.1	0.6	3.1	2.0	3.6	1.6	2.0	1.0	7.8	0.0	6.0	0.8	1.6	1.0	3.9	2.9	0.0	0.0	0.7	0.0	0.0
A24	7.6	10.3	9.8	7.2	6.8	6.6	10.0	14.2	10.7	9.0	9.7	6.1	18.1	13.6	18.4	17.2	7.8	9.6	11.3	16.7	9.6	35.1	22.8	12.6	19.9	16.7
A25	4.0	2.0	2.7	0.0	2.5	1.1	6.6	2.1	2.0	2.0	4.3	2.0	0.5	1.0	0.0	0.0	4.0	3.3	2.3	2.2	2.5	0.0	0.0	1.1	0.0	0.7
A26	6.0	4.6	3.8	1.0	4.5	2.6	2.4	3.3	8.7	7.1	3.4	5.0	6.6	1.0	3.4	3.0	4.8	3.9	5.3	3.5	5.3	10.9	8.1	2.8	0.7	3.6
A28	4.0	4.8	4.9	2.1	3.1	1.6	1.8	2.6	4.1	3.1	3.3	1.3	6.1	0.0	2.7	8.6	5.6	4.5	3.6	9.6	6.4	0.0	0.6	0.3	0.0	0.0

A29	1.0	4.8	1.6	0.5	9.3	1.1	0.0	1.1	1.5	3.3	0.8	1.3	3.0	0.0	0.0	3.0	4.0	2.6	1.8	3.9	4.6	0.0	0.4	1.0	0.0	0.0
A30	3.0	3.6	2.2	22.3	5.1	0.5	2.6	0.5	1.0	3.6	2.6	2.0	1.1	3.0	2.9	1.0	3.2	2.6	3.3	5.7	6.0	0.4	4.4	4.4	2.2	0.0
A31	1.0	3.8	1.9	0.5	2.4	4.2	1.8	4.8	1.5	2.6	3.3	0.0	2.5	1.0	5.5	2.0	0.0	2.0	2.2	4.4	4.8	8.0	3.9	3.1	2.9	0.7
A32	3.0	0.8	4.8	8.1	2.6	1.1	1.8	1.5	6.1	2.0	4.3	6.0	2.0	0.0	3.4	7.1	5.6	5.1	5.8	3.9	2.9	0.0	1.0	0.7	0.7	0.7
A33	2.0	5.0	0.5	3.4	2.8	0.0	1.2	1.0	1.5	1.0	0.0	0.7	9.2	5.0	17.5	7.5	0.0	1.0	1.8	3.5	2.6	7.7	14.9	3.4	6.5	2.9
A34	1.0	0.0	0.0	0.0	0.8	1.1	0.6	0.5	0.5	0.0	0.0	0.0	0.0	0.0	1.4	0.0	1.8	0.6	0.3	1.3	0.9	0.0	0.0	1.1	0.0	0.0
A36	0.0	0.0	0.0	0.0	0.2	0.0	0.6	1.0	0.0	0.0	0.0	0.0	0.0	0.0	0.7	0.0	0.0	0.2	0.0	0.0	0.4	0.0	0.0	0.0	0.0	0.0
A43	0.0	0.0	0.0	0.0	0.0	0.0	0.0	0.0	0.0	0.0	0.0	0.0	0.0	0.0	0.7	0.0	0.0	0.0	0.0	0.0	0.0	0.0	0.0	0.0	0.0	0.0
AX	0.0	0.0	0.0	0.0	0.8	1.6	0.0	0.0	0.0	0.0	0.0	0.0	0.0	0.0	0.7	0.0	0.8	0.2	0.0	0.9	0.4	0.0	0.0	0.4	0.0	0.0
ABL	2.4	2.2	1.3	1.4	1.5	4.0	4.0	6.7	0.2	1.7	6.9	2.5	2.9	6.0	3.4	3.1	6.0	0.0	0.8	0.0	3.3	1.7	1.4	4.2	0.0	0.0

Ethnic	40	40	40	40	40	40	40	40	40	40	40	40	40	40	40	40	40	40	40	40	40	41	41	41	41
code	207	210	211	212	213	216	218	219	220	301	302	303	401	402	403	414	416	421	602	800	803	003	00A	102	10A
N =	70	60	99	187	73	165	87	79	104	89	138	59	242	71	149	68	51	73	53	51	56	61	72	156	79
A1	1.4	7.0	3.0	4.6	3.4	0.0	4.8	1.3	6.0	8.4	8.3	11.4	2.3	2.8	3.7	1.6	5.3	0.7	0.9	4.9	3.6	0.8	6.9	1.6	12.7
A2	30.7	17.2	32.7	34.2	22.6	18.1	23.5	26.4	23.1	27.5	23.1	29.2	25.5	25.0	25.9	14.8	11.0	37.0	24.5	25.5	42.0	32.0	37.1	16.7	13.3
A3	0.7	17.6	2.0	5.2	1.4	0.0	7.6	4.4	7.8	4.1	4.8	8.0	1.5	2.1	1.7	1.5	0.0	0.7	0.9	2.9	2.7	0.9	3.9	1.6	9.5
A11	42.1	5.8	13.6	14.6	2.7	11.3	18.1	29.1	8.8	8.0	8.7	9.2	32.5	28.6	26.3	14.5	10.8	26.0	10.4	1.0	0.9	0.8	0.0	1.0	5.7
A23	0.7	0.8	0.0	1.1	0.0	0.0	0.0	4.4	2.0	5.1	2.2	0.8	0.0	0.0	0.3	0.0	0.0	0.0	0.0	1.0	1.8	0.0	0.0	0.3	2.5
A24	16.4	14.8	32.2	17.1	43.8	61.0	14.1	9.5	15.0	21.3	18.6	20.4	14.6	16.7	13.5	39.1	43.7	16.4	41.5	19.6	26.8	0.8	33.7	63.8	25.3
A25	0.0	1.7	1.5	0.3	0.0	0.3	0.0	1.6	0.5	0.6	0.7	0.8	0.4	0.0	0.0	0.0	0.0	0.0	0.9	0.0	0.9	0.0	0.7	0.0	0.0
A26	0.7	7.7	2.0	1.4	6.8	4.7	1.7	6.3	6.0	3.9	5.8	5.3	1.9	1.4	2.1	5.4	7.2	0.7	0.9	2.0	0.0	0.0	0.7	0.3	2.5
A28	0.7	1.7	0.0	0.8	0.0	0.0	0.0	1.3	0.5	1.7	0.7	1.7	0.8	0.0	0.7	0.0	0.0	0.7	1.9	6.9	12.5	2.6	6.8	12.5	1.9
A29	0.0	0.9	0.0	0.0	0.7	0.0	0.0	0.6	1.9	1.1	1.4	0.0	0.6	0.0	8.8	0.0	0.0	2.1	0.9	2.0	0.0	0.0	2.8	0.3	0.6
A30	1.4	2.5	3.5	5.6	1.4	0.3	7.2	5.1	3.6	0.0	4.3	0.8	1.1	1.4	3.0	0.0	0.0	0.0	0.0	2.0	0.0	0.8	0.0	0.0	0.6
A31	0.7	1.7	4.9	4.0	15.1	0.0	3.5	3.8	9.8	5.2	4.2	4.2	1.7	1.4	2.3	0.7	3.0	3.4	0.9	27.5	3.6	43.6	1.4	0.0	19.6
A32	0.0	4.2	0.0	1.4	0.7	0.0	0.0	0.0	0.0	0.0	1.8	0.0	0.2	0.0	0.0	0.0	0.0	0.0	0.0	2.0	2.7	0.0	0.7	1.0	0.6
A33	4.3	0.0	0.0	4.8	0.7	1.2	7.4	1.9	4.5	3.4	6.3	1.7	13.6	18.7	7.9	6.9	0.0	8.2	3.8	1.0	1.8	12.3	2.1	1.0	5.1
A34	0.0	0.0	0.0	0.0	0.0	0.3	0.0	3.0	0.5	0.0	1.4	0.0	1.3	0.7	0.0	2.3	1.0	4.1	12.3	0.0	0.9	0.0	0.7	0.0	0.0
A36	0.0	2.6	0.0	0.3	0.7	0.0	0.0	0.6	0.5	0.6	3.1	0.0	0.2	0.0	0.0	0.0	1.0	0.0	0.0	0.0	0.0	0.0	0.0	0.0	0.0
A43	0.0	0.0	0.5	0.5	0.0	0.0	0.0	0.0	0.0	0.0	0.0	0.0	0.4	0.0	0.0	0.0	0.0	0.0	0.0	0.0	0.0	0.0	0.0	0.0	0.0
AX	0.0	0.0	0.0	0.0	0.0	0.0	0.0	0.0	0.0	0.0	0.0	0.0	0.0	0.0	0.3	0.0	0.0	0.0	0.0	2.0	0.0	0.0	0.0	0.0	0.0
ABL	0.0	14.0	3.9	4.1	0.0	2.7	12.1	0.7	9.7	9.1	4.3	6.4	1.5	1.1	3.5	13.3	17.0	0.0	0.0	0.0	0.0	5.3	1.2	0.0	0.0

HLA-B

Ethnic	10	10	10	10	10	10	10	10	10	10	10	20	20	30	30	30	30	30	30	30	30	30	30	30	30	30
code	100	102	200	201	202	204	206	207	210	400	600	100	202	101	102	103	104	105	106	109	110	111	113	114	115	117
N =	59	99	103	103	65	90	61	70	105	447	116	99	51	208	141	98	60	64	117	101	96	150	321	295	231	551
B7	4.2	4.5	8.1	12.8	12.3	7.8	6.2	6.7	4.3	8.3	9.6	1.5	0.0	7.7	9.0	15.8	9.2	10.2	8.5	13.9	9.7	9.2	7.5	10.7	5.9	5.8
B8	3.4	6.3	9.8	13.3	0.0	3.5	2.8	2.9	5.7	3.2	2.2	1.0	0.0	4.7	2.9	9.2	5.8	5.5	13.7	11.4	8.9	8.9	6.8	9.8	3.6	6.8
B13	0.0	0.0	2.4	0.5	0.8	1.7	1.0	0.0	2.4	0.9	1.8	7.5	12.7	3.1	1.1	5.0	2.5	2.3	1.3	1.5	8.3	2.0	3.0	3.2	2.6	3.9

Table 6. (*continued*)

HLA-B

Ethnic	10	10	10	10	10	10	10	10	10	10	10	20	20	30	30	30	30	30	30	30	30	30	30	30	30	30
code	100	102	200	201	202	204	206	207	210	400	600	100	202	101	102	103	104	105	106	109	110	111	113	114	115	117
N =	59	99	103	103	65	90	61	70	105	447	116	99	51	208	141	98	60	64	117	101	96	150	321	295	231	551
B14	3.6	1.7	3.9	2.4	4.6	3.9	4.9	3.0	4.3	3.0	2.6	3.0	0.0	4.4	3.7	2.0	3.3	4.7	5.6	4.5	3.1	1.4	7.2	3.7	2.8	3.6
B18	6.7	4.7	1.9	7.1	4.6	7.0	4.1	2.3	5.7	3.3	3.4	0.0	0.0	10.8	3.7	6.1	16.7	7.0	3.4	5.4	4.9	4.0	4.8	3.4	6.3	9.6
B27	1.7	2.0	0.0	0.0	4.6	0.0	0.8	2.1	2.9	1.6	0.4	6.6	5.6	0.7	1.8	3.1	6.7	1.6	3.8	4.0	3.5	3.0	3.6	4.6	2.4	2.2
B35	4.8	13.1	0.5	0.0	0.8	2.2	3.3	13.6	4.8	7.7	10.0	3.5	0.0	10.3	11.6	11.7	5.0	6.3	8.5	5.4	8.1	6.0	8.4	7.6	18.3	15.4
B37	0.8	1.0	0.0	0.0	0.8	0.0	1.6	0.7	2.4	0.9	0.9	0.5	0.0	1.9	0.7	0.5	0.0	1.6	2.1	0.0	2.1	2.4	2.3	2.0	1.7	1.4
B38	0.8	0.0	0.0	0.0	0.0	0.0	0.0	0.0	1.9	0.2	0.9	2.0	0.0	6.0	4.6	0.5	0.0	3.1	1.3	0.0	4.2	3.2	3.0	1.9	2.4	3.5
B39	4.5	3.0	1.0	0.0	3.8	0.6	2.5	0.0	1.4	0.8	1.3	1.5	2.0	2.9	1.1	2.0	0.8	2.3	2.1	0.0	2.1	3.3	2.5	1.9	2.2	3.3
B41	0.0	0.0	1.5	4.1	10.3	2.2	4.9	0.0	1.9	1.8	0.0	0.5	0.0	1.2	2.1	1.5	0.0	0.8	1.3	2.0	1.0	2.0	1.6	1.7	3.1	1.6
B42	5.9	3.0	12.8	1.5	2.3	7.5	3.3	4.3	3.8	4.7	6.6	0.0	0.0	0.0	0.0	0.0	0.0	0.0	0.0	0.0	0.0	0.0	0.0	0.2	0.2	0.0
B44	5.1	5.6	7.1	2.4	6.9	5.8	5.7	2.9	8.6	6.2	7.0	6.6	1.0	10.9	5.4	9.6	21.7	10.2	10.3	20.8	12.5	12.9	10.8	9.2	9.4	9.2
B45	3.4	1.0	2.4	0.5	3.1	10.0	2.8	4.5	2.4	3.8	2.6	0.0	0.0	0.0	1.8	0.5	0.0	0.0	1.3	1.0	0.0	0.7	0.6	0.8	0.2	1.0
B46	0.0	0.0	0.0	0.0	0.0	0.0	0.0	0.0	0.5	0.0	0.0	0.0	0.0	0.0	0.0	0.0	0.0	0.0	0.0	0.0	0.0	0.0	0.0	0.2	0.0	0.1
B47	0.0	1.5	0.0	0.0	1.5	0.0	0.8	0.0	1.9	0.1	0.0	0.0	0.0	1.0	0.0	1.5	0.8	0.0	2.1	0.0	0.5	0.7	0.2	0.7	0.4	0.5
B48	0.8	0.0	1.0	0.0	0.0	0.0	0.0	0.0	0.5	0.2	0.9	0.0	0.0	0.0	0.0	0.0	0.0	0.0	0.0	0.0	0.0	0.7	0.0	0.2	0.0	0.1
B49	6.1	5.6	0.5	0.5	0.0	1.1	5.7	3.7	1.4	2.7	3.1	0.0	0.0	1.9	8.0	2.0	0.8	3.9	1.3	1.0	1.6	1.0	3.0	1.5	3.5	2.9
B50	1.7	0.5	0.0	0.0	0.0	0.0	0.0	0.7	0.5	1.0	1.7	0.0	1.0	1.4	3.9	0.0	0.8	0.0	1.3	2.0	1.6	0.7	1.5	1.0	0.4	1.8
B51	2.5	2.1	0.0	0.5	1.5	0.6	2.5	7.1	6.2	3.2	1.7	2.5	0.0	8.9	9.6	7.6	6.7	7.0	5.1	2.5	5.7	5.7	6.9	4.9	14.4	8.4
B52	2.5	5.1	0.0	0.0	0.8	0.0	0.0	3.9	1.4	1.1	3.6	0.0	0.0	4.2	8.0	1.0	1.7	3.9	0.9	0.5	1.0	0.7	2.5	1.0	2.5	2.6
B53	7.6	4.2	1.0	1.9	0.0	7.0	9.8	11.0	1.0	12.8	4.4	0.0	0.0	1.0	2.6	0.5	0.0	1.6	0.9	0.0	1.0	1.0	1.2	0.8	0.9	1.0
B54	0.0	0.0	0.0	0.0	0.0	0.0	0.0	0.0	0.0	0.0	0.0	0.0	0.0	0.2	0.0	0.0	0.0	1.6	0.0	0.0	0.0	0.0	0.0	0.0	0.2	0.0
B55	0.8	0.0	0.0	0.0	0.9	0.0	0.8	0.0	0.5	0.8	3.1	1.0	0.0	1.9	3.3	0.5	0.8	4.7	3.4	2.0	1.6	2.0	3.7	1.5	2.9	1.8
B56	0.0	4.0	0.0	0.0	0.0	0.0	0.0	0.0	0.5	0.0	0.4	17.2	22.5	0.5	0.0	1.0	0.0	2.3	1.3	0.5	1.0	1.7	0.8	1.5	1.1	0.4
B57	4.2	1.0	2.1	2.6	5.8	9.6	3.3	2.9	4.8	4.2	3.1	3.5	0.0	1.9	1.1	4.1	5.8	3.1	5.6	4.5	1.6	2.4	1.9	6.3	0.0	2.0
B58	6.3	8.6	17.2	35.6	11.1	11.8	4.9	7.9	11.0	6.9	1.7	1.0	0.0	1.4	0.7	1.0	2.5	0.8	0.9	0.5	1.0	1.3	1.9	3.4	1.8	1.8
B59	0.0	0.0	0.0	0.0	0.0	0.0	0.0	0.0	0.0	0.1	0.4	0.0	0.0	0.0	0.4	0.0	0.0	0.0	0.0	0.0	0.0	0.0	0.0	0.0	0.0	0.0
B60	0.0	0.0	0.0	0.0	0.0	0.0	0.0	0.0	2.9	1.3	0.9	10.0	15.5	1.7	1.4	4.5	0.8	3.9	3.8	5.0	5.1	8.6	3.0	4.6	1.7	0.5
B61	0.8	0.0	0.0	0.5	0.0	0.0	0.0	0.0	0.0	0.2	1.3	19.6	2.9	1.7	1.1	1.0	0.0	3.1	1.7	1.0	1.6	1.7	2.7	2.5	1.8	1.4
B62	1.8	0.0	0.0	0.5	0.8	0.0	0.8	0.7	2.4	1.4	1.7	7.1	28.3	1.9	0.7	5.1	5.0	4.7	6.0	9.9	2.1	6.8	5.3	4.9	1.7	2.8
B63	3.4	0.0	1.0	0.0	0.0	0.6	4.9	0.0	0.0	2.8	0.9	0.0	0.0	0.7	1.1	1.5	1.7	0.8	1.3	0.5	2.9	0.0	0.9	1.7	0.6	1.7
B67	0.0	0.0	0.5	0.0	0.0	0.0	0.0	0.7	0.0	0.7	0.4	0.0	0.0	0.0	0.0	0.0	0.0	0.0	0.0	0.0	0.0	0.0	0.0	0.0	0.0	0.2
B70	8.5	9.4	22.6	10.4	18.8	13.5	16.8	8.1	8.6	8.2	2.2	0.0	0.0	0.7	2.1	0.5	0.8	2.3	0.9	0.5	0.0	0.7	0.6	1.4	0.9	0.3
B73	0.0	0.0	0.0	0.0	0.0	0.0	0.0	0.7	0.0	0.2	0.0	0.0	0.0	0.0	0.4	0.0	0.0	0.0	0.0	0.0	0.0	0.3	0.3	0.2	0.0	0.1
B75	1.7	0.0	0.0	0.0	0.0	0.0	0.0	0.7	2.4	0.3	0.0	3.4	6.7	0.7	0.7	0.0	0.0	0.0	0.4	0.0	0.0	0.0	0.0	0.3	0.0	0.1
B76	0.0	0.0	0.0	0.0	0.0	0.0	0.0	0.0	0.0	0.0	0.0	0.0	0.0	0.0	0.0	0.0	0.0	0.0	0.0	0.0	0.0	0.0	0.0	0.0	0.0	0.1

B77	0.0	0.0	0.0	0.0	0.0	0.0	2.5	0.0	1.0	0.0	0.4	0.0	1.0	0.2	0.0	0.0	0.0	0.0	0.0	0.0	0.0	0.0	0.2	0.2	0.0	0.0
B5102	0.8	2.6	0.0	0.0	0.0	0.0	0.8	1.4	0.0	0.2	4.1	0.0	0.0	0.5	1.4	0.0	0.0	0.0	0.0	0.0	0.0	0.0	0.0	0.0	0.2	0.2
B7801	0.0	5.2	0.0	0.0	0.0	0.0	0.0	1.4	0.0	0.2	0.0	0.0	0.0	0.0	0.0	0.0	0.0	0.0	0.0	0.0	0.0	0.0	0.0	0.0	0.0	0.0
BX	1.7	0.0	0.0	0.5	1.5	0.6	0.0	0.0	0.5	1.1	5.8	0.0	0.0	0.0	0.0	0.0	0.0	0.8	0.0	0.0	0.0	1.7	0.3	0.5	0.0	0.4
BBL	4.4	4.0	2.8	2.4	2.5	3.0	2.5	6.0	0.0	3.6	8.7	0.4	0.9	2.7	4.2	0.4	0.0	0.0	0.0	0.0	3.2	3.4	0.9	0.0	4.0	1.9

Ethnic	30	50	30	30	30	30	30	30	30	30	30	30	30	30	30	30	30	30	30	30	30	40	40	40	40	40
code	119	120	121	122	124	125	127	128	129	130	131	141	400	401	406	413	601	701	702	801	902	101	104	201	202	204
N =	50	125	91	97	264	95	83	97	98	98	61	75	99	50	73	50	63	246	197	114	351	1023	261	145	138	69
B7	8.0	7.9	6.3	1.0	7.6	15.7	10.0	16.0	8.2	5.5	9.0	4.7	9.5	8.0	8.9	2.3	10.3	10.0	11.1	6.6	5.1	5.0	4.1	3.1	0.0	0.0
B8	6.0	7.5	4.9	1.5	6.4	16.0	8.5	4.6	10.7	13.7	13.3	1.3	3.8	0.0	3.4	9.6	12.8	10.0	11.1	3.9	4.9	0.0	0.6	0.0	0.0	0.0
B13	6.0	1.6	5.2	1.5	2.5	1.1	3.6	10.3	2.0	4.6	6.6	0.0	1.0	2.0	4.5	2.3	3.4	3.0	2.5	0.9	1.7	1.8	6.3	16.4	8.5	13.0
B14	2.0	5.6	2.7	5.9	6.9	1.1	3.0	2.6	0.0	2.9	0.0	1.3	2.0	0.0	0.0	0.0	4.0	4.1	6.5	6.1	5.0	0.1	0.8	0.0	0.4	0.0
B18	3.0	2.8	9.1	28.5	5.5	1.1	5.5	1.0	11.2	4.6	5.7	8.7	2.5	7.6	0.0	5.0	5.1	4.9	5.8	5.7	5.3	0.0	0.0	0.0	0.0	0.0
B27	5.6	2.8	4.4	0.5	2.8	8.8	4.2	7.2	7.1	5.6	4.9	7.3	2.8	4.0	0.7	0.0	3.2	4.1	4.3	1.8	3.1	0.4	3.0	1.4	1.8	2.2
B35	13.8	12.8	9.7	12.7	8.4	8.7	7.7	8.2	9.7	11.7	12.3	12.7	12.0	12.7	14.3	10.7	7.1	8.5	7.6	14.5	13.7	8.1	7.0	3.6	2.9	1.4
B37	0.0	0.8	4.4	1.0	1.4	1.1	0.6	0.0	2.0	1.5	0.0	4.0	3.8	1.0	3.4	4.0	0.8	2.2	1.8	0.4	0.9	0.7	1.9	2.1	0.0	0.0
B38	6.0	2.8	2.9	0.5	3.0	0.0	1.2	1.5	6.1	6.1	5.2	0.0	0.0	0.0	0.7	0.0	1.6	4.1	1.5	1.3	3.6	0.3	1.3	1.0	2.9	10.9
B39	1.0	2.0	3.3	3.6	0.6	0.5	2.5	6.2	0.0	1.0	0.9	4.0	1.5	0.0	0.0	0.0	0.0	1.8	2.0	6.6	4.9	4.5	1.7	1.7	2.9	3.6
B41	2.0	1.2	0.5	0.0	1.6	0.5	2.5	2.1	0.5	1.5	1.8	0.0	0.0	2.0	1.4	4.0	3.2	1.0	0.8	1.8	1.1	0.0	0.0	0.3	0.0	0.0
B42	0.0	0.4	0.0	0.0	0.0	0.0	0.0	0.0	0.0	0.0	0.0	0.0	0.0	2.0	0.0	0.0	0.8	0.0	0.0	0.4	0.6	0.0	0.0	0.0	0.0	0.0
B44	13.6	14.3	7.7	4.6	17.0	11.1	11.1	5.7	15.3	8.7	9.0	5.3	7.1	15.5	10.0	10.0	15.6	10.4	13.3	11.4	10.6	7.4	9.9	4.1	1.1	0.7
B45	1.3	1.2	0.5	3.1	1.4	1.1	0.6	0.0	0.0	1.5	1.6	0.7	0.0	0.0	0.0	0.0	0.8	0.6	0.3	1.3	2.1	0.0	0.0	0.7	0.0	0.0
B46	0.0	0.0	0.0	0.0	0.0	0.0	0.6	0.0	0.0	0.0	0.0	0.0	0.0	0.0	0.0	0.0	0.0	0.0	0.3	0.0	0.0	4.4	4.0	2.8	15.4	21.6
B47	0.0	0.4	1.6	0.0	0.8	0.5	0.0	0.0	0.0	0.5	0.8	0.0	0.0	0.0	1.4	0.0	0.0	0.6	0.0	0.4	0.0	0.1	0.0	0.0	1.1	0.0
B48	0.0	0.0	0.0	0.0	0.0	0.5	0.0	1.5	0.5	0.0	0.0	0.0	0.5	0.0	0.0	0.0	0.0	0.2	0.0	2.6	0.0	3.2	4.0	2.8	0.7	0.0
B49	1.0	3.2	3.3	6.5	3.7	0.0	1.8	2.1	1.0	2.0	0.8	4.0	1.5	2.0	2.1	0.0	0.0	2.2	2.0	2.2	3.0	0.0	0.2	0.3	0.0	0.0
B50	0.0	2.2	1.1	1.5	2.7	0.0	0.6	0.0	0.5	0.5	0.0	0.0	1.5	0.0	0.0	6.0	0.0	2.2	1.3	3.5	1.9	0.0	0.0	0.0	0.0	0.0
B51	7.0	11.4	12.6	5.9	7.8	2.6	5.0	3.1	9.7	6.0	3.3	9.3	7.6	2.0	10.1	6.0	2.4	3.7	4.0	3.9	7.1	9.3	7.8	11.8	4.3	5.8
B52	1.0	1.6	2.9	0.5	2.0	0.0	3.1	0.5	1.5	1.0	0.8	4.7	6.1	6.0	6.2	11.3	0.8	1.2	0.8	3.5	1.5	10.7	2.4	2.1	1.8	0.7
B53	0.0	0.4	0.5	0.0	0.8	0.5	1.2	2.6	1.0	1.5	0.0	0.0	1.0	4.5	0.7	1.0	0.8	0.8	0.8	1.3	1.5	0.0	0.0	0.7	0.0	0.0
B54	0.0	0.0	0.5	0.0	0.0	0.0	0.6	0.0	0.0	0.0	0.0	0.0	0.0	1.0	0.7	0.0	0.0	0.0	0.0	0.0	0.1	6.3	6.5	2.4	4.9	0.7
B55	1.0	1.2	2.2	1.5	2.0	0.0	1.2	0.5	1.5	1.5	1.6	2.0	2.5	3.0	4.1	0.0	2.4	2.2	1.3	0.9	2.0	2.9	1.8	3.2	4.0	7.9
B56	2.0	0.4	0.0	0.5	0.4	0.5	0.0	0.0	0.5	1.5	0.9	0.0	1.5	0.0	0.0	0.0	0.8	1.0	1.5	0.0	0.0	1.6	0.6	0.3	0.0	1.4
B57	1.0	3.1	1.1	2.2	2.0	0.5	0.6	1.5	3.1	4.1	3.3	4.7	3.5	1.2	7.5	3.0	4.5	3.9	3.6	2.2	2.2	0.0	0.8	2.8	1.1	1.8
B58	5.0	1.6	1.1	9.1	1.9	2.4	1.2	5.2	1.0	1.5	0.8	2.0	2.5	2.0	0.7	6.0	0.0	1.0	0.5	2.2	1.0	0.7	5.2	3.2	4.9	2.2
B59	0.0	0.0	0.0	0.0	0.0	0.0	0.6	0.0	0.0	0.0	0.0	0.0	0.0	0.0	0.0	0.0	0.0	0.0	0.0	0.0	0.0	1.9	1.0	0.0	0.0	0.0
B60	6.0	1.6	2.2	0.5	1.9	6.2	1.9	7.7	1.0	0.5	3.3	0.0	2.5	3.0	3.1	2.0	2.4	4.5	5.6	3.1	2.4	5.6	4.2	6.0	17.1	11.5
B61	0.0	0.8	1.8	2.2	1.7	0.0	1.2	1.5	2.0	3.1	0.0	21.3	12.9	4.0	4.1	8.0	4.0	3.3	0.5	4.4	1.9	10.7	9.2	4.2	5.4	0.7
B62	3.0	4.4	2.7	0.5	3.7	4.7	5.5	3.1	3.1	3.6	5.7	1.3	5.6	5.0	3.8	1.0	6.7	5.5	7.0	0.9	4.2	8.3	10.5	11.8	14.7	5.8
B63	3.0	0.8	0.5	0.5	1.1	4.2	1.8	0.5	0.5	2.4	0.8	0.0	1.5	4.0	4.1	1.0	0.8	0.4	0.1	0.4	0.7	0.0	0.0	2.6	0.0	0.7
B67	0.0	0.0	0.0	0.0	0.0	0.0	0.6	0.0	0.0	0.0	0.0	0.0	0.0	0.0	0.0	0.0	0.0	0.0	0.3	0.0	0.1	1.5	0.4	0.0	0.4	0.0

Table 6. (*continued*)

HLA-B

Ethnic	30	50	30	30	30	30	30	30	30	30	30	30	30	30	30	30	30	30	30	30	30	40	40	40	40	40
code	119	120	121	122	124	125	127	128	129	130	131	141	400	401	406	413	601	701	702	801	902	101	104	201	202	204
N =	50	125	91	97	264	95	83	97	98	98	61	75	99	50	73	50	63	246	197	114	351	1023	261	145	138	69
B70	0.0	2.4	0.0	0.0	0.4	0.0	0.0	0.0	0.0	0.0	0.8	0.7	0.0	4.0	0.7	2.0	0.8	1.6	0.3	3.1	0.1	1.6	1.9	0.4	0.4	0.0
B73	0.0	0.0	0.0	0.0	0.0	0.0	0.0	0.0	0.0	0.0	0.0	0.0	0.5	0.0	0.0	0.0	0.0	0.4	0.0	0.4	0.0	0.0	0.0	0.3	0.0	0.0
B75	0.0	0.0	0.0	0.5	0.0	0.5	0.6	4.1	0.0	0.0	1.6	0.0	0.5	0.0	1.4	1.0	0.0	0.0	0.0	1.8	1.0	1.1	1.7	2.4	1.1	6.4
B76	0.0	0.0	0.0	0.0	0.0	0.0	0.0	0.0	0.0	0.0	0.0	0.0	0.0	1.0	0.0	0.0	0.0	0.2	0.0	0.0	0.0	0.0	0.0	0.0	0.0	0.0
B77	0.0	0.0	0.0	0.0	0.2	0.0	0.0	0.5	0.0	0.0	0.0	0.0	0.0	0.0	0.0	0.0	0.0	0.0	0.0	0.0	0.0	0.0	0.0	0.0	0.0	0.0
B5102	0.0	0.0	0.0	0.0	0.0	0.0	0.0	0.0	0.0	0.5	0.0	0.0	0.0	0.0	0.7	1.0	0.0	0.0	0.0	0.0	0.9	0.3	0.0	0.3	0.4	0.0
B7801	0.0	0.0	0.0	0.0	0.0	0.0	0.0	0.0	0.0	0.0	0.0	0.0	0.0	0.0	0.0	0.0	0.0	0.0	0.0	0.0	0.0	0.0	0.0	0.0	0.0	0.0
BX	0.0	0.0	0.5	0.5	0.0	8.9	0.0	0.0	0.0	0.0	0.0	0.0	0.5	0.0	0.0	0.0	0.8	0.4	0.3	0.4	2.0	0.0	0.0	0.0	0.4	0.0
BBL	1.7	0.8	3.3	2.7	2.0	1.1	10.7	0.0	0.0	0.7	5.0	0.0	1.6	2.5	1.5	2.8	4.4	0.0	0.6	0.0	3.0	1.3	1.3	4.8	1.6	0.8

HLA-B

Ethnic	40	40	40	40	40	40	40	40	40	40	40	40	40	40	40	40	40	40	40	40	40	41	41	41	41
code	207	210	211	212	213	216	218	219	220	301	302	303	401	402	403	414	416	421	602	800	803	003	00A	102	15A
N =	70	60	99	187	73	165	87	79	104	89	138	59	242	71	149	68	51	73	53	51	56	61	72	156	79
B7	0.8	2.6	0.0	3.0	2.1	0.3	3.4	8.0	5.3	4.6	3.5	4.2	2.7	3.5	12.2	1.5	2.0	2.1	3.8	3.9	5.8	0.8	3.3	2.2	3.2
B8	0.7	7.5	1.0	1.1	0.7	0.0	0.6	0.0	4.1	0.0	6.4	4.2	0.2	0.0	0.3	0.0	0.0	0.7	0.9	5.6	1.8	0.8	1.4	2.2	0.0
B13	10.7	12.7	0.5	11.4	1.4	9.4	13.1	18.7	7.2	4.7	6.2	1.7	9.3	9.9	7.4	2.3	10.5	6.8	1.9	1.0	0.9	0.8	0.0	0.0	3.8
B14	0.0	4.3	0.0	1.6	0.0	0.0	1.8	0.0	0.5	0.0	0.7	1.7	0.4	0.0	0.7	0.0	0.0	0.0	0.9	2.9	0.9	5.9	0.7	0.0	0.0
B18	0.0	0.0	0.5	0.0	0.0	0.0	1.1	0.0	1.9	1.1	1.1	0.0	2.5	1.4	1.3	6.4	2.0	0.0	1.9	1.0	1.8	0.0	1.4	0.0	0.6
B27	0.7	3.3	10.5	2.4	2.1	3.4	4.0	2.5	2.4	2.8	5.1	4.4	6.0	4.9	3.2	3.8	6.2	3.7	0.9	2.9	10.1	0.0	0.0	10.7	9.2
B35	0.7	1.7	5.2	5.4	6.8	0.0	4.6	2.5	4.3	5.2	6.6	1.7	2.5	3.5	5.0	12.0	5.9	2.1	0.9	18.6	12.7	29.0	29.8	4.7	6.6
B37	1.5	0.0	1.0	3.8	2.1	0.3	2.9	0.0	3.5	2.3	4.3	0.0	1.4	0.7	0.7	0.0	0.0	0.0	0.0	0.0	2.7	0.0	0.0	0.6	7.6
B38	2.9	4.3	4.7	0.3	0.7	0.0	4.0	0.6	2.5	0.0	2.2	0.0	3.5	1.4	4.2	0.0	2.0	4.1	0.9	0.0	0.0	0.8	5.6	0.0	1.9
B39	1.4	0.0	1.0	1.4	0.7	15.8	0.0	1.3	2.0	0.0	2.2	1.7	1.7	1.4	2.7	0.0	0.0	2.9	9.4	17.4	0.0	18.3	17.9	1.0	0.0
B41	0.0	0.0	0.0	0.5	0.0	0.0	0.0	0.6	0.0	2.8	1.4	0.8	0.0	0.0	0.3	0.0	2.0	0.0	0.0	0.0	0.0	0.0	0.0	0.3	0.0
B42	0.0	0.0	0.5	0.0	0.0	0.0	0.0	0.0	0.0	0.0	0.0	0.8	0.2	0.0	0.0	0.0	0.0	0.0	0.0	0.0	0.0	0.0	0.0	0.0	0.0
B44	0.0	4.3	1.0	6.5	1.4	0.3	4.8	8.2	4.3	11.4	1.4	2.5	5.4	2.8	3.4	6.9	0.0	0.7	0.9	3.9	0.0	0.0	2.8	1.3	4.4
B45	0.7	0.0	0.0	0.0	0.7	0.0	0.0	0.0	1.4	1.8	0.0	0.8	0.0	0.0	0.7	0.0	0.0	0.0	0.0	0.0	0.0	0.0	0.0	0.3	0.0
B46	18.4	3.5	1.0	5.6	0.7	1.2	4.0	3.8	2.4	1.7	1.1	0.8	14.0	7.7	13.2	2.9	2.0	15.1	0.0	0.0	0.0	0.0	0.0	0.0	0.0
B47	1.4	0.8	0.5	0.0	0.0	0.3	0.0	0.0	0.0	0.6	0.0	0.0	0.0	0.0	0.0	0.0	0.0	1.5	0.0	0.0	0.9	0.0	0.0	1.3	0.6
B48	0.7	0.8	5.9	5.1	19.8	14.5	1.7	0.6	1.4	3.9	4.7	0.8	1.0	1.4	1.3	0.0	3.9	6.2	3.8	6.6	4.5	2.5	3.5	9.9	6.3

B49	0.7	1.7	0.0	0.0	0.0	0.0	0.0	1.5	0.5	0.0	0.0	0.8	0.0	0.0	0.0	0.0	1.0	0.0	0.0	2.0	0.0	0.0	2.1	0.0	0.0	
B50	0.0	3.5	1.0	1.9	1.4	0.0	0.6	0.0	2.4	0.0	2.9	0.0	0.0	0.0	0.3	0.0	0.0	0.0	0.0	0.0	0.0	0.0	0.0	0.3	3.2	
B51	6.4	6.9	13.4	6.5	17.8	0.3	3.8	5.7	9.4	11.4	8.7	8.5	6.4	4.2	1.3	4.4	5.0	4.1	0.0	11.8	1.8	13.3	3.5	14.8	5.9	
B52	0.7	7.9	1.0	2.7	0.0	0.0	3.4	0.0	1.4	2.8	5.8	3.5	3.1	2.1	1.0	9.8	1.0	0.7	0.0	2.9	0.0	0.8	3.5	0.0	0.0	
B53	0.0	1.7	0.5	0.0	0.0	0.0	1.1	1.3	0.0	1.1	0.4	13.3	0.0	0.0	0.0	0.0	0.0	0.0	0.0	0.0	4.5	0.0	1.4	0.6	0.6	
B54	1.4	1.7	1.5	2.9	2.1	0.0	1.7	3.2	2.4	2.8	1.4	1.7	0.6	2.1	1.7	0.0	0.0	2.9	0.0	0.0	0.0	0.0	0.0	0.0	0.0	
B55	5.9	0.8	5.6	1.9	0.7	9.3	4.0	8.2	4.3	2.2	2.2	0.8	2.5	1.4	1.0	0.7	0.0	1.4	22.6	0.0	0.0	0.0	0.0	0.0	0.6	
B56	1.4	0.0	1.0	0.3	0.0	0.6	0.0	5.5	1.9	0.0	0.7	0.8	1.4	0.7	1.7	1.5	0.0	2.1	17.0	0.0	0.9	0.0	0.0	0.0	0.0	
B57	1.5	0.0	0.5	1.3	0.0	0.0	0.0	1.3	1.0	3.9	1.8	4.3	5.2	1.4	2.7	2.2	0.0	0.0	0.0	1.0	0.0	0.0	0.0	0.0	2.7	
B58	3.6	2.5	0.0	2.9	0.0	0.9	4.8	8.2	5.5	5.8	5.1	3.4	2.9	15.5	3.0	1.5	3.1	7.2	1.9	0.0	0.9	0.0	0.7	0.0	5.9	
B59	0.0	0.0	0.0	0.3	0.0	0.0	0.0	0.0	0.0	0.0	0.4	0.0	0.0	0.7	0.0	0.0	0.0	0.0	0.0	0.0	0.0	0.0	0.0	0.0	0.0	
B60	11.8	2.5	6.4	5.4	11.6	32.3	3.6	5.1	12.8	3.4	4.0	6.2	8.3	16.2	3.2	1.5	1.0	11.4	15.1	2.9	25.7	1.7	0.0	1.3	7.5	
B61	2.3	0.9	9.0	4.8	9.6	2.8	5.2	0.6	2.4	4.7	6.9	4.3	4.3	2.8	1.7	0.0	1.0	2.1	8.5	9.4	6.7	1.7	13.0	27.3	6.2	
B62	4.5	0.9	12.6	9.8	11.6	4.2	3.8	8.2	3.4	2.8	5.4	4.3	5.0	7.0	10.1	26.7	32.4	4.1	2.8	4.9	3.7	6.9	4.2	16.9	16.0	
B63	0.0	3.3	5.2	0.3	0.0	0.0	8.3	0.6	0.0	1.8	3.3	1.7	0.4	0.7	1.1	0.0	1.0	7.9	0.0	0.0	0.9	0.0	0.0	1.6	0.6	
B67	0.7	0.0	0.0	0.5	0.0	0.3	2.3	0.0	0.0	0.0	0.7	0.0	0.0	0.0	0.0	0.0	0.0	0.7	1.9	0.0	0.0	0.0	0.0	0.0	0.6	
B70	0.0	0.0	0.5	0.5	1.4	0.0	1.7	0.0	0.5	0.0	0.4	0.8	0.0	0.7	0.7	0.0	0.0	0.7	0.0	0.0	0.0	0.8	3.5	0.0	0.6	
B73	0.0	0.0	0.0	0.0	0.0	0.0	0.0	0.0	0.0	0.0	0.0	0.0	0.0	0.0	0.3	0.0	0.0	0.0	0.9	0.0	0.0	0.0	0.0	0.0	0.0	
B75	10.0	0.9	0.5	4.2	3.4	0.3	4.6	0.0	3.0	1.7	0.0	0.9	8.3	4.2	6.7	0.7	2.0	0.0	2.8	0.0	3.6	0.8	0.7	0.0	0.6	
B76	0.7	0.0	0.0	0.0	0.0	0.0	0.0	0.7	0.0	0.0	0.0	0.0	0.0	1.4	2.1	0.0	0.0	2.1	0.0	0.0	0.0	0.0	0.0	0.3	0.0	
B77	0.7	0.0	0.0	0.0	0.7	0.0	0.0	0.0	0.0	0.0	0.0	0.0	0.4	0.0	0.0	0.7	0.0	0.0	0.0	0.0	0.0	0.0	0.0	0.0	0.0	
B5102	0.0	0.8	1.0	0.5	0.7	0.0	0.0	1.3	0.5	0.6	0.4	0.0	0.0	0.0	1.3	0.0	0.0	3.4	0.0	0.0	0.0	0.0	0.7	0.0	0.0	
B7801	0.0	0.0	0.0	0.0	0.0	0.0	0.0	0.0	0.0	0.0	0.0	0.8	0.0	0.0	0.0	0.0	0.0	0.0	0.0	0.0	0.0	0.0	0.0	0.0	0.0	
BX	0.7	0.0	0.0	0.3	0.0	0.0	0.0	0.0	0.0	0.0	0.0	0.0	0.4	0.0	1.7	0.0	0.0	0.0	0.0	0.0	0.0	0.0	0.0	0.6	0.0	
BBL	6.2	18.3	7.0	4.8	0.2	3.5	4.9	1.7	5.4	12.2	2.7	17.2	0.0	0.0	1.8	14.4	16.1	3.6	0.0	1.2	9.4	14.9	0.6	1.7	4.5	

HLA-C

Ethnic	10	10	10	10	10	10	10	10	10	10	10	20	20	30	30	30	30	30	30	30	30	30	30	30	30	30
code	100	102	200	201	202	204	206	207	210	400	600	100	202	101	102	103	104	105	106	109	110	111	113	114	115	117
N =	59	99	103	103	65	90	61	70	105	447	116	99	51	208	141	98	60	64	117	101	96	150	321	295	231	551
Cw1	1.7	5.1	0.0	0.0	0.8	0.0	1.7	0.7	1.4	1.1	2.2	22.9	27.3	2.9	4.4	3.6	5.9	5.6	3.5	4.2	4.7	4.0	4.2	3.3	2.6	3.1
Cw2	6.1	7.3	13.8	14.6	11.5	10.2	17.0	14.8	8.7	9.7	6.8	8.7	0.0	5.5	2.5	5.6	4.3	1.6	4.3	2.0	4.8	6.9	5.2	7.2	6.5	4.0
Cw4	11.8	17.8	7.0	14.6	20.6	15.3	17.1	15.1	11.6	21.0	21.5	17.6	37.1	14.2	9.6	14.1	8.7	11.5	11.4	7.8	13.1	9.2	11.1	11.2	16.2	17.1
Cw5	6.0	3.5	0.5	0.0	0.8	0.0	1.7	0.0	1.4	2.4	2.2	4.1	0.0	2.2	1.4	5.3	6.8	5.5	3.1	12.4	4.9	4.4	5.8	4.7	2.4	5.7
Cw6	8.0	5.7	19.4	33.0	19.7	18.4	11.2	11.0	23.2	7.4	9.0	16.1	2.0	5.4	10.2	13.1	6.1	5.6	12.1	8.8	12.9	9.9	8.9	14.2	6.7	10.1
Cw7	13.7	13.8	25.8	33.5	31.4	19.2	19.8	13.8	29.9	15.6	17.8	14.1	2.0	18.4	19.7	38.8	24.5	28.3	25.5	35.9	24.3	20.0	20.8	28.0	15.1	22.1
Cw9	7.1	1.0	2.0	1.0	0.0	5.6	5.0	3.6	4.0	2.3	9.0	5.8	11.0	2.9	3.3	4.6	5.3	5.7	6.6	10.0	7.1	7.0	7.4	5.3	4.5	2.8
Cw10	1.7	15.1	4.4	1.0	0.0	4.0	4.2	4.4	5.0	4.1	2.6	0.5	2.9	2.9	0.7	5.3	0.8	2.4	4.5	6.6	4.2	9.6	2.7	4.8	2.6	0.7
Cw11	0.0	0.0	0.0	0.0	0.0	0.0	0.8	0.0	0.5	0.0	0.4	0.0	0.0	0.5	0.4	0.0	0.0	0.0	0.0	0.0	0.0	0.0	0.0	0.3	0.0	0.1
CwX	3.4	1.0	0.0	2.4	3.3	1.7	0.0	0.0	5.1	1.8	0.0	0.5	2.0	0.2	2.5	0.0	0.0	1.6	1.7	0.0	0.0	0.7	3.8	1.7	0.0	1.7
CBL	40.5	29.8	27.1	0.0	11.9	25.5	21.6	36.5	9.3	34.6	28.4	9.7	15.6	45.0	45.3	9.6	37.5	32.2	27.3	12.4	24.0	28.3	30.2	19.3	43.5	32.7

Table 6. *(continued)*

HLA-C

Ethnic code	30 119	30 120	30 121	30 122	30 124	30 125	30 127	30 128	30 129	30 130	30 131	30 141	30 400	30 401	30 406	30 413	30 601	30 701	30 702	30 801	30 902	40 101	40 104	40 201	40 202	40 204
N =	50	125	91	97	264	95	83	97	98	98	61	75	99	50	73	50	63	246	197	114	351	1023	261	145	138	69
Cw1	2.1	1.2	5.1	3.1	3.1	3.7	1.8	11.4	3.1	3.1	5.0	0.0	5.7	9.4	5.9	5.1	3.2	4.3	5.5	1.8	2.7	11.8	12.7	12.9	16.3	6.0
Cw2	8.2	7.0	6.8	5.8	4.1	7.7	4.9	8.2	8.0	7.9	3.3	11.5	2.5	10.4	2.1	0.0	4.8	5.4	4.4	3.1	6.4	0.1	0.8	1.4	1.1	1.4
Cw4	18.1	15.7	11.7	12.1	12.8	9.3	6.1	7.3	12.7	14.3	14.5	9.7	13.4	11.2	19.3	9.6	12.2	9.8	10.0	19.0	16.4	4.2	5.9	4.2	8.0	3.7
Cw5	1.0	4.1	1.1	22.0	9.7	3.7	6.9	2.6	6.4	3.7	3.3	4.1	0.5	2.0	0.0	0.0	9.8	3.7	6.4	3.1	3.9	0.4	0.8	0.3	0.0	0.0
Cw6	10.8	5.4	7.5	7.3	9.6	3.7	10.8	11.2	4.7	13.9	7.6	16.3	11.2	8.0	17.8	18.9	7.6	9.5	9.4	5.4	6.6	1.0	7.5	7.3	4.4	2.2
Cw7	16.6	17.0	17.8	25.9	24.3	26.3	15.5	20.6	33.2	21.1	20.2	12.9	23.7	16.2	24.1	19.5	27.6	21.5	28.9	22.0	22.1	15.3	12.8	6.5	16.8	22.9
Cw9	6.2	5.3	2.8	0.5	4.5	5.3	3.7	17.4	3.1	5.7	9.3	2.7	2.5	10.0	4.9	3.0	3.2	3.9	7.2	2.7	7.4	13.9	13.8	8.8	14.8	11.1
Cw10	3.1	1.6	4.5	0.5	2.3	7.6	1.8	3.6	8.7	1.0	5.0	0.0	3.6	4.1	2.8	3.1	2.4	3.9	5.7	7.7	2.8	8.8	11.0	4.9	9.0	9.6
Cw11	0.0	0.0	0.0	0.0	0.0	0.0	0.0	7.1	2.1	0.5	0.0	0.0	0.5	9.4	0.7	0.0	0.0	0.2	0.5	0.0	0.0	4.2	3.9	2.5	3.6	19.1
CwX	0.0	0.0	1.1	0.5	1.5	6.1	0.0	0.0	0.0	0.0	0.0	0.0	0.5	0.0	0.0	0.0	5.0	3.3	5.4	0.9	1.0	0.6	0.2	0.0	0.0	0.0
CBL	34.0	42.8	41.7	22.2	28.1	26.5	48.4	10.7	18.1	28.8	31.7	43.0	35.7	19.2	22.5	40.7	24.2	34.4	16.6	34.4	30.6	39.5	30.7	51.1	26.0	24.4

HLA-C

Ethnic code	40 207	40 210	40 211	40 212	40 213	40 216	40 218	40 219	40 220	40 301	40 302	40 303	40 401	40 402	40 403	40 414	40 416	40 421	40 602	40 800	40 803	41 003	41 00A	41 102	41 10A
N =	70	60	99	187	73	165	87	79	104	89	138	59	242	71	149	68	51	73	53	51	56	61	72	156	70
Cw1	14.3	2.5	9.7	7.5	4.2	11.1	8.5	13.2	12.5	5.7	9.1	6.9	3.3	4.3	3.4	3.7	5.0	11.8	38.7	2.9	3.6	7.5	14.9	0.6	3.2
Cw2	1.4	3.4	1.5	0.8	2.1	0.6	1.2	28.9	1.4	1.7	2.9	4.3	2.7	0.0	0.0	0.7	4.9	2.1	0.9	1.0	10.5	2.5	0.7	12.1	0.8
Cw4	0.7	9.6	16.1	4.4	6.2	7.4	5.3	14.6	5.0	11.6	4.3	8.7	3.6	6.5	7.3	10.0	9.2	4.8	10.8	22.9	11.3	16.8	17.1	5.6	12.7
Cw5	0.0	4.3	0.0	2.4	0.0	0.0	1.2	5.1	1.5	0.0	0.0	8.8	0.6	0.0	1.0	0.0	2.0	0.0	0.0	1.0	0.0	0.0	0.0	0.3	1.3
Cw6	3.6	9.7	2.5	13.3	4.2	0.0	15.8	5.9	16.4	8.8	13.6	4.3	7.2	6.0	6.9	0.7	4.0	6.2	9.8	2.9	7.5	0.0	11.0	1.0	14.5
Cw7	15.9	7.9	16.8	15.1	15.2	21.5	5.3	10.3	13.1	8.1	17.2	5.2	25.5	24.0	18.8	14.1	10.5	18.6	21.6	30.7	2.7	34.3	26.9	4.5	6.8
Cw9	18.7	0.8	8.9	15.2	20.9	14.2	9.7	6.3	22.1	8.1	7.6	9.1	14.2	10.3	2.0	2.2	8.1	1.5	6.2	6.3	8.5	8.4	3.5	33.1	14.8
Cw10	6.6	4.2	7.9	8.4	15.3	17.2	7.3	6.1	5.4	13.0	17.1	18.6	6.9	19.5	8.0	0.0	2.0	34.8	4.0	13.1	10.5	4.3	7.8	19.8	16.7
Cw11	5.2	1.7	2.6	3.9	0.7	1.2	2.9	2.5	2.4	3.4	1.8	0.0	13.2	7.3	11.8	0.0	2.0	15.0	0.0	0.0	1.8	0.8	0.0	0.0	1.3
CwX	0.0	0.0	0.0	0.0	0.0	1.5	0.0	0.0	0.0	0.0	0.0	0.0	0.6	7.9	4.8	0.0	0.0	1.4	0.0	6.1	5.7	0.0	0.0	0.3	0.0
CBL	33.5	55.8	34.0	29.1	31.1	25.3	43.0	7.1	20.2	39.6	26.4	34.2	22.1	14.2	36.0	68.5	52.4	4.0	8.1	13.1	38.0	25.4	18.1	22.6	18.9

HLA-DR

Ethnic code	10	10	10	10	10	10	10	30	30	30	30	30	30	30	30	30	30	30	30	30	30	30	30	30	30	30
	100	200	201	204	210	400	600	101	102	103	104	105	106	109	110	111	111	114	115	117	120	121	122	124	128	129
N =	56	89	54	79	96	318	112	153	105	94	57	55	101	101	79	124	285	256	181	507	85	82	76	219	48	100
DR1	6.3	2.8	0.0	9.2	4.7	6.3	8.9	5.8	8.6	10.1	10.5	10.9	10.7	5.0	12.7	8.9	10.8	9.5	12.7	8.5	10.5	8.5	9.1	10.9	8.3	9.3
DR3	14.0	29.2	1.9	15.3	10.4	14.7	12.4	8.5	11.7	12.2	21.9	8.2	12.4	11.4	7.0	8.9	11.0	10.1	8.2	10.2	10.3	11.0	26.2	11.6	10.4	11.5
DR4	5.4	2.8	43.5	1.9	10.4	3.4	6.2	10.5	17.8	10.1	7.9	12.7	13.9	24.8	8.2	17.2	10.8	13.4	5.2	7.1	8.2	8.7	11.8	16.0	16.4	10.3
DR7	10.0	6.8	7.3	7.6	14.5	9.5	8.9	10.1	7.3	13.3	28.9	12.7	20.8	19.3	14.6	11.9	13.5	11.3	6.8	13.7	18.0	9.1	5.3	18.6	13.8	8.8
DR8	8.9	1.1	5.6	3.8	1.9	8.3	4.0	2.0	1.0	3.2	3.5	1.8	3.7	2.0	5.7	3.6	5.2	4.2	4.4	3.2	5.9	1.2	1.3	3.0	7.3	4.5
DR9	1.8	1.2	1.9	0.6	1.0	2.9	4.3	2.9	0.0	2.1	0.0	1.8	2.0	2.5	0.6	2.0	1.4	1.0	0.0	0.1	0.0	0.6	0.7	1.8	7.9	0.5
DR10	0.9	1.7	0.0	3.8	2.6	1.1	4.0	0.0	1.4	0.5	0.9	0.9	1.5	0.0	0.6	2.0	2.5	1.2	2.9	1.5	1.2	1.8	0.7	0.9	0.0	1.5
DR11	11.6	24.7	6.3	22.7	18.2	13.5	13.2	19.1	26.4	13.8	8.8	10.0	6.4	7.9	17.7	6.9	9.8	18.1	27.0	26.1	9.9	13.9	14.3	11.1	13.5	15.9
DR12	4.5	5.1	1.9	4.8	8.3	3.8	2.7	3.3	2.1	1.1	0.9	2.7	4.0	2.0	3.2	4.0	0.4	3.9	1.9	1.1	2.4	3.7	0.7	0.9	4.2	2.0
DR13	14.9	14.2	24.5	12.1	7.8	13.8	12.7	10.3	7.1	11.7	5.3	15.5	9.8	12.9	10.1	14.0	13.2	10.9	5.1	9.8	19.0	11.4	2.0	9.3	3.1	12.4
DR14	2.9	2.8	0.0	0.0	2.1	2.4	0.9	5.6	3.3	6.4	0.9	3.6	3.2	1.0	2.5	3.0	3.8	3.3	3.0	2.3	0.0	2.7	3.3	2.1	3.6	2.0
DR15	12.8	5.1	1.9	11.8	9.9	11.7	16.2	9.5	11.2	14.9	9.6	17.3	10.2	11.4	10.1	11.6	11.2	8.8	6.3	7.4	5.7	10.0	7.7	8.4	9.0	8.3
DR16	1.8	0.0	0.0	0.0	2.1	2.0	3.4	6.7	0.5	0.5	0.9	1.8	0.0	0.0	6.3	1.2	2.1	2.3	10.4	3.6	2.9	14.9	16.3	1.0	0.0	8.3
DR125	0.0	0.0	0.0	0.0	0.0	0.5	0.4	0.0	0.0	0.0	0.0	0.0	0.0	0.0	0.0	0.8	0.0	0.6	0.0	0.1	0.0	0.0	0.0	0.0	0.0	0.0
DR2L0	0.0	0.0	0.0	0.0	3.1	0.3	0.4	0.0	0.0	0.0	0.0	0.0	0.0	0.0	0.0	0.8	0.0	0.2	0.0	0.0	0.0	0.0	0.0	0.0	0.0	1.0
DRX	0.9	0.0	0.9	1.3	2.6	2.2	0.0	1.0	0.0	0.0	0.0	0.0	0.0	0.0	0.6	0.8	1.2	0.6	0.0	0.5	0.0	0.0	0.0	2.1	0.0	1.0
DRBL	3.4	2.5	4.5	5.1	0.4	3.7	1.2	4.8	1.6	0.0	0.0	0.0	1.6	0.0	0.0	2.4	3.1	0.7	6.0	4.9	6.1	2.4	0.8	2.3	2.4	2.6

Ethnic code	30	30	30	30	30	30	30	40	40	40	40	40	40	40	40	40	40	40	40	40	40	40	40	40	41	41
	130	400	601	701	702	801	902	101	104	201	202	212	216	219	220	302	401	402	403	414	416	421	602	803	00A	102
N =	80	93	59	232	146	86	291	898	237	87	79	90	93	72	54	101	238	70	142	61	49	77	51	57	65	145
DR1	10.9	5.6	11.4	10.1	12.0	5.2	8.5	5.5	5.1	0.6	1.3	3.3	0.0	16.1	7.7	4.0	0.2	0.0	1.4	0.0	1.0	0.6	2.9	0.9	2.3	0.7
DR3	15.1	4.5	9.6	10.1	11.3	6.1	9.5	0.2	2.3	2.9	2.7	5.7	1.1	21.5	8.9	9.1	4.8	7.1	4.0	0.8	5.2	5.2	2.9	0.9	2.3	3.9
DR4	10.3	10.6	22.5	12.8	18.2	16.8	11.0	22.8	21.4	14.4	15.8	9.1	8.7	9.0	12.4	13.4	9.0	16.4	8.7	4.1	5.2	19.5	20.0	10.9	41.7	38.3
DR7	9.9	13.3	7.4	15.1	14.4	12.4	12.2	0.4	9.2	16.3	2.0	15.2	0.0	2.1	11.8	13.4	8.3	6.4	5.8	3.5	0.0	1.9	0.0	1.8	0.8	0.7
DR8	5.0	1.6	1.7	3.0	2.7	9.5	5.5	13.3	9.4	4.3	3.2	8.5	7.9	2.8	4.6	7.9	4.0	4.3	6.7	6.9	3.2	6.5	9.8	7.0	4.7	3.8
DR9	0.0	1.1	0.8	1.5	2.1	0.0	1.9	13.0	8.3	12.3	15.9	7.6	6.0	17.0	7.4	5.9	11.9	9.3	11.5	2.5	0.0	16.9	11.3	3.5	13.7	5.3
DR10	1.3	5.6	1.7	1.7	0.7	0.6	2.1	0.6	2.1	1.2	1.9	0.6	0.0	1.4	2.9	2.0	2.5	2.1	5.7	0.0	1.0	1.3	0.0	0.9	0.0	0.0
DR11	12.8	8.1	8.0	9.7	7.5	8.9	15.8	2.6	2.7	8.3	11.4	6.4	19.3	8.1	12.9	8.0	4.2	6.4	2.1	3.5	12.3	13.6	9.8	0.0	12.1	11.2
DR12	3.1	1.6	3.4	1.6	0.7	1.8	3.7	7.0	7.0	7.7	13.6	7.8	23.6	2.1	2.8	8.4	13.7	12.1	29.7	49.7	7.5	13.6	21.2	1.8	0.8	0.7
DR13	12.2	12.3	0.2	12.7	11.6	6.7	11.5	7.8	11.8	3.0	13.2	6.8	0.5	2.1	5.8	5.4	3.6	5.7	4.2	3.3	3.1	1.3	1.0	4.4	4.6	7.5
DR14	0.0	6.5	5.3	1.5	2.1	9.1	2.1	5.5	5.6	0.0	1.3	3.5	12.2	3.5	0.9	2.0	7.1	6.4	5.6	0.8	3.1	3.2	9.8	60.2	1.5	10.8
DR15	7.3	25.6	14.9	10.8	14.4	8.9	8.5	17.4	11.3	19.5	9.1	16.7	7.1	4.3	11.5	13.0	20.6	17.1	9.5	20.6	42.0	9.7	9.6	4.9	2.3	1.4
DR16	4.7	0.0	0.8	1.3	2.1	2.9	2.8	0.8	0.2	2.4	1.3	0.0	0.0	2.9	0.0	2.8	4.5	2.9	2.1	0.0	0.0	3.2	0.0	0.0	0.8	1.4
DR125	1.3	0.0	0.0	0.0	0.0	2.3	0.7	1.9	2.1	0.6	0.0	0.6	0.5	0.0	0.0	1.5	0.2	0.0	0.4	0.0	0.0	0.0	0.0	0.9	0.8	5.2
BR2L6	0.0	0.0	0.0	0.2	0.0	0.0	0.0	0.0	0.0	0.0	0.0	0.0	0.0	0.0	0.0	0.0	1.7	2.1	0.0	0.0	0.0	3.2	0.0	0.0	0.0	0.3
DRX	0.0	0.0	0.0	5.3	0.3	0.0	0.3	0.2	0.0	0.0	0.0	0.0	0.5	0.0	0.0	0.0	0.2	1.4	0.4	0.0	0.0	0.0	0.0	0.0	0.0	0.3
DRBL	6.1	3.6	6.2	2.5	0.0	8.6	3.8	1.0	1.5	6.6	7.4	8.3	12.4	7.1	10.6	3.3	3.4	0.1	2.1	4.2	16.2	0.0	1.6	2.1	11.5	8.4

Table 6. (*continued*)

HLA-DQ

Ethnic	10	10	10	10	10	10	10	30	30	30	30	30	30	30	30	30	30	30	30	30	30	30	30	30	30	30
code	100	200	201	204	210	400	600	101	102	103	104	105	106	109	110	111	113	114	115	117	120	121	122	124	128	129
N =	56	89	54	79	96	318	112	153	105	94	57	55	101	101	79	124	285	256	181	507	85	82	76	219	48	100
DQ1	50.9	42.7	31.3	59.1	40.8	45.1	47.8	40.9	31.5	39.9	28.9	49.2	36.9	28.7	40.8	43.9	45.9	36.3	45.5	35.4	43.0	50.6	38.0	34.3	32.1	44.5
DQ2	21.4	16.9	10.2	15.5	19.3	19.8	17.1	16.8	16.7	19.7	50.0	16.1	26.2	25.7	16.3	16.8	21.4	19.1	14.3	22.0	20.2	19.5	27.3	25.4	17.5	19.0
DQ3	2.7	1.7	28.3	3.8	6.9	5.0	4.4	5.2	8.3	7.2	9.6	11.6	11.0	20.8	9.1	11.4	10.0	11.1	10.1	7.3	7.6	9.8	12.9	13.9	2.1	11.5
DQ4	5.4	22.5	5.8	7.3	4.7	10.0	5.8	2.3	1.0	5.2	4.4	2.7	3.1	2.0	4.4	4.8	4.5	6.1	3.9	2.3	6.6	0.6	0.7	3.9	5.5	4.0
DQ7	19.6	16.3	15.0	11.6	25.5	18.6	18.3	29.6	38.2	24.6	7.0	18.8	19.4	22.8	23.9	21.0	16.2	27.3	19.9	29.1	14.4	19.5	15.4	19.0	30.7	21.0
DQX	0.0	0.0	0.0	0.0	0.5	0.0	0.0	0.0	0.0	0.0	0.0	0.0	0.0	0.0	0.0	0.0	0.0	0.0	0.0	0.0	0.0	0.0	0.0	0.2	0.0	0.0
DQBL	0.0	0.0	9.4	2.7	2.4	1.5	6.5	5.3	4.2	3.4	0.0	1.6	3.2	0.0	5.5	2.2	2.1	0.0	6.2	3.9	8.2	0.0	5.7	3.4	6.2	0.0

Ethnic	30	30	30	30	30	30	30	40	40	40	40	40	40	40	40	40	40	40	40	40	40	40	40	40	41	40
code	130	400	601	701	702	801	902	101	104	201	202	212	216	219	220	302	401	402	403	414	416	421	602	803	00A	102
N =	80	93	59	232	146	86	291	898	237	87	79	90	93	72	54	101	238	70	142	61	49	77	51	57	65	145
DQ1	35.9	56.1	37.8	43.7	41.5	26.6	38.1	45.6	40.6	35.3	39.4	34.4	33.5	39.2	31.1	36.3	48.4	37.6	33.7	33.5	59.0	26.6	31.8	11.0	6.6	13.4
DQ2	22.6	13.2	14.4	22.9	21.8	16.7	18.2	0.6	10.2	12.6	5.4	17.2	1.1	16.6	14.0	20.8	11.5	10.0	9.8	4.9	11.1	8.4	4.9	1.8	3.1	1.7
DQ3	8.8	13.4	15.9	12.1	11.6	11.8	7.1	18.8	17.0	9.2	8.2	11.1	1.6	21.4	12.4	2.5	18.6	22.9	16.9	9.8	6.1	27.3	20.2	6.0	31.8	5.3
DQ4	2.5	1.6	3.4	2.4	3.1	7.7	6.6	14.9	8.3	4.1	6.3	3.3	1.6	5.6	3.7	1.5	3.7	4.6	4.2	9.8	2.0	7.1	10.8	8.8	12.0	5.7
DQ7	17.4	12.0	21.8	16.3	21.4	25.1	24.3	15.2	15.8	19.6	17.8	19.4	48.2	8.6	27.2	25.7	17.0	21.2	35.1	41.8	21.4	30.5	30.9	67.9	31.7	70.4
DQX	0.0	0.0	0.0	0.0	0.0	0.0	0.0	0.1	0.0	0.0	0.0	0.0	0.5	0.0	0.0	0.0	0.0	0.0	0.0	0.0	0.0	0.0	0.0	0.0	0.0	0.0
DQBL	12.8	3.7	6.7	2.7	0.6	12.2	5.7	4.8	8.1	19.3	2.9	14.5	13.4	8.6	11.5	13.3	0.8	3.7	0.3	0.2	0.3	0.0	1.4	4.6	14.9	3.5

Table 7. Frequencies and linkage disequilibrium for five-locus haplotypes in different ethnic groups (HLA-A, -C, -B, -DR, -DQ). (Reproduced with permission from *HLA 1991*, © Oxford University Press and the Committee of the 11th International Histocompatibility Workshop and Conference, 1992)

10100 North African Negroid (*N* = 56)

A	C	B	DR	DQ	HF (%)	LD
A30	CBL	B42	DR3	DQ4	2.7	2.7

10200 South African Negroid (*N* = 89)

A	C	B	DR	DQ	HF (%)	LD
A23	CBL	B70	DR11	DQ7	2.9	2.9
A30	CBL	B42	DR3	DQ4	2.8	2.8
A29	Cw7	B44	DR11	DQ1	2.8	2.8
ABL	Cw6	B58	DR12	DQ4	2.2	2.2
A3	Cw6	B58	DR7	DQ2	2.2	2.2
A2	CBL	B58	DR13	DQ1	2.2	2.2
A30	Cw7	B42	DR3	DQ4	2.0	2.2

10201 Bushman (San) (*N* = 54)

A	C	B	DR	DQ	HF (%)	LD
A30	Cw4	B58	DR13	DQ1	8.2	8.1
A3	Cw6	B58	DR11	DQ7	4.6	4.6
A23	Cw6	B58	DR4	DQ7	4.6	4.6
A43	Cw7	B7	DR4	DQ7	3.7	3.7
A28	Cw6	B58	DR8	DQ1	3.7	3.7
A23	Cw6	B58	DR4	DQ3	3.7	3.6
A32	Cw4	B70	DR7	DQ2	2.8	2.8
A2	Cw7	B18	DR13	DQ3	2.8	2.8
A28	Cw7	B41	DR4	DQ3	2.8	2.8
A2	Cw7	B57	DR4	DQ3	2.5	2.4

10204 Zimbabwean (*N* = 61)

A	C	B	DR	DQ	HF (%)	LD
A30	CBL	B45	DR1	DQ1	4.8	4.7
A30	Cw6	B58	DR15	DQ1	2.5	2.4
A28	Cw7	B7	DR15	DQ1	2.5	2.4
A2	CBL	B45	DR13	DQ1	2.5	2.4
A30	Cw6	B57	DRBL	DQ1	2.1	2.1

10210 South African Negroid (Cape) (*N* = 95)

A	C	B	DR	DQ	HF (%)	LD
A30	C Cw7	B8	DR7	DQ2	2.1	2.1

10400 North American Negroid (*N* = 312)

A	C	B	DR	DQ	HF (%)	LD
A36	Cw4	B53	DR11	DQ1	1.1	1.1
A2	CBL	B45	DR15	DQ1	1.1	1.1
A2	Cw4	B53	DR11	DQ2	1.0	0.9
A34	Cw4	B53	DR15	DQ1	0.8	0.8

Table 7. (*continued*)

10400 North American Negroid (*continued*)
(N = 312)

A	C	B	DR	DQ	HF (%)	LD
A30	Cw7	B42	DR3	DQ4	0.8	0.8
A2	Cw7	B7	DR15	DQ1	0.8	0.8
A2	CBL	B53	DR13	DQ1	0.8	0.8
A34	CBL	B51	DR3	DQ2	0.6	0.6
A33	CBL	B63	DR1	DQ1	0.6	0.6
A33	CBL	B42	DR3	DQ4	0.6	0.6
A30	Cw7	B18	DR11	DQ1	0.6	0.6
A3	Cw4	B35	DR11	DQ7	0.6	0.6
A23	Cw2	B70	DR11	DQ7	0.6	0.6

10600 South American Negroid
(N = 111)

A	C	B	DR	DQ	HF (%)	LD
A23	Cw7	B7	DR15	DQ1	2.3	2.2

30101 Albanian
(N = 148)

A	C	B	DR	DQ	HF (%)	LD
A2	Cw7	B18	DR11	DQ7	3.7	3.7
A3	CBL	B38	DR13	DQ1	1.7	1.7
A2	Cw4	B35	DR12	DQ7	1.7	1.7

30102 Armenian
(N = 97)

A	C	B	DR	DQ	HF (%)	LD
A33	CwX	B14	DR1	DQ1	3.1	3.1
A24	CBL	B18	DR11	DQ7	2.1	2.0

30103 Austrian
(N = 93)

A	C	B	DR	DQ	HF (%)	LD
A1	Cw7	B8	DR3	DQ2	5.1	5.1
A3	Cw7	B7	DR15	DQ1	3.2	3.2
A24	Cw4	B35	DR11	DQ7	3.2	3.2
A2	Cw7	B7	DR15	DQ1	2.6	2.5
A2	Cw7	B8	DR3	DQ2	2.2	2.2
A2	Cw5	B44	DR11	DQ7	2.2	2.1

30104 Basque
(N = 56)

A	C	B	DR	DQ	HF (%)	LD
A29	CBL	B44	DR7	DQ2	5.4	5.2
A30	CBL	B18	DR3	DQ2	4.7	4.7
A11	Cw1	B27	DR1	DQ1	4.5	4.5
A2	Cw7	B7	DR15	DQ1	3.6	3.6
A24	CBL	B18	DR3	DQ2	3.6	3.5
A1	Cw7	B8	DR3	DQ2	3.6	3.5
A30	Cw5	B18	DR3	DQ2	3.0	3.0
A1	Cw7	B57	DR7	DQ2	2.7	2.7
A24	CBL	B44	DR7	DQ2	2.7	2.6

Table 7. *(continued)*

A	C	B	DR	DQ	HF (%)	LD
A2	Cw7	B7	DR7	DQ2	2.7	2.6
A2	CBL	B44	DR15	DQ1	2.7	2.6
ABL	CBL	B18	DR3	DQ2	2.1	2.0
30105	Belgian (N = 54)					
A1	Cw7	B8	DR11	DQ7	3.7	3.7
A11	CBL	B52	DR15	DQ1	2.8	2.8
30106	British (N = 77)					
A1	CBL	B8	DR3	DQ2	3.3	3.3
A1	Cw7	B8	DR3	DQ2	2.9	2.9
A1	Cw4	B35	DR1	DQ1	2.2	2.2
30109	Cornish (N = 101)					
A1	Cw7	B8	DR3	DQ2	8.4	8.4
A2	Cw5	B44	DR4	DQ7	5.0	4.9
A29	CBL	B44	DR7	DQ2	4.4	4.4
A2	Cw7	B7	DR15	DQ1	3.6	3.6
A2	Cw9	B62	DR4	DQ3	2.8	2.8
A2	Cw10	B60	DR4	DQ3	2.5	2.5
30110	Czech (N = 53)					
A3	Cw4	B35	DR1	DQBL	2.8	2.8
A2	Cw6	B13	DR7	DQ2	2.8	2.8
A2	Cw10	B60	DR11	DQ7	2.8	2.8
A1	Cw7	B8	DR3	DQ1	2.8	2.8
A3	Cw7	B7	DR15	DQ1	2.5	2.5
40414	Javanese (N = 57)					
ABL	CBL	B62	DR12	DQ7	8.2	7.7
A24	CBL	BBL	DR12	DQ7	7.2	6.4
A11	CBL	B62	DR12	DQ7	4.7	4.2
A24	CBL	B62	DR15	DQ1	3.9	3.4
A24	CBL	B18	DR12	DQ7	3.5	3.0
A24	Cw4	B35	DR12	DQ7	3.2	3.0
A2	Cw7	BBL	DR15	DQ1	3.0	3.0
A24	CBL	B35	DR12	DQ7	2.7	1.9
A24	Cw4	B35	DR12	DQ1	2.6	2.5
A11	CBL	B62	DR12	DQ3	2.6	2.5
ABL	CBL	BBL	DR15	DQ1	2.5	2.5

Table 7. *(continued)*

40416 Timor
(*N* = 48)

A	C	B	DR	DQ	HF (%)	LD
A24	CBL	B62	DR15	DQ1	8.6	6.8
A24	CBL	BBL	DR15	DQ1	5.8	4.8
A1	CBL	B58	DR3	DQ2	4.2	4.2
ABL	CBL	B62	DRBL	DQ1	4.2	4.0
A24	CBL	B27	DR15	DQ1	2.2	1.9
A11	CBL	B51	DRBL	DQ1	2.1	2.1
ABL	Cw2	B62	DR15	DQ1	2.1	2.0
A24	CBL	BBL	DR12	DQ1	2.1	1.9
A24	Cw9	B62	DR15	DQ1	2.1	1.8
A24	CBL	B48	DR15	DQ1	2.1	1.8

40421 Singapore Chinese
(*N* = 68)

A	C	B	DR	DQ	HF (%)	LD
A2	Cw11	B46	DR9	DQ3	7.2	7.2
A33	Cw10	B58	DR3	DQ2	3.5	3.5
A11	Cw10	B13	DR11	DQ7	2.9	2.9
A24	Cw4	B63	DR4	DQ3	2.2	2.2
A2	Cw1	B56	DR4	DQ3	2.2	2.2
A11	Cw10	B60	DR15	DQ1	2.2	2.2

40602 Maori
(*N* = 50)

A	C	B	DR	DQ	HF (%)	LD
A2	Cw1	B55	DR12	DQ7	8.1	8.0
A24	Cw1	B56	DR4	DQ3	6.0	5.9
A24	Cw1	B56	DR12	DQ7	5.0	4.8
A24	Cw7	B39	DR14	DQ1	3.0	3.0
A24	Cw4	B60	DR4	DQ4	3.0	3.0
A24	Cw1	B55	DR12	DQ7	3.0	2.8
A24	Cw7	B39	DR4	DQ1	2.7	2.6
A34	Cw6	B61	DR14	DQ1	2.0	2.0
A33	Cw6	B56	DR8	DQ1	2.0	2.0
A24	Cw6	B60	DR4	DQ3	2.0	2.0
A11	Cw4	B60	DR4	DQ4	2.0	2.0
A24	Cw1	B56	DR11	DQ7	2.0	1.9
A24	Cw1	B55	DR9	DQ3	2.0	1.9
A24	Cw1	B55	DR9	DQ1	2.0	1.9

40803 Tlingit
(*N* = 50)

A	C	B	DR	DQ	HF (%)	LD
A2	CBL	BBL	DR14	DQ7	10.4	9.7
A24	Cw2	B27	DR14	DQ7	6.0	5.9
A2	Cw4	B35	DR14	DQ7	4.7	4.5
A24	CBL	B60	DR14	DQ7	4.5	3.3
A24	CBL	B48	DR14	DQ7	3.0	2.8
A2	CBL	B60	DR14	DQ7	3.0	1.0

Table 7. *(continued)*

A	C	B	DR	DQ	HF (%)	LD
A2	CBL	B27	DR14	DQ7	2.2	1.4
A2	Cw10	B60	DR14	DQ7	2.1	1.6
A24	CBL	B62	DRBL	DQ7	2.0	2.0
A2	Cw6	B53	DR14	DQ7	2.0	2.0
A2	Cw1	B75	DR14	DQ7	2.0	2.0
A2	CBL	B37	DR8	DQ4	2.0	2.0
A2	Cw4	B53	DR14	DQ7	2.0	1.9
A28	CBL	BBL	DR14	DQ7	2.0	1.8
4100A	South American Indian ($N = 65$)					
A24	Cw1	B35	DR4	DQBL	4.5	4.5
A24	Cw4	B35	DR4	DQ7	4.5	4.3
A2	Cw7	B39	DRBL	DQBL	4.0	3.9
A2	CBL	B35	DR8	DQ4	3.0	3.0
A1	Cw6	B38	DR11	DQ7	3.0	3.0
A24	Cw7	B39	DRBL	DQ3	2.5	2.4
A24	CBL	B39	DR4	DQ3	2.5	2.3
A2	Cw7	B39	DR4	DQ3	2.4	2.2
A2	Cw10	B61	DR9	DQ3	2.3	2.2
A24	Cw10	B61	DR4	DQ7	2.1	2.0
A24	Cw4	B70	DR4	DQ3	2.0	2.0
41102	Inuit ($N = 144$)					
A24	CBL	B48	DR4	DQ7	9.4	9.0
A24	Cw10	B61	DR4	DQ7	5.5	4.6
A24	Cw9	B62	DR4	DQ7	4.8	4.0
A24	Cw9	B61	DR4	DQ7	3.2	1.8
A24	CBL	B61	DR14	DQ7	3.0	2.7
A2	Cw9	B62	DR4	DQ7	2.8	2.6
A2	Cw2	B27	DR4	DQ7	2.5	2.4
A24	Cw10	B51	DR11	DQ7	2.2	2.1
A28	CBL	B61	DR4	DQ7	2.2	2.0
A28	CBL	B51	DR4	DQ7	2.2	2.0
A24	Cw10	B61	DRBL	DQ1	1.7	1.7
A24	Cw10	B51	DRBL	DQ7	1.6	1.5
A24	Cw9	B51	DR11	DQ7	1.6	1.3
A24	CBL	BBL	DRBL	DQ7	1.5	1.5
A24	Cw4	B35	DRBL	DQ7	1.4	1.4
A24	CBL	B51	DR11	DQ7	1.4	1.2
A24	CBL	B52	DR8	DQ1	0.3	0.2
A24	CBL	B51	DR4	DQBL	0.3	0.2
A2	Cw9	B35	DR8	DQBL	0.3	0.2
A2	Cw10	B61	DR4	DQBL	0.3	0.2
A2	CBL	B52	DR15	DQ1	0.3	0.2
A24	Cw10	B61	DR14	DQ1	0.2	0.2
A24	Cw10	B60	DR4	DQ4	0.2	0.2

Table 7. (*continued*)

41102 Inuit (*continued*)
(*N* = 144)

A	C	B	DR	DQ	HF (%)	LD
A24	Cw1	B55	DR4	DQBL	0.2	0.2
A24	CBL	B52	DR14	DQ1	0.2	0.2
A24	CBL	B48	DR4	DQ4	0.2	0.2
A24	CBL	B51	DRJ25	DQ7	0.2	0.2
A24	CBL	B35	DRJ25	DQ7	0.2	0.2
A2	Cw9	B62	DR15	DQ1	0.2	0.2
A2	Cw9	B35	DRBL	DQ1	0.2	0.2
A2	Cw1	B54	DR8	DQ1	0.2	0.2
A2	CBL	B51	DR9	DQ3	0.2	0.2
A11	Cw4	B62	DR15	DQ1	0.2	0.2

40104 Korean
(*N* = 235)

A	C	B	DR	DQ	HF (%)	LD
A33	CBL	B44	DR13	DQ1	4.5	4.4
A33	Cw7	B44	DR7	DQ2	3.8	3.8
A30	Cw6	B13	DR7	DQ2	3.2	3.2
A24	CBL	B52	DR15	DQ1	2.3	2.3
A33	Cw10	B58	DR13	DQ1	2.1	2.1
A2	Cw1	B27	DR1	DQ1	1.6	1.6
A24	Cw4	B62	DR4	DQ3	1.3	1.3
A1	Cw6	B37	DR10	DQ1	1.3	1.3
A24	Cw7	B7	DR1	DQ1	1.2	1.2
A2	Cw11	B46	DR8	DQ1	1.1	1.1
A2	Cw10	B13	DR12	DQ7	1.1	1.1
A2	Cw1	B54	DR4	DQ4	1.1	1.1
A26	CBL	B61	DR9	DQ3	1.0	1.0
A24	CBL	B51	DR15	DQ1	1.0	1.0
A2	Cw1	B54	DR4	DQBL	1.0	1.0
A11	CBL	B62	DR14	DQ1	1.0	1.0
A24	CBL	B61	DR9	DQ3	0.9	0.8
A2	Cw9	B48	DR14	DQ1	0.9	0.8

40201 Northern Han
(*N* = 77)

A	C	B	DR	DQ	HF (%)	LD
A2	CBL	B62	DR15	DQ1	3.9	3.7
A2	CBL	B51	DR7	DQ2	2.6	2.5

40202 Southern Han
(*N* = 79)

A	C	B	DR	DQ	HF (%)	LD
A24	Cw1	B54	DR4	DQ7	3.5	3.5
A11	CBL	B60	DR11	DQ7	2.7	2.6
A11	Cw9	B60	DR9	DQ7	2.5	2.5
A11	CBL	B61	DR4	DQ4	2.5	2.5

Table 7. (*continued*)

40212 Manchu
(N = 89)

A	C	B	DR	DQ	HF (%)	LD
A30	Cw6	B13	DR7	DQ2	5.0	5.0
A24	Cw9	B62	DR15	DQ1	2.2	2.2

40216 Taiwan Aborigine
(N = 85)

A	C	B	DR	DQ	HF (%)	LD
A24	Cw7	B39	DR12	DQ7	10.4	10.0
A2	CBL	B48	DRBL	DQBL	6.1	6.1
A2	Cw1	B55	DR11	DQ7	5.2	5.1
A24	Cw7	B60	DR14	DQ1	4.1	3.8
A24	Cw9	B60	DR11	DQ7	3.8	3.6
ABL	Cw10	B13	DR12	DQ7	3.3	3.3
A24	Cw4	B60	DR4	DQ7	2.9	2.9
A24	Cw9	B55	DR11	DQ7	2.4	2.3
A24	Cw7	B60	DRBL	DQ1	2.3	2.0
A24	CBL	B60	DR15	DQ1	2.2	2.1

40219 Li
(N = 72)

A	C	B	DR	DQ	HF (%)	LD
A11	Cw4	B13	DRBL	DQ1	2.6	2.6
A2	Cw2	B62	DR1	BQBL	2.1	2.1
A11	Cw2	B55	DR3	DQ3	2.1	2.1
A11	Cw2	B13	DR3	DQ1	2.1	1.9

40220 Inner Mongolian
(N = 54)

A	C	B	DR	DQ :	HF (%) :	LD

40302 Mongolian
(N = 101)

A	C	B	DR	DQ	HF (%)	LD
A30	Cw6	B13	DR7	DQ2	4.0	4.0
A24	CBL	B52	DR15	DQ1	3.0	3.0
A2	Cw6	B50	DR7	DQ2	3.0	3.0
A2	Cw7	B8	DR3	DQ2	2.8	2.8

40401 Thais
(N = 238)

A	C	B	DR	DQ	HF (%)	LD
A2	Cw11	B46	DR9	DQ3	4.5	4.5
A33	Cw7	B44	DR7	DQ2	2.6	2.6
A11	Cw11	B46	DR15	DQ1	1.6	1.5
A11	CBL	B75	DR15	DQ1	1.6	1.5
A33	Cw9	B58	DR3	DQ2	1.5	1.5
A33	Cw9	B57	DR3	DQ2	1.5	1.5
A11	Cw11	B46	DR9	DQ3	1.5	1.5
A2	Cw9	B13	DR15	DQ1	1.5	1.4
A11	Cw11	B46	DR14	DQ1	1.3	1.3

Table 7. (*continued*)

40401 Thais (*continued*)
(*N* = 144)

A	C	B	DR	DQ	HF (%)	LD
A11	CBL	B51	DR8	DQ1	1.3	1.3
A24	Cw7	B38	DR15	DQ1	1.3	1.2
A2	CBL	B61	DR15	DQ1	1.3	1.2
A2	Cw9	B60	DR15	DQ1	1.1	1.0
A11	CBL	B75	DR12	DQ7	1.1	1.0
A11	Cw9	B13	DR15	DQ1	0.9	0.9

40402 Thai Chinese
(*N* = 70)

A	C	B	DR	DQ	HF (%)	LD
A33	Cw10	B58	DR3	DQ2	4.3	4.3
A2	Cw11	B46	DR9	DQ3	2.9	2.9
A24	Cw6	B13	DR7	DQ2	2.1	2.1
A11	Cw7	B60	DR4	DQ3	2.1	2.1
A11	Cw10	B13	DR15	DQ1	2.1	2.1
A11	CBL	B27	DR12	DQ7	2.1	2.1

40403 Vietnamese
(*N* = 140)

A	C	B	DR	DQ	HF (%)	LD
A29	CBL	B7	DR10	DQ1	4.6	4.6
A2	Cw11	B46	DR9	DQ3	4.5	4.5
A24	Cw4	B35	DR12	DQ7	2.1	2.1
A11	CBL	B62	DR12	DQ7	2.1	2.0
A11	CBL	B75	DR12	DQ7	2.0	2.0
A2	CBL	B62	DR12	DQ7	1.8	1.7
ABL	Cw11	B46	DR12	DQ7	1.6	1.6
A2	CBL	B46	DR12	DQ7	1.6	1.5

30129 Yugoslavian
(*N* = 98)

A	C	B	DR	DQ	HF (%)	LD
A1	Cw7	B8	DR3	DQ2	7.7	7.6
A2	Cw7	B18	DR11	DQ7	3.6	3.5
A3	Cw7	B7	DR15	DQ1	2.4	2.4
A2	Cw7	B18	DR4	DQ1	2.0	2.0

30130 Hungarian
(*N* = 80)

A	C	B	DR	DQ	HF (%)	LD
A1	Cw7	B8	DR3	DQ2	4.8	4.8
A1	CBL	B8	DR3	DQ2	3.8	3.8
A3	Cw4	B35	DR1	DQ1	3.1	3.1

30400 Indian
(*N* = 93)

A	C	B	DR	DQ	HF (%)	LD
A24	CBL	B61	DR15	DQ1	4.1	4.0
A1	Cw6	B37	DR10	DQ1	2.6	2.6
A11	Cw7	B44	DR7	DQ2	2.2	2.1

Table 7. (*continued*)

30601 Australian
(N = 59)

A	C	B	DR	DQ	HF (%)	LD
A1	Cw7	B8	DR3	DQ2	7.6	7.6
A1	Cw7	B7	DR15	DQ1	3.4	3.4
ABL	CBL	BBL	DRBL	DQBL	2.8	2.8
A2	CwX	B62	DR4	DQ3	2.8	2.8
A2	Cw5	B44	DR4	DQ7	2.8	2.8
A3	Cw5	B44	DR15	DQ1	2.5	2.5
A24	Cw4	B35	DR11	DQ7	2.5	2.5
A2	Cw7	B7	DR15	DQ1	2.5	2.5
A1	Cw7	B8	DR15	DQ1	2.5	2.5

30701 USA
(N = 226)

A	C	B	DR	DQ	HF (%)	LD
A1	Cw7	B8	DR3	DQ2	4.5	4.5
A1	CBL	B8	DR3	DQ2	1.9	1.9
A2	Cw7	B7	DR15	DQ1	1.6	1.6
A2	CBL	B38	DR13	DQ1	1.5	1.5
A3	Cw4	B35	DR1	DQ1	1.3	1.3
A2	Cw10	B60	DR13	DQ1	1.1	1.1
A1	CBL	B44	DR4	DQ7	1.1	1.1
A3	Cw7	B7	DR15	DQ1	1.0	1.0
A2	CBL	B14	DR7	DQ2	1.0	1.0
A3	Cw1	B27	DR1	DQ1	0.9	0.9
A29	CBL	B44	DR7	DQ2	0.9	0.9
A25	CBL	B18	DR15	DQ1	0.9	0.9
A24	Cw4	B35	DR11	DQ7	0.9	0.9
A24	Cw4	B35	DR1	DQ1	0.9	0.9
A2	CBL	B44	DR4	DQ3	0.9	0.9

30702 Canadian
(N = 142)

A	C	B	DR	DQ	HF (%)	LD
A1	Cw7	B8	DR3	DQ2	5.2	5.1
A3	Cw7	B7	DR15	DQ1	4.3	4.3
A2	Cw5	B44	DR4	DQ7	3.2	3.2
A32	CwX	B14	DR7	DQ2	2.1	2.1
A2	Cw2	B44	DR1	DQ1	2.1	2.1
A24	Cw9	B62	DR13	DQ1	1.4	1.4
A1	Cw6	B57	DR7	DQ3	1.4	1.4
A2	Cw7	B7	DR15	DQ1	1.4	1.3

30801 Mexican
(N = 86)

A	C	B	DR	DQ	HF (%)	LD
A2	CBL	B35	DR14	DQ1	2.8	2.8
A29	CBL	B44	DR7	DQ2	2.3	2.3

Table 7. (*continued*)

30902 Brazilian
(*N* = 286)

A	C	B	DR	DQ	HF (%)	LD
A24	Cw4	B35	DR11	DQ7	1.4	1.4
A2	Cw4	B35	DR7	DQ2	1.4	1.4
A2	CBL	B44	DR7	DQ2	1.4	1.4
A1	Cw7	B8	DR3	DQ2	1.3	1.3
A2	CBL	B51	DR1	DQ1	0.9	0.9
A24	Cw4	B35	DRBL	DQ1	0.8	0.8
A33	CBL	B14	DR1	DQ1	0.7	0.7
A3	Cw7	B7	DR7	DQ2	0.7	0.7
A2	Cw7	B39	DR8	DQ4	0.7	0.7
A2	CBL	B51	DR4	DQ7	0.7	0.7

40101 Japanese (Wajin)
(*N* = 893)

A	C	B	DR	DQ	HF (%)	LD
A24	CBL	B52	DR15	DQ1	8.3	8.2
A33	CBL	B44	DR13	DQ1	4.9	4.9
A24	Cw7	B7	DR1	DQ1	3.6	3.6
A2	Cw11	B46	DR8	DQ1	2.0	2.0
A24	Cw1	B54	DR4	DQ4	1.6	1.6
A11	Cw4	B62	DR4	DQ3	1.2	1.2
A24	CBL	B51	DR9	DQ3	1.0	0.9
A2	Cw9	B35	DR4	DQ3	0.8	0.8
A11	Cw7	B67	DR15	DQ1	0.7	0.7
A33	CBL	B44	DR8	DQ1	0.6	0.6
A31	CBL	B51	DR12	DQ7	0.6	0.6
A26	Cw9	B35	DR15	DQ1	0.6	0.6
A24	Cw1	B59	DR4	DQ4	0.6	0.6
A2	Cw7	B7	DR1	DQ1	0.6	0.6
A2	Cw7	B39	DR15	DQ1	0.6	0.6
A2	Cw11	B46	DR9	DQ3	0.6	0.6
A2	Cw10	B13	DR12	DQ7	0.6	0.6
A2	Cw1	B59	DR4	DQ4	0.6	0.6
A11	Cw7	B39	DR8	DQ1	0.6	0.6
A31	CBL	B51	DR8	DQ4	0.5	0.5
A26	Cw9	B62	DR1	DQ1	0.5	0.5
A26	Cw10	B61	DR9	DQ3	0.5	0.5
A24	Cw11	B46	DR8	DQ1	0.5	0.5
A24	Cw1	B54	DR4	DQBL	0.5	0.5
A24	CBL	B61	DR9	DQ3	0.5	0.5
A2	Cw1	B54	DR4	DQ4	0.5	0.5
A2	CBL	B61	DR9	DQ3	0.5	0.5
A11	Cw1	B54	DR4	DQ4	0.5	0.5
A26	CBL	B61	DR9	DQ3	0.5	0.4
A24	CBL	B61	DR12	DQ7	0.5	0.4
A31	CBL	B51	DR8	DQ1	0.4	0.4
A26	CBL	B52	DR15	DQ1	0.4	0.4
A24	Cw1	B54	DR8	DQ1	0.4	0.4

Table 7. (*continued*)

40101 Japanese (Wajin) (*continued*)
(N = 144)

A	C	B	DR	DQ	HF (%)	LD
A24	CBL	B54	DR4	DQ4	0.4	0.4
A24	CBL	B51	DR8	DQ3	0.4	0.4
A2	CBL	B48	DR15	DQ1	0.4	0.4
A1	Cw6	B37	DR10	DQ1	0.4	0.4
A31	CBL	B51	DR4	DQ4	0.4	0.3
A33	Cw10	B58	DR13	DQ1	0.3	0.3
ABL	Cw1	B55	DR4	DQ4	0.3	0.3
ABL	CBL	B61	DR12	DQ7	0.3	0.3
A31	Cw10	B60	DR8	DQ1	0.3	0.3
A31	CBL	B51	DRJ25	DQ7	0.3	0.3
A26	Cw9	B62	DR9	DQ7	0.3	0.3
A26	Cw9	B35	DR14	DQ1	0.3	0.3
A26	Cw9	B35	DR4	DQ3	0.3	0.3
A24	Cw9	B35	DR11	DQ7	0.3	0.3
A24	Cw9	B35	DR4	DQBL	0.3	0.3
A24	Cw7	B70	DR4	DQ7	0.3	0.3
A24	Cw7	B7	DR4	DQ4	0.3	0.3
A24	Cw4	B62	DR4	DQ3	0.3	0.3
A24	Cw10	B61	DR9	DQ3	0.3	0.3
A24	Cw10	B60	DR11	DQ7	0.3	0.3
A24	Cw1	B54	DR14	DQ1	0.3	0.3
A24	CBL	B51	DR11	DQ7	0.3	0.3
A24	CBL	B35	DR4	DQ4	0.3	0.3
A2	Cw9	B61	DR9	DQ3	0.3	0.3
A2	Cw9	B35	DR15	DQ1	0.3	0.3
A2	Cw7	B39	DR9	DQ3	0.3	0.3
A2	Cw11	B46	DR12	DQ7	0.3	0.3
A2	Cw10	B61	DR14	DQ1	0.3	0.3
A2	Cw10	B61	DR9	DQ3	0.3	0.3
A2	Cw10	B60	DR4	DQ4	0.3	0.3
A2	CBL	B48	DR4	DQ3	0.3	0.3
A2	CBL	B51	DRJ25	DQ7	0.3	0.3
A2	CBL	B13	DR12	DQ7	0.3	0.3
A11	Cw7	B67	DR16	DQ1	0.3	0.3
A11	Cw1	B55	DR4	DQ4	0.3	0.3
A11	CBL	B51	DR4	DQ4	0.3	0.3
A24	Cw9	B62	DR4	DQ4	0.3	0.2

30111 Dane
(N = 122)

A	C	B	DR	DQ	HF (%)	LD
A3	Cw7	B7	DR15	DQ1	3.6	3.6
A1	Cw7	B8	DR3	DQ2	3.4	3.3
A1	CBL	B8	DR3	DQ2	2.8	2.8
A29	CBL	B44	DR7	DQ2	2.0	2.0

Table 7. (*continued*)

30113 French
(*N* = 244)

A	C	B	DR	DQ	HF (%)	LD
A1	Cw7	B8	DR3	DQ2	2.5	2.5
A29	CBL	B44	DR7	DQ2	2.2	2.2
A3	Cw7	B7	DR15	DQ1	1.8	1.8
A33	CBL	B14	DR1	DQ1	1.4	1.4
A1	CBL	B8	DR3	DQ2	1.4	1.4
A2	CBL	B51	DR13	DQ1	1.2	1.1
A2	Cw4	B35	DR3	DQ2	1.0	1.0

30114 German
(*N* = 203)

A	C	B	DR	DQ	HF (%)	LD
A1	Cw7	B8	DR3	DQ2	4.8	4.8
A3	Cw7	B7	DR15	DQ1	2.5	2.5
A33	CBL	B14	DR1	DQ1	1.2	1.2
A2	Cw7	B8	DR3	DQ2	1.2	1.2
A2	Cw5	B44	DR4	DQ7	1.2	1.2
A3	Cw4	B35	DR11	DQ7	1.0	1.0
A3	Cw4	B35	DR1	DQ1	1.0	1.0
A3	CBL	B7	DR4	DQ3	1.0	1.0
A28	Cw7	B44	DR11	DQ7	1.0	1.0
A24	Cw4	B35	DR11	DQ7	1.0	1.0
A23	Cw4	B44	DR7	DQ2	1.0	1.0
A1	CBL	B8	DR3	DQ2	1.0	1.0

30115 Greek
(*N* = 176)

A	C	B	DR	DQ	HF (%)	LD
A2	Cw4	B35	DR11	DQ7	2.0	1.9
A1	Cw7	B8	DR3	DQ2	1.9	1.9
A2	CBL	B51	DR1	DQ1	1.7	1.6
A2	CBL	B51	DR11	DQ7	1.6	1.5
A11	Cw4	B35	DR1	DQ1	1.3	1.2
A33	CBL	B14	DR1	DQ1	1.1	1.1
A3	Cw7	B7	DR16	DQ1	1.1	1.1
A3	Cw4	B35	DR3	DQ2	1.1	1.1
A2	Cw7	B8	DR3	DQ1	1.1	1.1
A2	Cw7	B18	DR11	DQ7	1.1	1.1
A11	Cw4	B35	DR14	DQ1	1.1	1.1
A2	CBL	B35	DR11	DQ1	1.1	0.9

30117 Italian
(*N* = 483)

A	C	B	DR	DQ	HF (%)	LD
A2	Cw4	B35	DR11	DQ7	2.3	2.3
A1	Cw7	B8	DR3	DQ2	2.3	2.3
A3	Cw4	B35	DR11	DQ7	2.2	2.1
A3	Cw4	B35	DR1	DQ1	1.6	1.6
A32	Cw4	B35	DR11	DQ7	1.3	1.3

Table 7. (*continued*)

30117 Italian (*continued*)
(*N* = 483)

A	C	B	DR	DQ	HF (%)	LD
A29	CBL	B44	DR7	DQ2	1.3	1.3
A1	CBL	B8	DR3	DQ2	1.3	1.3
A2	CBL	B51	DR11	DQ7	1.3	1.2
A30	Cw6	B13	DR7	DQ2	1.1	1.1
A2	Cw7	B18	DR11	DQ7	1.0	1.0
A2	CBL	B18	DR11	DQ7	1.0	1.0
A2	Cw5	B44	DR11	DQ7	0.9	0.9
A24	CBL	B18	DR11	DQ7	0.8	0.8
A33	CBL	B14	DR1	DQ1	0.7	0.7
A24	Cw4	B35	DR11	DQ7	0.7	0.7
A2	Cw7	B8	DR3	DQ2	0.7	0.7
A2	Cw5	B44	DR13	DQ1	0.7	0.7
A3	Cw7	B7	DR15	DQ1	0.6	0.6
A2	CBL	B44	DR7	DQ2	0.6	0.5
A2	Cw6	B13	DR7	DQ2	0.5	0.5
A2	Cw5	B44	DR4	DQ7	0.5	0.5
A2	Cw5	B18	DR3	DQ2	0.5	0.5
A2	Cw4	B35	DR13	DQ1	0.5	0.5
A2	CBL	B52	DR15	DQ1	0.5	0.5
A11	Cw4	B35	DR1	DQ1	0.5	0.5
A1	Cw7	B63	DR13	DQ1	0.5	0.5
ABL	Cw7	B7	DR15	DQ1	0.4	0.4
A32	Cw9	B62	DR11	DQ7	0.4	0.4
A30	Cw5	B18	DR3	DQ2	0.4	0.4
A3	CBL	B38	DR13	DQ1	0.4	0.4
A3	CBL	B18	DR11	DQ7	0.4	0.4
A3	CBL	B14	DR1	DQ1	0.4	0.4
A26	Cw4	B35	DR1	DQ1	0.4	0.4
A11	Cw4	B35	DR7	DQ2	0.4	0.4
A1	Cw7	B8	DR7	DQ2	0.4	0.4

30120 Portuguese
(*N* = 85)

A	C	B	DR	DQ	HF (%)	LD
A2	CBL	B44	DR11	DQ7	2.4	2.3
A29	CBL	B44	DR7	DQ2	2.2	2.2

30121 Romanian
(*N* = 82)

A	C	B	DR	DQ	HF (%)	LD
A2	CBL	B51	DR16	DQ1	3.0	2.9
A1	Cw7	B8	DR3	DQ2	2.8	2.8
A3	Cw4	B35	DR1	DQ1	2.4	2.4

Table 7. (*continued*)

30122 Sardinian
(N = 74)

A	C	B	DR	DQ	HF (%)	LD
A30	Cw5	B18	DR3	DQ2	11.4	11.3
A2	Cw7	B58	DR16	DQ1	3.9	3.8
A33	CBL	B14	DR1	DQ1	2.7	2.7
A2	Cw7	B49	DR4	DQ3	2.7	2.7
A2	Cw4	B35	DR3	DQ2	2.7	2.7
A2	Cw5	B18	DR3	DQ2	2.7	2.5
A30	Cw5	B18	DR3	DQBL	2.6	2.6
A32	Cw7	B18	DR16	DQ1	2.0	2.0

30124 Spanish
(N = 192)

A	C	B	DR	DQ	HF (%)	LD
A29	CBL	B44	DR7	DQ2	3.0	3.0
A1	Cw7	B8	DR3	DQ2	2.0	2.0
A2	Cw7	B7	DR15	DQ1	1.6	1.6
A30	Cw5	B18	DR3	DQ2	1.5	1.5
A2	CBL	B51	DR11	DQ7	1.5	1.5
A11	Cw4	B35	DR1	DQ1	1.5	1.5
ABL	CBL	B14	DR7	DQ2	1.3	1.3
A30	Cw6	B13	DR7	DQ2	1.3	1.3
A2	Cw5	B44	DR13	DQ1	1.3	1.3
A33	CBL	B14	DR1	DQ1	1.3	1.2
A2	CBL	BBL	DR11	DQ7	1.0	1.0

30128 Uralic
(N = 48)

A	C	B	DR	DQ	HF (%)	LD
A11	Cw4	B35	DR3	DQ2	3.1	3.1
A31	CBL	B41	DR8	DQ2	2.1	2.1
A3	Cw4	B7	DR1	DQ1	2.1	2.1
A3	Cw4	B35	DR8	DQ7	2.1	2.1
A3	Cw1	B39	DR15	DQ1	2.1	2.1
A28	Cw6	B44	DR3	DQ7	2.1	2.1
A24	Cw7	B39	DR4	DQ7	2.1	2.1
A2	Cw9	B14	DR11	DQ7	2.1	2.1
A2	Cw7	B7	DR14	DQ1	2.1	2.1
A2	Cw6	B58	DR4	DQ1	2.1	2.1
A11	Cw11	B39	DR11	DQ7	2.1	2.1
A11	Cw1	B27	DR12	DQBL	2.1	2.1
A26	Cw2	B60	DRBL	DQ1	2.0	2.0

Table 8. Frequencies of DRB1–DPB1 haplotypes in different ethnic groups. (Reproduced with permission from *HLA 1991*, © Oxford University Press and the Committee of the 11th International Histocompatibility Workshop and Conference, 1992)

10102 Senegal
($N = 52$)

DRB1	DPB1	HF (%)	LD	RLD	ChiSQ
1304	1701	10.5	7.2	0.55	24.6
0301	0402	5.8	2.0	0.17	1.7
1101	0402	5.2	1.4	0.12	0.9
07	0101	4.8	3.3	0.46	10.1
1101	1701	4.8	1.7	0.14	1.5
1101	0101	4.4	1.8	0.14	1.7
1304	1301	3.0	1.4	0.18	1.7
1102	BL	2.9	2.9	5.95	187.4
1302	0101	2.9	2.1	0.54	7.1
1301	0402	2.9	1.5	0.34	2.2
0301	1301	2.8	1.3	0.16	1.6
1102	0402	2.8	0.4	0.05	0.1

10200 South African Negroid
($N = 84$)

DRB1	DPB1	HF (%)	LD	RLD	ChiSQ
0302	0101	10.4	4.4	0.35	10.1
1101	0402	7.0	3.0	0.19	5.7
0302	0402	6.2	2.6	0.17	4.7
1101	0101	6.1	−0.4	−0.07	0.1
1302	0101	5.0	2.0	0.31	3.6
1101	1301	3.6	2.2	0.43	8.5
07	0401	3.0	2.2	0.32	12.4
1302	0201	2.9	1.7	0.21	5.4
07	0402	2.7	1.1	0.18	1.9
0401	0101	2.4	1.6	1.00	8.7
1501	0201	2.4	1.6	0.31	7.4
1201	1801	2.3	2.1	0.47	36.8
1101	0301	2.1	1.0	0.23	2.0

10201 Bushman (San)
($N = 53$)

DRB1	DPB1	HF (%)	LD	RLD	ChiSQ
0401	0101	15.1	2.4	0.12	1.2
0401	0401	10.4	2.2	0.16	1.3
1301	BL	7.3	4.7	0.51	13.0
1302	BL	6.4	3.4	0.32	6.0
07	0401	5.1	3.4	0.55	10.0
1301	0401	4.8	2.3	0.24	3.1
0802	0101	4.7	3.1	1.00	10.5
1302	0101	4.7	0.1	0.01	0.0
0401	0402	4.0	−0.7	−0.14	0.2
BL	0101	3.3	1.4	0.39	1.9
0404	BL	2.9	1.8	0.50	4.7
0401	0201	2.7	0.5	0.13	0.2
0401	BL	2.5	−5.9	−0.70	9.0
07	BL	2.4	0.7	0.11	0.4

Table 8. (*continued*)

10201	Bushman (San) (*continued*) (*N* = 53)				
DRB1	DPB1	HF (%)	LD	RLD	ChiSQ
1102	0402	2.2	1.9	1.00	17.0
BL	BL	2.2	1.0	0.23	1.1
0401	0301	2.1	1.3	1.11	4.2
10202	Hottentot (Khoi) (*N* = 90)				
1501	0402	10.4	2.6	0.13	2.9
0401	0402	9.3	5.2	0.52	20.0
1302	0201	6.3	4.4	0.39	24.0
0404	0402	4.9	3.2	0.72	15.5
1501	BL	4.9	1.6	0.19	2.2
1501	0401	4.7	1.6	0.20	2.5
0401	0101	3.0	0.6	0.05	0.3
1501	0101	2.9	−1.7	−0.37	1.9
0301	0101	2.8	1.5	0.24	4.2
1501	1101	2.8	1.1	0.25	1.9
0301	0402	2.8	0.6	0.12	0.5
0102	BL	2.7	2.2	0.56	19.7
1302	0301	2.6	1.8	0.37	10.0
10400	North American Negroid (*N* = 124)				
0302	0101	5.8	3.8	0.78	28.2
1501	0401	3.7	2.7	0.37	24.6
1302	0101	3.2	1.0	0.18	1.8
07	0201	2.9	1.6	0.17	5.7
1501	0101	2.9	0.5	0.09	0.5
07	0402	2.4	1.5	0.18	6.9
0301	0101	2.4	0.2	0.05	0.1
1302	1701	2.3	1.6	0.22	10.4
0102	0101	2.3	1.3	0.50	5.9
1101	1301	2.0	1.7	0.45	23.8
1301	1801	2.0	1.7	0.40	26.8
0804	BL	2.0	1.6	0.24	14.2
1503	0101	2.0	0.9	0.31	2.6
0804	0101	1.8	−0.2	−0.10	0.1
1101	BL	1.7	1.1	0.18	6.6
1101	1701	1.7	1.0	0.13	3.9
0301	0401	1.7	0.9	0.13	2.8
0804	0201	1.7	0.8	0.12	1.9
20202	Highlanders (*N* = 83)				
1101	0401	16.2	4.9	0.35	8.6
BL	0201	12.2	6.1	0.36	18.0

Table 8. (*continued*)

20202 Highlanders (*continued*)
(N = 83)

DRB1	DPB1	HF (%)	LD	RLD	ChiSQ
BL	0401	9.6	−2.1	−0.18	1.6
1405	0401	4.6	−0.7	−0.13	0.3
1101	0201	4.4	−1.5	−0.25	1.1
0803	0401	4.0	1.6	0.52	3.2
1407	0401	3.1	0.6	0.18	0.4
1405	0201	3.1	0.4	0.05	0.2
1501	0401	3.0	−0.6	−0.17	0.3
1501	0501	2.6	1.3	0.19	2.8
1101	0301	2.6	0.9	0.19	1.2
1407	0501	2.5	1.6	0.33	5.7
0405	0501	2.4	1.9	0.76	15.7
BL	0301	2.4	0.7	0.14	0.7
1405	0501	2.4	0.5	0.06	0.4
other	0501	2.2	1.4	0.35	5.5

30113 French
(N = 171)

DRB1	DPB1	HF (%)	LD	RLD	ChiSQ
1501	0401	5.6	2.5	0.59	13.0
0301	0101	5.1	4.3	0.57	90.9
1101	0401	4.9	1.1	0.22	2.2
07	0401	4.8	−0.6	−0.11	0.5
0301	0401	3.9	−0.4	−0.09	0.2
other	0401	3.9	1.3	0.36	4.1
0401	0401	3.6	1.5	0.49	6.2
0101	0401	3.1	−0.6	−0.17	0.7
1301	0401	3.0	0.7	0.21	1.2
07	0402	2.2	0.9	0.11	2.9
07	1101	1.9	1.6	0.68	28.9
0101	0402	1.8	0.9	0.11	3.7
07	0301	1.7	0.5	0.06	0.8
1601	0401	1.6	0.4	0.21	0.6
1101	0301	1.5	0.6	0.07	1.8
1101	0201	1.5	0.4	0.05	0.6
0801	0301	1.4	1.1	0.34	13.9
0101	BL	1.3	1.0	0.47	19.4
BL	0201	1.2	1.0	0.69	19.3
0103	0201	1.2	1.0	0.62	17.2
1302	0401	1.2	0.4	0.42	1.5

30114 German
(N = 90)

DRB1	DPB1	HF (%)	LD	RLD	ChiSQ
07	0401	6.5	5.5	0.63	66.0
BL	BL	5.6	2.6	0.55	7.4
1501	0401	5.0	1.9	0.38	3.7
0401	0401	4.0	2.5	1.01	12.3
0801	0401	3.8	0.2	0.04	0.0

Table 8. (*continued*)

30114 German (*continued*)
(*N* = 90)

DRB1	DPB1	HF (%)	LD	RLD	ChiSQ
0301	0401	7.2	2.5	0.33	4.6
1101	0301	3.7	2.6	0.32	13.0
1104	0401	2.9	1.6	0.74	5.7
0301	0402	2.9	1.5	0.18	3.4
0101	0201	2.8	2.2	0.36	18.2
1301	0401	2.4	0.7	0.25	0.9
1101	0401	2.2	−1.3	−0.37	1.5
0101	BL	2.2	1.6	0.26	8.2
1101	0402	2.1	0.7	0.09	0.7

30117 Italian
(*N* = 268)

DRB1	DPB1	HF (%)	LD	RLD	ChiSQ
1101	0401	6.3	2.2	0.27	10.3
07	0401	3.5	−0.8	−0.19	1.4
0301	0101	3.2	2.7	0.62	90.8
1101	0402	3.2	1.4	0.13	7.2
1104	0402	3.1	2.0	0.35	26.9
0101	0401	3.1	0.9	0.22	3.4
BL	0401	2.9	1.4	0.48	10.9
1501	0401	2.3	0.4	0.11	0.7
07	0402	2.1	0.1	0.01	0.1
07	0301	2.0	0.6	0.06	1.9
1601	0401	2.0	0.3	0.10	0.5
0301	0401	1.9	−1.7	−0.48	7.3
1104	0201	1.9	1.0	0.17	7.8
1401	0401	1.9	0.1	0.02	0.0
0101	0402	1.7	0.7	0.13	3.3
1104	0401	1.4	−1.0	−0.41	3.5
0301	0402	1.4	−0.2	−0.12	0.2
1302	0301	1.4	1.0	0.27	13.0
1101	0201	1.3	−0.3	−0.19	0.4
1301	0201	1.3	0.7	0.17	5.0
0301	BL	1.1	1.0	1.47	79.6
07	1101	1.1	0.9	0.54	20.9
1401	1001	1.1	0.9	0.35	33.7
1401	0301	1.1	0.5	0.11	2.8
07	1001	1.0	0.6	0.25	6.5
1601	0201	1.0	0.4	0.09	1.4
0102	0401	1.0	0.0	0.01	0.0
1302	0201	0.9	0.4	0.11	1.7
1301	0301	0.9	0.4	0.10	2.0
1301	0402	0.9	0.3	0.07	0.7
1501	0301	0.9	0.3	0.05	0.7
0401	0401	0.9	0.1	0.04	0.0
07	0201	0.8	−0.8	−0.49	2.7
0801	0401	0.8	−0.1	−0.11	0.1

Table 8. (*continued*)

10102 Italian (*continued*)
(N = 268)

DRB1	DPB1	HF (%)	LD	RLD	ChiSQ
0402	0201	0.8	0.6	0.46	11.4
07	0901	0.8	0.6	0.44	11.1
1502	0402	0.8	0.5	0.23	3.9
0102	0301	0.8	0.5	0.21	5.5
0801	0201	0.8	0.5	0.20	3.9
07	1701	0.7	0.6	0.51	12.8
1501	1401	0.7	0.6	0.24	14.8
0403	0301	0.7	0.5	0.26	6.5

30122 Sardinian
(N = 83)

DRB1	DPB1	HF (%)	LD	RLD	ChiSQ
0301	BL	9.8	2.6	0.15	2.8
1601	BL	9.1	3.7	0.30	7.5
0301	0301	6.7	1.4	0.08	1.0
1101	0401	5.8	3.5	0.48	12.8
0405	0301	5.7	3.6	0.48	14.2
0301	0401	5.5	−0.3	−0.06	0.1
1601	0201	5.2	3.6	0.48	17.5
1104	0401	3.9	2.5	0.58	10.7
1601	0301	2.8	−1.2	−0.29	0.9
0405	BL	2.6	−0.3	−0.11	0.1
0102	0401	2.4	1.4	0.43	4.3
0101	BL	2.3	0.8	0.24	1.1

30124 Spanish
(N = 148)

DRB1	DPB1	HF (%)	LD	RLD	ChiSQ
07	0401	5.3	−0.9	−0.15	0.7
1501	0401	4.5	1.5	0.24	3.6
07	0201	4.4	1.1	0.08	1.8
BL	0401	4.1	1.3	0.21	2.7
0101	0401	3.4	1.6	0.39	6.1
07	1101	3.3	2.5	0.72	28.3
BL	BL	2.8	2.3	0.40	32.0
1104	0401	2.7	1.4	0.48	6.3
0102	0201	2.6	1.6	0.32	9.8
0401	0401	2.6	1.0	0.30	3.0
07	0101	2.5	1.2	0.22	4.2
07	0301	2.3	0.8	0.13	1.8
1301	0401	2.2	0.8	0.26	2.0
0301	0301	1.7	1.2	0.19	9.7
0402	0201	1.7	1.1	0.45	9.5
1501	0402	1.5	0.7	0.09	2.3

Table 8. (*continued*)

30141 Gypsy (Spanish) (N = 70)

DRB1	DPB1	HF (%)	LD	RLD	ChiSQ
1401	0201	9.9	3.8	0.27	5.9
1401	0401	9.1	−0.6	−0.06	0.1
1401	BL	7.1	1.6	0.12	1.1
1601	BL	4.9	3.1	0.36	9.5
1101	0401	4.4	2.4	0.54	6.1
1502	0401	4.1	1.6	0.31	2.4
1401	0402	4.0	0.0	0.00	0.0
0101	0201	2.9	2.3	1.00	16.3
BL	0401	2.7	0.9	0.22	0.9
1601	0402	2.6	1.3	0.15	2.3
BL	0201	2.3	1.2	0.25	2.2
07	0401	2.2	0.5	0.14	0.3
1101	BL	2.0	0.8	0.16	1.1

30701 USA (N = 143)

DRB1	DPB1	HF (%)	LD	RLD	ChiSQ
1501	0401	6.9	4.0	0.61	25.3
BL	BL	4.1	3.1	0.38	33.7
0401	0401	3.5	1.4	0.28	3.9
07	0402	3.2	1.6	0.17	5.6
07	0401	3.1	−0.2	−0.06	0.0
0101	0402	3.0	2.0	0.37	14.1
BL	0201	2.7	1.5	0.16	7.3
BL	0401	1.8	−1.5	−0.45	3.0
07	1301	1.7	1.4	0.74	29.2
0301	0101	1.6	1.3	0.37	19.1
0401	0402	1.6	0.5	0.08	0.8
0102	BL	1.5	1.1	0.26	9.3
0301	0401	1.4	−0.9	−0.39	1.5
1302	0401	1.4	−0.1	−0.07	0.0
0301	0402	1.4	0.3	0.04	0.2

30702 Canadian (N = 166)

DRB1	DPB1	HF (%)	LD	RLD	ChiSQ
1501	0401	5.9	1.4	0.21	2.8
0401	0401	5.8	1.8	0.31	5.2
07	0401	3.9	−0.6	−0.14	0.6
0301	0101	3.5	2.6	0.36	27.2
BL	0401	3.5	1.7	0.67	10.1
0301	0401	3.1	−1.8	−0.36	4.1
1501	0402	2.5	1.1	0.11	3.7
1301	0201	2.3	1.8	0.41	26.5
0101	0401	2.0	−0.3	−0.13	0.2
0301	0301	2.0	0.7	0.08	1.7
0403	0401	1.9	0.7	0.37	2.2
0101	0201	1.8	1.2	0.23	9.9

Table 8. (*continued*)

30702 Canadian (*continued*)
(N = 166)

DRB1	DPB1	HF (%)	LD	RLD	ChiSQ
0101	0402	1.8	1.1	0.21	6.3
0404	0401	1.8	0.8	0.56	3.9
07	0301	1.8	0.7	0.07	1.6
07	0402	1.8	0.4	0.04	0.4
1104	0402	1.7	1.4	0.59	22.5
A-UND	0401	1.6	0.9	0.74	5.4
0407	0301	1.5	1.4	0.19	54.3
0301	1301	1.5	1.2	0.49	16.0
07	0201	1.4	0.3	0.03	0.3
0401	0402	1.4	0.2	0.02	0.1
1301	0401	1.3	−0.7	−0.34	1.3
1101	0401	1.3	0.2	0.12	0.2

40101 Japanese (Wajin)
(N = 415)

DRB1	DPB1	HF (%)	LD	RLD	ChiSQ
0901	0501	7.8	2.4	0.28	16.5
1502	0901	7.2	6.4	0.82	526.1
0405	0501	6.1	1.3	0.18	5.8
0803	0501	4.6	1.7	0.36	14.1
0901	0201	4.4	1.5	0.13	9.3
1302	0401	3.5	3.1	0.70	270.5
0101	0402	3.2	2.7	0.59	155.7
1501	0201	2.8	1.4	0.25	14.4
1201	0501	2.7	1.4	0.66	19.6
0405	0201	2.6	0.1	0.01	0.0
0802	0501	2.4	0.7	0.29	4.8
1401	0501	2.1	0.5	0.21	2.5
BL	0501	1.9	0.9	0.54	10.5
0406	0201	1.6	0.8	0.30	10.7
1302	0201	1.5	0.0	0.00	0.0
0403	0201	1.4	0.8	0.34	10.9
1405	0501	1.4	0.6	0.51	6.7
1101	0501	1.4	0.5	0.40	4.7
1501	0501	1.3	−1.4	−0.52	10.6
0803	BL	1.3	1.1	0.39	47.0
0406	0501	1.0	−0.4	−0.28	1.5
0803	0202	1.0	0.7	0.16	12.3
1202	0501	1.0	0.3	0.25	1.5
0405	0402	1.0	0.0	−0.04	0.0
1502	0201	0.9	−1.0	−0.51	5.6
0401	0201	0.9	0.5	0.34	7.4
0802	0402	0.9	0.5	0.13	6.6
1401	0202	0.8	0.7	0.17	21.6
0405	BL	0.8	0.5	0.19	6.2
1302	0501	0.7	−2.1	−0.76	23.0
0101	0201	0.7	−0.4	−0.35	1.5

Table 8. (*continued*)

40101 Japanese (Wajin) (*continued*)
(*N* = 415)

DRB1	DPB1	HF (%)	LD	RLD	ChiSQ
1403	0201	0.7	0.4	0.33	5.1
0901	0402	0.6	−0.6	−0.52	3.6
1403	0501	0.6	0.1	0.09	0.2
1501	0402	0.5	−0.1	−0.19	0.2
1302	0402	0.5	−0.1	−0.16	0.2
1406	1301	0.5	0.4	0.31	40.9
0401	0401	0.5	0.4	0.22	15.6
0101	BL	0.5	0.4	0.14	8.7
1301	0501	0.5	0.3	0.58	3.2
1001	0201	0.5	0.3	0.40	4.9
0405	0301	0.5	0.1	0.04	0.4
1406	0501	0.5	0.0	−0.06	0.0
0405	0202	0.5	0.0	−0.02	0.0

40104 Korean
(*N* = 99)

DRB1	DPB1	HF (%)	LD	RLD	ChiSQ
0901	0501	4.7	2.5	0.39	8.1
07	1301	4.5	3.8	0.48	44.9
0803	0501	3.7	1.7	0.31	4.5
0405	0501	3.3	1.5	0.28	3.4
1302	0401	3.1	1.7	0.17	5.1
0405	BL	2.8	2.4	0.57	41.0
0101	0201	2.7	1.7	0.34	6.9
1302	0501	2.6	−0.7	−0.20	0.4
0101	0501	2.2	0.6	0.14	0.7
1501	0401	2.1	1.1	0.16	3.5
0301	0401	2.0	1.7	0.62	18.6
0406	0501	2.0	1.4	0.73	7.8

40204 Buyi
(*N* = 70)

DRB1	DPB1	HF (%)	LD	RLD	ChiSQ
1602	0501	12.3	4.8	0.66	10.3
1401	0501	8.7	0.2	0.02	0.0
1501	0501	6.5	−1.3	−0.16	0.7
1202	0501	4.6	0.4	0.10	0.1
0901	0501	4.5	0.3	0.07	0.1
1501	0202	4.1	1.4	0.11	1.4
1303	1301	4.0	2.8	0.39	12.5
1401	0202	3.7	0.7	0.05	0.3
1303	0501	3.6	−0.6	−0.14	0.3
1401	0201	3.6	2.1	0.30	5.8
0301	0501	3.3	0.9	0.39	1.0
1501	0201	2.9	1.5	0.21	3.2
1202	0202	2.9	1.4	0.20	2.3
1502	1301	2.8	2.2	0.60	14.2
BL	0501	2.4	1.3	1.26	4.8

Table 8. (*continued*)

40204 Buyi (*continued*) (*N* = 70)

DRB1	DPB1	HF (%)	LD	RLD	ChiSQ
0901	1301	2.4	1.2	0.17	2.4
1101	0202	2.1	1.2	0.27	2.6

40421 Singapore Chinese (*N* = 70)

DRB1	DPB1	HF (%)	LD	RLD	ChiSQ
0901	0501	10.1	2.6	0.32	2.8
1101	0501	7.3	1.8	0.31	1.8
1501	0201	6.9	5.1	0.45	27.2
1602	0501	6.2	2.9	0.80	7.1
0803	0501	4.3	1.5	0.51	2.3
0301	0401	4.2	3.8	0.71	64.3
0901	1301	4.1	2.7	0.35	9.4
1202	0501	4.1	0.0	−0.01	0.0
1501	0501	3.7	−2.8	−0.43	3.7
0406	0501	2.1	−0.3	−0.13	0.1
1202	1301	2.1	1.4	0.18	4.1
D-UND	0501	2.1	0.8	0.51	1.1
1201	0501	2.1	0.2	0.10	0.1

40802 Zuni (*N* = 50)

DRB1	DPB1	HF (%)	LD	RLD	ChiSQ
0802	0402	20.0	7.2	0.88	13.1
1402	0401	17.4	6.9	0.32	9.8
1602	0402	15.7	4.1	0.55	4.5
1402	0402	11.6	−8.0	−0.41	12.2
1406	0401	5.9	2.9	0.49	4.7
0407	0402	4.7	1.0	0.44	0.8
0403	0402	4.0	1.6	1.00	2.7
1602	0401	3.3	−2.9	−0.47	2.5
1406	0402	3.1	−2.4	−0.44	2.9
1402	0501	3.0	2.0	1.00	6.6
0411	0401	3.0	1.3	0.40	1.7
0411	0301	2.0	1.9	0.65	24.8
0101	0401	2.0	1.3	1.00	4.1
0901	0402	2.0	0.8	1.00	1.3

Table 9. List of ethnic groups in Table 6. (Reproduced with permission from *HLA 1991*, © Oxford University Press and the Committee of the 11th International Histocompatibility Workshop and Conference, 1992)

Ethnic code	Ethnic group
10 100	North African Black
10 102	Senegal
10 200	South African Black
10 201	Bushman (San)
10 202	Hottentot (Khoi)
10 204	Zimbabwean
10 206	Zaire
10 207	West African
10 210	South African Black (Cape)
10 400	North American Black
10 600	South American Black
20 100	Australian Aborigine
20 201	Coastals (Papua New Guinea)
20 202	Highlanders (Papua New Guinea)
30 101	Albanian
30 102	Armenian
30 103	Austrian
30 104	Basque
30 105	Belgian
30 106	British
30 109	Cornish
30 110	Czech
30 111	Dane
30 113	French
30 114	German
30 115	Greek
30 117	Italian
30 119	Polish
30 102	Portuguese
30 121	Romanian
30 122	Sardinian
30 124	Spanish
30 125	Swedish
30 127	Ukrainian
30 128	Uralic
30 129	Yugoslavian
30 130	Hungarian
30 131	Slovak
30 141	Gypsy (Spanish)
30 210	South African Caucasoid
30 400	Indian
30 401	Bhargavas
30 406	Iyers
30 413	Tribal

Table 9. (*continued*)

30 502	Usbekistan
30 601	Australian
30 701	USA
30 702	Canadian
30 801	Mexican
30 902	Brazilian
40 101	Japanese (Wajin)
40 104	Korean
40 201	Northern Han
40 202	Southern Han
40 204	Buyi
40 207	Miao
40 209	Yi
40 210	Uygur
40 211	Tibetan
40 212	Manchu
40 213	Orochon
40 216	Taiwan Aborigine
40 217	Dai
40 218	Hui
40 219	Li
40 220	Inner Mongolian
40 301	Buriat
40 302	Mongolian
40 303	Kazakhstan
40 401	Thais
40 402	Thai Chinese
40 403	Vietnamese
40 404	Filipino
40 414	Javanese
40 416	Timor
40 421	Singapore Chinese
40 602	Maori
40 800	Native North American
40 802	Zuni
40 803	Tlingit
41 003	Native Brazilian
41 00A	Native South American
41 102	Inuit
41 10A	Yakut

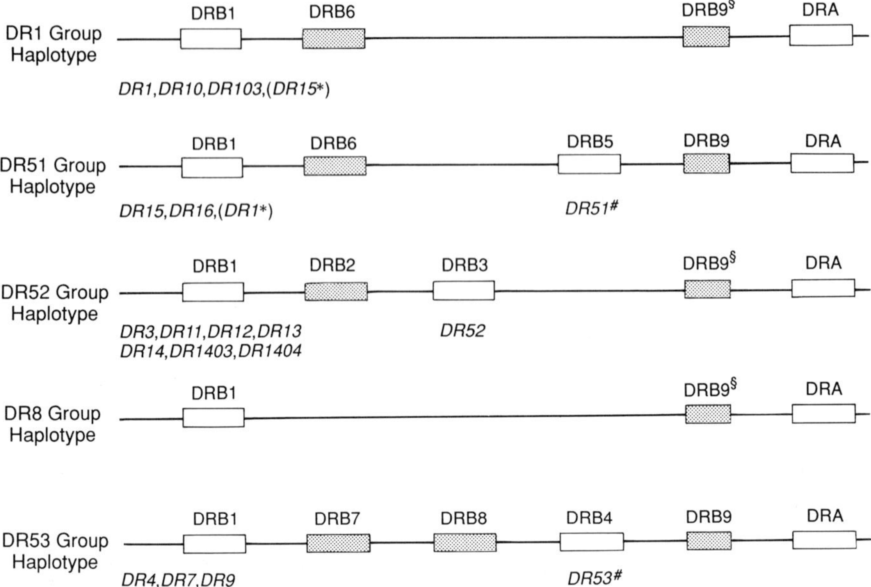

Figure 1. Genomic organization of the HLA-DR region and encoded products (specificities). Pseudogenes are indicated by shaded boxes, expressed genes by open boxes. The serological specificity encoded by a gene is shown underneath in italics.
*rarely observed haplotypes
#DR51 and DR53 may not be expressed on certain haplotypes
§the presence of DRB9 in these haplotypes needs confirmation

References

1. Imanishi, T., Akaza, T., Kimura, A., Tokugawa, K., and Gojobori, T. (1992). Estimation of allele and haplotype frequencies for HLA and complement loci. *HLA 1991* (ed. K. Tsuji *et al.*), Vol. 1. Oxford University Press.
2. Imanishi, T., Akaza, T., Kimura, A., Kimura, A., Tokugawa, K., and Gojobori, T. (1992). Allele frequencies and haplotype frequencies for HLA and complement loci in various ethnic groups. *HLA 1991* (ed. K. Tsuji *et al.*) Vol. 1 Oxford University Press.

A1

List of suppliers

Agfa Gevaert Ltd., French's Avenue, Dunstable, Bedfordshire LU6 1DF, UK.

Aldrich, New Road, Gillingham, Dorset SP8 4JL, UK.

Alpha Laboratories, 40 Parham Drive, Eastleigh, Hampshire SO5 4NU, UK.

American Society of Histocompatibility and Immunogenetics, PO Box 15804, Lenexa, KS 66285-5804, USA.

American Type Culture Collection (ATCC), 12301 Parklawn Drive, Rockville, Maryland 20852, USA.

Amersham International, Lincoln Place, Green End, Aylesbury, Buckinghamshire HP20 2TP, UK; 2636 South Clearbrook Drive, Arlington Heights, IL 60005, USA.

Applied Biosystems Ltd., Kelvin Close, Birchwood Science Park North, Warrington, Cheshire WA3 7PB, UK; 850 Lincoln Center Drive, Foster City, CA 94404, USA.

AstroScan, Bradden, Isle of Man, UK.

At Biochem, 30 Spring Mill Drive, Malvern, PA 19355, USA.

Baker Company, Inc., Sanford, Maine, USA.

BDH Chemicals, Broom Road, Poole, Dorset BH12 4NN, UK.

Becton Dickinson Ltd., Between Towns Road, Cowley, Oxford OX4 3LY, UK; 2350 Qume Drive, San Jose, California 95131-1807, USA.

Bellco, PO Box B, 340 Eldrudo Road, Vineland, NJ 08360, USA.

Bethesda Research Laboratories, PO Box 35, Trident House, Renfrew Road, Paisley PA3 4EF, UK.

Bibby Sterilin, Lampton House, Lampton Road, Hounslow, Middlesex TW3 4EE, UK.

Bio 101, Inc., PO Box 2284, La Jolla, CA 92038-2284, USA.

BioRad, 1414 Harbour Way South, Richmond, CA 98404, USA; Marylands Avenue, Hemel Hempstead, Hertfordshire HP2 7TD, UK.

Biotest Diagnostics, Unit 21A, Monkspath Business Park, Shirley, Solihull, West Midlands B90 4NY, UK; 6072 Dreieich, Germany.

Boehringer Mannheim, Bell Lane, Lewes, East Sussex BN7 1LG, UK; PO Box 50414, Indianapolis, IN 26250, USA.

Burroughs Wellcome Company, 3030 Cornwallis Road, Research Triangle Park, NC 27709, USA.

Calbiochem, PO Box 12087, San Diego, CA 92112-4180, USA; 3 Heathcoat

Building, Highfields Science Park, University Boulevard, Nottingham NG7 2QJ, UK.

Camlab, Nuffield Road, Cambridge CB4 1TH, UK.

Carnation, Los Angeles, CA 90036, USA.

Cellmark Diagnostics, 20271 Goldenrod Lane, Germantown, MD 20874, USA; Blacklands Way, Abingdon Business Park, Abingdon, Oxon OX14 1DY, UK.

CNTS, 91943 Les Ulis Cedex, France.

Collaborative Research Incorporated, 128 Spring Street, Lexington, Massachusetts 02173, USA.

Corning Inc., Corning, NY 14831, USA.

Co-Star UK Ltd., Victoria House, 28/38 Desborough Street, High Wycombe, Bucks HP11 2NF, UK.

CryoMed, 51529 Birch Street, New Baltimore, MI 48047, USA.

C-Six, 9653 N. Granville Road, Mequon, WI 53092, USA.

Dade, Aguada, Puerto Rico 00602.

Dako Ltd., 16 Manor Courtyard, Hughenden Avenue, High Wycombe, Bucks HP13 5RE, UK.

Dakopatts a/s, PO Box 1359, DK-2600 Glostrup, Denmark.

Dia Med AG, 3280 Murtin, Switzerland.

Difco Laboratories, PO Box 14B, Central Avenue, East Molesey, Surrey KT8 0SE, UK; PO Box 331058, Detroit, MI 48232-7058, USA.

Du Pont (UK) Ltd., Wedgwood Way, Stevenage, Herts SG1 4QN, UK.

Dynal (UK) Ltd., Station House, 26 Grove Street, New Ferry, Wirral, Merseyside L62 5AZ, UK.

Dynatech, Daux Road, Billinghurst, Sussex RH14 95J, UK.

EM Science, PO Box 70, 480 Democrat Road, Gibbstown, NJ 08027, USA.

European Collection of Animal Cell Cultures, Porton Down, Salisbury, Wiltshire SP4 0JG, UK.

Fisher Biotech, 1600 W. Glenlake Avenue, Itasca, IL 60143, USA.

Fisher Scientific Co., St Helens Street, New Enterprise House, Derby DE1 3GY, UK.

Flowgen Instruments Ltd., Broad Oak Enterprise Village, Broad Road, Sittingbourne, Kent ME9 8AQ, UK.

FMC Corp. Bioproducts, 5 Maple Street, Rockland, ME 04841, USA.

Fresenius Diagnostics, Daimlerstr 22, 6380 Bad Homberg, Germany.

Gallenkamp, Belton Road West, Loughborough, Leicestershire LE11 0TR, UK.

Gen Probe, San Diego, CA, USA.

Genetic Research Instrumentation Ltd., Gene House, Station Road, Takeley, Bishop's Stortford, Herts CM22 6SG, UK.

Gibco, PO Box 35, Trident House, Renfrew Road, Paisley PA3 4EF, UK; PO Box 68, Grand Island, NY 14072-0068, USA.

Grand Island Biological Company, 3175 Staley Road, Grand Island, NY 14072, USA.

Greiner BV, PO Box 280, 2400 AG Alphen a/d Rijn, The Netherlands.

Hamilton, PO Box 10030, Reno, Nevada 89520, USA.

Hellma England (Ltd.), 23 Station Road, Westcliff-on-Sea, Essex SS0 7RA, UK.

Hendley Ltd., Oakwood Hill Industrial Estate, Loughton, Essex, UK.

Hillcross Pharmaceuticals, Burnley, Lancs BB10 2HP, UK.

Hoefer Scientific Instruments, 654 Minnesota Street, PO Box 77387, San Francisco, CA 94107-9885, USA; Unit 12, Croft Road Workshops, Newcastle under Lyme, Staffordshire ST5 0TW, UK.

Hybaid Ltd., 111–113 Waldegrave Road, Teddington, Middlesex TW11 8LL, UK.

ICN/FLOW, Eagle House, Peregrine Business Park, Gomm Road, High Wycombe, Bucks HP13 7DL, UK; 3300 Hyland Avenue, Costa Mesa, California 92626, USA.

Kent Laboratories Inc., 23404 N-E 8th St, Redmond, Washington 98052, USA.

Kodak Ltd., PO Box 66, Kodak House, Station Road, Hemel Hempstead, UK.

Koh-I-Nor, Bloomsbury, NJ, USA.

Labo-Systems, 4 PL.C1. Monet, 91400 Saclay, France.

Lederle Labs Division, Gosport, Hampshire, UK.

Leica, Postfach 2040, W-6300 Wetzlar 1, Germany.

Luckham Ltd., Victoria Gardens, Burgess Hill, Sussex, UK.

Medfor Products, 15 King's Road, Fleet, Hants, UK.

Merck, E Merck, D-6100 Darmstadt, Germany.

Micron Separations Incorporated, PO Box 1046, 135 Flanders Road, Westborough, MA 01581, USA.

Miles Labs Ltd., PO Box 37, Stoke Poges, Slough, Bucks SL2, UK.

Millipore Ltd., The Boulevard, Blackmoor Lane, Watford, Herts WD1 8YW, UK; PO Box 255, Bedford, MA 01730, USA.

MSE Scientific Instruments, Bishop Meadow Road, Loughborough, Leicester LE11 0RG, UK.

New England Biolabs, 32 Tozer Road, Beverly, MA 01915-5599, USA.

New England Nuclear, 549 Albany Street, Boston, MA 02118, USA.

Northumbria Biologicals Ltd., South Nelson Industrial Estate, Cramlington, Northumberland NE23 9HL, UK.

Nunc, Kamstrupvej 90, Kamstrup, DK-4000 Roskilde, Denmark.

Nycomed UK Ltd., Nycomed House, 2111 Coventry Road, Sheldon, Birmingham B26 3EA, UK.

Omega, 1 Omega Drive, Box 4047, Stanford, CT 06907, USA.

One Lambda Inc., 21001 Kittridge St, Canoga Park, California 91303-2801, USA.

Organon Teknika Ltd., Science Park, Milton Road, Cambridge CB4 4FL, UK; D-6904 Eppelheim, Germany.

Ortho Diagnostic Systems Ltd., Enterprise House, Station Road, Loudwater, High Wycombe, Bucks HP10 9UF, UK.

Oxoid Ltd., Wade Road, Basingstoke, Hants RG24 0PW, UK.

Perkin Elmer Cetus, 761 Main Avenue, Norwalk, CT 06859-0156, USA.

Pharmacia, Biotechnology Division, Davy Avenue, Knowlhill, Milton Keynes MK5 8PH, UK; 800 Centennial Avenue, PO Box 1327, Piscataway, NJ 08855, USA.

Philip Harris Scientific, 618 Western Avenue, Park Royal, London W3 0TE, UK.

Pierce, PO Box 117, Rockford, Illinois 61105, USA.

Polaroid (UK) Ltd., Ashley Road, St. Albans, UK.

Promega Corporation, 2800 Woods Hollow Road, Madison, Wisconsin 53711-5399, USA.

Robbins Scientific Corporation, 814 San Aleso Avenue, Sunnyvale, California 94086-1411, USA.

Ross Labs, Unit 13, Fence Ave Industrial Estate, Macclesfield SK10 1LT, UK.

Rubbermaid, Wooster, OH 44691, USA.

Sandoz Research Institute, Frimley Business Park, Frimley, Camberley, Surrey GU16 58G, UK.

Sarstedt Ltd., 68 Boston Road, Leicester LE4 1AW, UK.

Savant Instruments, Inc., 110-103 Bi-County Boulevard, Farmingdale, NY 11735, USA.

Sera Lab Ltd., Crawley Down, Sussex RH10 4FF, UK.

Serotec, 22 Bankside, Station Approach, Kidlington, Oxon OX5 1JE, UK.

Sherwood Medical Industries Ltd., London Road, County Oak, Crawley RH11 7YQ, UK.

Sigma Chemical Company Ltd., Fancy Road, Poole, Dorset BH17 7NH, UK; PO Box 14508, St Louis, MO 63178, USA.

Stratagene Ltd., Cambridge Innovation Centre, Cambridge Science Park, Milton Road, Cambridge CB4 4GF, UK.

Sweetheart Plastics, Chipping Sodbury, Avon, UK.

Techmate Ltd., 10 Bridgeturn Avenue, Old Wolverton, Milton Keynes MK12 5QL, UK.

Travenol Laboratories, Morton Grove, Illinois, USA.

Tribiotics, Crawley, Oxon, UK.

Unipath Ltd., Norse Road, Bedford MK41 0QG, UK.

United States Biochemical Corporation, PO Box 22400, Cleveland, Ohio 44122, USA.

U.V. Products, Science Park, Milton Road, Cambridge CB4 4BN, UK.

VWR Scientific, Box 7900, San Francisco, California 94120, USA.

Whatman Ltd., Springfield Mill, Maidstone, Kent, UK.

Index

ABO blood group 101, 103, 211, 214, 219
acceptable mismatching 103
accreditation 79, 212
acridine orange 87
alkaline phosphatase 194
alleles 2
allelic association 73
alloantibodies 51
alloantigens 1–3, 5
alloantisera 13
American Society for Histocompatibility and Immunogenetics 91, 94
animals
- UK guidelines on use 35

ankylosing spondylitis 160
annealing 176
ante-natal clinics 15, 21
antibody
- classes 88
- dependent cellular cytotoxicity 98
- screening 81, 85

antigenic epitopes 5
anti-globulin 96, 214
anti-idiotype 93
anti-lymphocyte globulin 88
anti-thymocyte globulin 88
APAAP 194
asceptic culture technique 153
atypical DR/DQ associations 125
autoantibody 89–90
automation 65–6
autoradiography 122
avidin 194

bacterial
- stab culture 108–10
- competent cells 108

BAT genes 247
Bayes' theorem 224
beta-2 microglobulin 4
beta-galactosidase 194
Bf gene 235
biotin 194
blocking antibody 97
blood transfusion 81
- Centre 14, 15, 17
- HLA antibody induction 14

branchoalveolor lavage 154–5
broad specificities 70
bromophenol blue 164
buffy coat 55

C2 235
C4 235
- genotyping 237–8

calmodulin 246
carbonyl iron 55
carboxyfluoroscein diacetate 58, 98
carboxypepsidase 165
CD3 100
cDNA
- library plating 239
- library screening 241

cell suspension concentration 52
chain termination sequencing 177
chi-square test 77
chromosome
- map
  - of Mhc 235
  - HLA class I 239
  - HLA class II 236
  - HLA class III 237
  - walking 228

chronic lymphocytic leukaemia 64, 84, 86
chymotrypsin 165
COL11A2 244
Collaborative Transplant Study 2
complement 63
- evaluation 64

computer analysis 25–6
- 2 × 2 contingency table 26
- correlation coefficient 26
- tail analysis 26

controls in cytotoxicity test 63, 102
cosmid genomic insert plating 240
CpG rich islands 227, 243
crossover 219
crossmatch 1, 94
crossreacting groups 5, 69
crossreactivity 68
cumulative power of exclusion 221
cryostat 192
CYP21B 235
cytokines 191
cytotoxic
- antibodies 2–3
- T lymphocytes 159

cytotoxicity 52, 60
  effect of drugs on assay 68
  false positive 62
  interpreting results 68
  reading tests 67

DAB 198
Daudi cell line 32
disease
  association 247
  susceptibility 5–6
dithioerythritol 89
dithiothreitol 89, 178
DNA
  assay 113
  digest 115–17
  isolation 113, 128, 229
  plasmid 108–13
  probes
    concensus 130–3
    digoxigenin labelled 135
    pRTV1 108–9, 111
    pIIβ-I 108–9, 111
    pDCH1 108–9, 111
    radiolabelled 117, 119–22
  purification 172
  storage 113
DNAse 56
dot blot 132–4
double normalized value 148
Duchenne muscular dystrophy 207
Duffy blood group 214
dynabeads 23, 57

eosin 52
electroblotting 167
electrophoresis 164
epithelial cell antibodies 91
ethidium bromide 52, 87, 112, 116, 119, 233
ethnic groups 292
Eurotransplant organization 101

family study 71
  tree 74
fast blue 196
fast red TR 196
fine needle aspirates 155
flowcytometry 18, 99
fluorobeads 56, 61
fluorochromes 52
fluorography 165
4E 169

frequency
  allele 75
  antigen 75
  gene 76
Freund's adjuvant 33
frozen cell panels 86
frozen tissue sections 192

gel
  agarose 216
  low melting point 111, 119
  mini 112, 116, 130–1
  loading buffer 112, 117
  polyacrylamide 216
  starch 216
  transfer 116–18
glucose oxidase 194
graft infiltrating cells 154–6
graft-versus-host disease 156

HIT trial 103
HLA
  alleles 254–8
  allele frequencies 260–8
  alloantibodies 13–27, 81
    blood donors 14
    class I 14, 19
    class II 18–19
    commercial companies 27
    concentration 24
    dilution 24
    exchange 26
    extra reactions 23
    placental extraction 21
    plasmapheresis 21
    screening 14–21, 51
    storage 22–3
    volunteer immunization 14
  antigen
    distribution
      clinical application 208
      heart 199–203
      kidney 205
      liver 205
      lungs 204
      tisues 200–1, 206
    expression 191
    structure 4–6
    racial variations 76
  -Bw4 70, 259
  -Bw6 70, 259
  class I
    locus specific nucleotides 175
  polymorphism 159

polypeptide backbone ribbon diagram 186
crystallographic structure 5
disease 3–6, 77
DRB region 293
genes 251
gene classes 4–6
gene sequence 2–6
genetic map 4
haplotype 6
haplotype frequencies 269–91
locus
-A 4–5
-B 4
-DP 4–6
-DQ 6
-DR 4–6
-E 5
-H 5
-J 5
nomenclature 1–4
specificities 252–3
typing 51
Hamilton dispenser 61, 65
haplotype 73
Hardy–Weinberg equilibrium 218
heparin 53
hepatitis B 26
HIV 26
Homozygous typing cells 147–8
horseraddish peroxidase 194
HSP70 243, 247
hybridization 120–1, 131–3, 136–41
blocking 120
cross 122, 132
dehybridization 122, 141
incubator 120
prehybridization 120–1
rehybridization 122
sequence specific 125, 131
hybridomas 29–30, 35–8
cloning 38
hydrolink 179
hyperacute rejection 3
hypervariable region 5–6

ICE5 235
IDDM 207
immunocytochemistry 193
immunomagnetic beads 57, 84
immunoprecipitation 161
immunoprobing 167
International Cell Exchange 23
International Histocompatibility Workshop 1–2, 27, 67, 117, 168
cytotoxicity test scoring 76
isoelectric focusing 159–62, 216
banding pattern analysis 168

Kell blood group 214

leucoagglutination 3, 100
ligation 173
likelihood ratio 211, 223
estimate 151–2
limiting dilution assay 151
linkage disequilibrium 71, 73, 108, 218
loading buffer 164
LMP 244, 251
lymphnode cells
isolation 54
storage 59
lymphobeads 56
lymphocyte
freezing 16–18
isolation 15, 84
B lymphocytes 19
separation medium 53
storage 58
thawing 16
lymphocytotoxicity 3, 51, 82
DTT modification 89
lymphoid cell lines 30–2, 34–5, 91, 150–4
Epstein–Barr transformation 152–3
European collection 35
lysis of cells 166

M13 181
MNSs blood group 214
major histocompatibility complex 2–4
genomic structure 239
magnet 58
matching
transplantation 101
heirachy 102
highly sensitised patients 102
membranes 117–23, 131–41
Mendel's law of inheritance 219
mitomycin-C 144
mixed lymphocyte culture 143–7
autologous 145
controls 145
frozen cells 146
HLA class I molecules 143
HLA class II molecules 143
pooled human sera 144, 146
radiation source 143, 144
reactivity to DR, DQ, DP 144

monoclonal antibody 3, 27–50
ascities 41
carry-over prevention 42–3
complement for cytotoxicity 43
dilution 42
Elisa screening 39–40
fusion 29, 31, 34–7
immunization schedules 29–34
purified immunoglobulin 41
sequence epitopes 43–50
storage 193
monocyte antibodies 91
mycoplasma 153

napthol AS-MX phosphate 196
non-radioactive assays 132
nuclei lysis buffer 113
nucleotide sequence analysis 183
neuraminidase 164
non-HLA antibodies 90
northern blotting 228
novel Mhc genes 227, 244–5
nylon wool 56

obliterative bronchiolitis 156
odds ratio 78
OKT3 88
oligonucleotide 4
synthesizer 129
typing 125, 127, 131–41
chemiluminescence 140
high resolution 127
low resolution 127, 135–6
organ
donation 1–3
donor typing 54
transplantation 2–5
antibodies 82
orthogonal field alteration gel electrophoresis 228

panel cells 15–19, 23, 83
PAP 194
paraffin sections 192
Park–Terasaki medium 59
paternity 75, 211–25
chain of custody 212
DNA testing 211–12, 217–18
exclusion 220, 222–3
haplotype frequencies 219–20, 225
index 223–4
probability 211–12, 219, 223–4
proof of identity 212

peptide binding
groove 5, 189
site pockets 185, 188
peripheral blood lymphocytes 51
phase contrast microscopy 67
phenol extraction of DNA 172
phycoerythrin 99
platelet absorption 19–20, 23–4
polymerase chain reaction 4–6, 125, 127–31, 169
controls 131
DNA electrophoresis 130
HLA-DQB1 alleles 130
HLA-DRB1 alleles 129
primers 125, 127, 129–30
Taq polymerase 130
thermal cycling 127, 130
plasma protein 211, 213, 216, 220
plasma units 18
polymorphism 3–6
preclearing 163
precondensation 166
pregnancy 13–15, 81
primed lymphocyte test 148–50, 156
feeder cells 149–51
HLA class I 148
HLA class II 148, 150
transplant recipients 149
protein sequence analysis 184
pseudogenes 5–6, 251
publication of data 79
pulsed field gel electrophoresis 227–8
standards 233

rabbit complement 14, 43, 63
radiolabelling 162
rare cutting restriction endo-nucleases 229
RD gene 243
recombination 75
recurrent absortion 14
red cell
enzymes 211, 213, 216, 220
lysis buffer 113
regression analysis 152
relative response 144, 146, 151
relative risk 78
reporting results 78
responder
cells 143–52
normalized values 148
restriction
enzyme digestion 107–8, 111–12, 114–15, 172, 231
fragment length polymorphism 3–4, 107–8, 114–26, 217
comparison to serology 108

controls 115
HLA-DQA 124
HLA-DQB 124
HLA-DRB 108
indistinguishable DRB combinations 108
interpretation 123
novel 125
paternity 217
stringency of washing 122–3
troubleshooting 125–6
Rh blood group 214–20
RING genes 244

segregation distortion 75
sequence specific oligonucleotide probes 4
sequencing 168, 176
nucleotide 125
polymorphic 125
serum
dispenser 86
samples 85
sheep red blood cells 18, 56
size markers 112, 117–19, 132, 217
Southern blotting 232
spleen cell
isolation 54
transport 60
Staphylococcus aureaus 162
statistical analysis 3, 76
stimulation index 146
stimulating cells 143–52

T cell
cloning 149
precursor analysis 151, 156
receptor 5
recognised epitopes 156
TAP 244, 251
Taq 1 107, 115–16
Terasaki trays 60, 86
tetramethylammonium chloride 134–7
tissue freezing 192
TNF 235
transfectants 33, 45
transformation 173
transplantation
bone marrow 104
history 8
HLA antibody induction 14
kidney 3
organ 82
recipient 1, 3
rejection 2, 3
tritiated thymidine 143–7
trypan blue 52

UKTS 103
ultraviolet transilluminator 112, 116–19, 131

variable number tandem repeat sequences 217–18

W6/32 162
western blotting 166

X-ray film 122–3

yeast artificial chromosomes 228

zoo blots 243, 246

ORDER OTHER TITLES OF INTEREST TODAY

Price list for: UK, Europe, Rest of World (excluding US and Canada)

Forthcoming Titles

124. Human Genetic Disease Analysis Davies, K.E. (Ed)
.... Spiralbound hardback 0-19-963309-6 **£30.00**
.... Paperback 0-19-963308-8 **£18.50**
123. Protein Phosphorylation Hardie, G. (Ed)
.... Spiralbound hardback 0-19-963306-1 **£32.50**
.... Paperback 0-19-963305-3 **£22.50**
122. Immunocytochemistry Beesley, J. (Ed)
.... Spiralbound hardback 0-19-963270-7 **£32.50**
.... Paperback 0-19-963269-3 **£22.50**
121. Tumour Immunobiology Gallagher, G., Rees, R.C. & others (Eds)
.... Spiralbound hardback 0-19-963370-3 **£35.00**
.... Paperback 0-19-963369-X **£25.00**
120. Transcription Factors Latchman, D.S. (Ed)
.... Spiralbound hardback 0-19-963342-8 **£30.00**
.... Paperback 0-19-963341-X **£19.50**
119. Growth Factors McKay, I.A. & Leigh, I. (Eds)
.... Spiralbound hardback 0-19-963360-6 **£30.00**
.... Paperback 0-19-963359-2 **£19.50**
118. Histocompatibility Testing Dyer, P. & Middleton, D. (Eds)
.... Spiralbound hardback 0-19-963364-9 **£32.50**
.... Paperback 0-19-963363-0 **£22.50**
117. Gene Transcription Hames, D.B. & Higgins, S.J. (Eds)
.... Spiralbound hardback 0-19-963292-8 **£35.00**
.... Paperback 0-19-963291-X **£25.00**
116. Electrophysiology Wallis, D.I. (Ed)
.... Spiralbound hardback 0-19-963348-7 **£32.50**
.... Paperback 0-19-963347-9 **£22.50**
115. Biological Data Analysis Fry, J.C. (Ed)
.... Spiralbound hardback 0-19-963340-1 **£50.00**
.... Paperback 0-19-963339-8 **£27.50**
114. Experimental Neuroanatomy Bolam, J.P. (Ed)
.... Spiralbound hardback 0-19-963326-6 **£32.50**
.... Paperback 0-19-963325-8 **£22.50**
112. Lipid Analysis Hamilton, R.J. & Hamilton, S.J. (Eds)
.... Spiralbound hardback 0-19-963098-4 **£35.00**
.... Paperback 0-19-963099-2 **£25.00**
111. Haemopoiesis Testa, N.G. & Molineux, G. (Eds)
.... Spiralbound hardback 0-19-963366-5 **£32.50**
.... Paperback 0-19-963365-7 **£22.50**

Published Titles

113. Preparative Centrifugation Rickwood, D. (Ed)
.... Spiralbound hardback 0-19-963208-1 **£45.00**
.... Paperback 0-19-963211-1 **£25.00**
110. Pollination Ecology Dafni, A.
.... Spiralbound hardback 0-19-963299-5 **£32.50**
.... Paperback 0-19-963298-7 **£22.50**
109. In Situ Hybridization Wilkinson, D.G. (Ed)
.... Spiralbound hardback 0-19-963328-2 **£30.00**
.... Paperback 0-19-963327-4 **£18.50**
108. Protein Engineering Rees, A.R., Sternberg, M.J.E. & others (Eds)
.... Spiralbound hardback 0-19-963139-5 **£35.00**
.... Paperback 0-19-963138-7 **£25.00**
107. Cell-Cell Interactions Stevenson, B.R., Gallin, W.J. & others (Eds)
.... Spiralbound hardback 0-19-963319-3 **£32.50**
.... Paperback 0-19-963318-5 **£22.50**
106. Diagnostic Molecular Pathology: Volume I Herrington, C.S. & McGee, J. O'D. (Eds)
.... Spiralbound hardback 0-19-963237-5 **£30.00**
.... Paperback 0-19-963236-7 **£19.50**
105. Biomechanics-Materials Vincent, J.F.V. (Ed)
.... Spiralbound hardback 0-19-963223-5 **£35.00**
.... Paperback 0-19-963222-7 **£25.00**
104. Animal Cell Culture (2/e) Freshney, R.I. (Ed)
.... Spiralbound hardback 0-19-963212-X **£30.00**
.... Paperback 0-19-963213-8 **£19.50**
103. Molecular Plant Pathology: Volume II Gurr, S.J., McPherson, M.J. & others (Eds)
.... Spiralbound hardback 0-19-963352-5 **£32.50**
.... Paperback 0-19-963351-7 **£22.50**
101. Protein Targeting Magee, A.I. & Wileman, T. (Eds)
.... Spiralbound hardback 0-19-963206-5 **£32.50**
.... Paperback 0-19-963210-3 **£22.50**
100. Diagnostic Molecular Pathology: Volume II: Cell and Tissue Genotyping Herrington, C.S. & McGee, J.O'D. (Eds)
.... Spiralbound hardback 0-19-963239-1 **£30.00**
.... Paperback 0-19-963238-3 **£19.50**
99. Neuronal Cell Lines Wood, J.N. (Ed)
.... Spiralbound hardback 0-19-963346-0 **£32.50**
.... Paperback 0-19-963345-2 **£22.50**
98. Neural Transplantation Dunnett, S.B. & Björklund, A. (Eds)
.... Spiralbound hardback 0-19-963286-3 **£30.00**
.... Paperback 0-19-963285-5 **£19.50**
97. Human Cytogenetics: Volume II: Malignancy and Acquired Abnormalities (2/e) Rooney, D.E. & Czepulkowski, B.H. (Eds)
.... Spiralbound hardback 0-19-963290-1 **£30.00**
.... Paperback 0-19-963289-8 **£22.50**
96. Human Cytogenetics: Volume I: Constitutional Analysis (2/e) Rooney, D.E. & Czepulkowski, B.H. (Eds)
.... Spiralbound hardback 0-19-963288-X **£30.00**
.... Paperback 0-19-963287-1 **£22.50**
95. Lipid Modification of Proteins Hooper, N.M. & Turner, A.J. (Eds)
.... Spiralbound hardback 0-19-963274-X **£32.50**
.... Paperback 0-19-963273-1 **£22.50**
94. Biomechanics-Structures and Systems Biewener, A.A. (Ed)
.... Spiralbound hardback 0-19-963268-5 **£42.50**
.... Paperback 0-19-963267-7 **£25.00**
93. Lipoprotein Analysis Converse, C.A. & Skinner, E.R. (Eds)
.... Spiralbound hardback 0-19-963192-1 **£30.00**
.... Paperback 0-19-963231-6 **£19.50**
92. Receptor-Ligand Interactions Hulme, E.C. (Ed)
.... Spiralbound hardback 0-19-963090-9 **£35.00**
.... Paperback 0-19-963091-7 **£25.00**
91. Molecular Genetic Analysis of Populations Hoelzel, A.R. (Ed)
.... Spiralbound hardback 0-19-963278-2 **£32.50**
.... Paperback 0-19-963277-4 **£22.50**

90. **Enzyme Assays** Eisenthal, R. & Danson, M.J. (Eds)
.... Spiralbound hardback 0-19-963142-5 **£35.00**
.... Paperback 0-19-963143-3 **£25.00**

89. **Microcomputers in Biochemistry** Bryce, C.F.A. (Ed)
.... Spiralbound hardback 0-19-963253-7 **£30.00**
.... Paperback 0-19-963252-9 **£19.50**

88. **The Cytoskeleton** Carraway, K.L. & Carraway, C.A.C. (Eds)
.... Spiralbound hardback 0-19-963257-X **£30.00**
.... Paperback 0-19-963256-1 **£19.50**

87. **Monitoring Neuronal Activity** Stamford, J.A. (Ed)
.... Spiralbound hardback 0-19-963244-8 **£30.00**
.... Paperback 0-19-963243-X **£19.50**

86. **Crystallization of Nucleic Acids and Proteins** Ducruix, A. & Gieg‹130›, R. (Eds)
.... Spiralbound hardback 0-19-963245-6 **£35.00**
.... Paperback 0-19-963246-4 **£25.00**

85. **Molecular Plant Pathology: Volume I** Gurr, S.J., McPherson, M.J. & others (Eds)
.... Spiralbound hardback 0-19-963103-4 **£30.00**
.... Paperback 0-19-963102-6 **£19.50**

84. **Anaerobic Microbiology** Levett, P.N. (Ed)
.... Spiralbound hardback 0-19-963204-9 **£32.50**
.... Paperback 0-19-963262-6 **£22.50**

83. **Oligonucleotides and Analogues** Eckstein, F. (Ed)
.... Spiralbound hardback 0-19-963280-4 **£32.50**
.... Paperback 0-19-963279-0 **£22.50**

82. **Electron Microscopy in Biology** Harris, R. (Ed)
.... Spiralbound hardback 0-19-963219-7 **£32.50**
.... Paperback 0-19-963215-4 **£22.50**

81. **Essential Molecular Biology: Volume II** Brown, T.A. (Ed)
.... Spiralbound hardback 0-19-963112-3 **£32.50**
.... Paperback 0-19-963113-1 **£22.50**

80. **Cellular Calcium** McCormack, J.G. & Cobbold, P.H. (Eds)
.... Spiralbound hardback 0-19-963131-X **£35.00**
.... Paperback 0-19-963130-1 **£25.00**

79. **Protein Architecture** Lesk, A.M.
.... Spiralbound hardback 0-19-963054-2 **£32.50**
.... Paperback 0-19-963055-0 **£22.50**

78. **Cellular Neurobiology** Chad, J. & Wheal, H. (Eds)
.... Spiralbound hardback 0-19-963106-9 **£32.50**
.... Paperback 0-19-963107-7 **£22.50**

77. **PCR** McPherson, M.J., Quirke, P. & others (Eds)
.... Spiralbound hardback 0-19-963226-X **£30.00**
.... Paperback 0-19-963196-4 **£19.50**

76. **Mammalian Cell Biotechnology** Butler, M. (Ed)
.... Spiralbound hardback 0-19-963207-3 **£30.00**
.... Paperback 0-19-963209-X **£19.50**

75. **Cytokines** Balkwill, F.R. (Ed)
.... Spiralbound hardback 0-19-963218-9 **£35.00**
.... Paperback 0-19-963214-6 **£25.00**

74. **Molecular Neurobiology** Chad, J. & Wheal, H. (Eds)
.... Spiralbound hardback 0-19-963108-5 **£30.00**
.... Paperback 0-19-963109-3 **£19.50**

73. **Directed Mutagenesis** McPherson, M.J. (Ed)
.... Spiralbound hardback 0-19-963141-7 **£30.00**
.... Paperback 0-19-963140-9 **£19.50**

72. **Essential Molecular Biology: Volume I** Brown, T.A. (Ed)
.... Spiralbound hardback 0-19-963110-7 **£32.50**
.... Paperback 0-19-963111-5 **£22.50**

71. **Peptide Hormone Action** Siddle, K. & Hutton, J.C.
.... Spiralbound hardback 0-19-963070-4 **£32.50**
.... Paperback 0-19-963071-2 **£22.50**

70. **Peptide Hormone Secretion** Hutton, J.C. & Siddle, K. (Eds)
.... Spiralbound hardback 0-19-963068-2 **£35.00**
.... Paperback 0-19-963069-0 **£25.00**

69. **Postimplantation Mammalian Embryos** Copp, A.J. & Cockroft, D.L. (Eds)
.... Spiralbound hardback 0-19-963088-7 **£35.00**
.... Paperback 0-19-963089-5 **£25.00**

68. **Receptor-Effector Coupling** Hulme, E.C. (Ed)
.... Spiralbound hardback 0-19-963094-1 **£30.00**
.... Paperback 0-19-963095-X **£19.50**

67. **Gel Electrophoresis of Proteins (2/e)** Hames, B.D. & Rickwood, D. (Eds)
.... Spiralbound hardback 0-19-963074-7 **£35.00**
.... Paperback 0-19-963075-5 **£25.00**

66. **Clinical Immunology** Gooi, H.C. & Chapel, H. (Eds)
.... Spiralbound hardback 0-19-963086-0 **£32.50**
.... Paperback 0-19-963087-9 **£22.50**

65. **Receptor Biochemistry** Hulme, E.C. (Ed)
.... Spiralbound hardback 0-19-963092-5 **£35.00**
.... Paperback 0-19-963093-3 **£25.00**

64. **Gel Electrophoresis of Nucleic Acids (2/e)** Rickwood, D. & Hames, B.D. (Eds)
.... Spiralbound hardback 0-19-963082-8 **£32.50**
.... Paperback 0-19-963083-6 **£22.50**

63. **Animal Virus Pathogenesis** Oldstone, M.B.A. (Ed)
.... Spiralbound hardback 0-19-963100-X **£30.00**
.... Paperback 0-19-963101-8 **£18.50**

62. **Flow Cytometry** Ormerod, M.G. (Ed)
.... Paperback 0-19-963053-4 **£22.50**

61. **Radioisotopes in Biology** Slater, R.J. (Ed)
.... Spiralbound hardback 0-19-963080-1 **£32.50**
.... Paperback 0-19-963081-X **£22.50**

60. **Biosensors** Cass, A.E.G. (Ed)
.... Spiralbound hardback 0-19-963046-1 **£30.00**
.... Paperback 0-19-963047-X **£19.50**

59. **Ribosomes and Protein Synthesis** Spedding, G. (Ed)
.... Spiralbound hardback 0-19-963104-2 **£32.50**
.... Paperback 0-19-963105-0 **£22.50**

58. **Liposomes** New, R.R.C. (Ed)
.... Spiralbound hardback 0-19-963076-3 **£35.00**
.... Paperback 0-19-963077-1 **£22.50**

57. **Fermentation** McNeil, B. & Harvey, L.M. (Eds)
.... Spiralbound hardback 0-19-963044-5 **£30.00**
.... Paperback 0-19-963045-3 **£19.50**

56. **Protein Purification Applications** Harris, E.L.V. & Angal, S. (Eds)
.... Spiralbound hardback 0-19-963022-4 **£30.00**
.... Paperback 0-19-963023-2 **£18.50**

55. **Nucleic Acids Sequencing** Howe, C.J. & Ward, E.S. (Eds)
.... Spiralbound hardback 0-19-963056-9 **£30.00**
.... Paperback 0-19-963057-7 **£19.50**

54. **Protein Purification Methods** Harris, E.L.V. & Angal, S. (Eds)
.... Spiralbound hardback 0-19-963002-X **£30.00**
.... Paperback 0-19-963003-8 **£20.00**

53. **Solid Phase Peptide Synthesis** Atherton, E. & Sheppard, R.C.
.... Spiralbound hardback 0-19-963066-6 **£30.00**
.... Paperback 0-19-963067-4 **£18.50**

52. **Medical Bacteriology** Hawkey, P.M. & Lewis, D.A. (Eds)
.... Spiralbound hardback 0-19-963008-9 **£38.00**
.... Paperback 0-19-963009-7 **£25.00**

51. **Proteolytic Enzymes** Beynon, R.J. & Bond, J.S. (Eds)
.... Spiralbound hardback 0-19-963058-5 **£30.00**
.... Paperback 0-19-963059-3 **£19.50**

50. **Medical Mycology** Evans, E.G.V. & Richardson, M.D. (Eds)
.... Spiralbound hardback 0-19-963010-0 **£37.50**
.... Paperback 0-19-963011-9 **£25.00**

49. **Computers in Microbiology** Bryant, T.N. & Wimpenny, J.W.T. (Eds)
.... Paperback 0-19-963015-1 **£19.50**

48. **Protein Sequencing** Findlay, J.B.C. & Geisow, M.J. (Eds)
.... Spiralbound hardback 0-19-963012-7 **£30.00**
.... Paperback 0-19-963013-5 **£18.50**

47. **Cell Growth and Division** Baserga, R. (Ed)
.... Spiralbound hardback 0-19-963026-7 **£30.00**
.... Paperback 0-19-963027-5 **£18.50**

46. **Protein Function** Creighton, T.E. (Ed)
.... Spiralbound hardback 0-19-963006-2 **£32.50**
.... Paperback 0-19-963007-0 **£22.50**

45. **Protein Structure** Creighton, T.E. (Ed)
.... Spiralbound hardback 0-19-963000-3 **£32.50**
.... Paperback 0-19-963001-1 **£22.50**

44. **Antibodies: Volume II** Catty, D. (Ed)
.... Spiralbound hardback 0-19-963018-6 **£30.00**
.... Paperback 0-19-963019-4 **£19.50**

43. **HPLC of Macromolecules** Oliver, R.W.A. (Ed)
.... Spiralbound hardback 0-19-963020-8 **£30.00**
.... Paperback 0-19-963021-6 **£19.50**
42. **Light Microscopy in Biology** Lacey, A.J. (Ed)
.... Spiralbound hardback 0-19-963036-4 **£30.00**
.... Paperback 0-19-963037-2 **£19.50**
41. **Plant Molecular Biology** Shaw, C.H. (Ed)
.... Paperback 1-85221-056-7 **£22.50**
40. **Microcomputers in Physiology** Fraser, P.J. (Ed)
.... Spiralbound hardback 1-85221-129-6 **£30.00**
.... Paperback 1-85221-130-X **£19.50**
39. **Genome Analysis** Davies, K.E. (Ed)
.... Spiralbound hardback 1-85221-109-1 **£30.00**
.... Paperback 1-85221-110-5 **£18.50**
38. **Antibodies: Volume I** Catty, D. (Ed)
.... Paperback 0-947946-85-3 **£19.50**
37. **Yeast** Campbell, I. & Duffus, J.H. (Eds)
.... Paperback 0-947946-79-9 **£19.50**
36. **Mammalian Development** Monk, M. (Ed)
.... Hardback 1-85221-030-3 **£30.50**
.... Paperback 1-85221-029-X **£22.50**
35. **Lymphocytes** Klaus, G.G.B. (Ed)
.... Hardback 1-85221-018-4 **£30.00**
34. **Lymphokines and Interferons** Clemens, M.J., Morris, A.G. & others (Eds)
.... Paperback 1-85221-035-4 **£22.50**
33. **Mitochondria** Darley-Usmar, V.M., Rickwood, D. & others (Eds)
.... Hardback 1-85221-034-6 **£32.50**
.... Paperback 1-85221-033-8 **£22.50**
32. **Prostaglandins and Related Substances** Benedetto, C., McDonald-Gibson, R.G. & others (Eds)
.... Hardback 1-85221-032-X **£32.50**
.... Paperback 1-85221-031-1 **£22.50**
31. **DNA Cloning: Volume III** Glover, D.M. (Ed)
.... Hardback 1-85221-049-4 **£30.00**
.... Paperback 1-85221-048-6 **£19.50**
30. **Steroid Hormones** Green, B. & Leake, R.E. (Eds)
.... Paperback 0-947946-53-5 **£19.50**
29. **Neurochemistry** Turner, A.J. & Bachelard, H.S. (Eds)
.... Hardback 1-85221-028-1 **£30.00**
.... Paperback 1-85221-027-3 **£19.50**
28. **Biological Membranes** Findlay, J.B.C. & Evans, W.H. (Eds)
.... Hardback 0-947946-84-5 **£32.50**
.... Paperback 0-947946-83-7 **£22.50**
27. **Nucleic Acid and Protein Sequence Analysis** Bishop, M.J. & Rawlings, C.J. (Eds)
.... Hardback 1-85221-007-9 **£35.00**
.... Paperback 1-85221-006-0 **£25.00**
26. **Electron Microscopy in Molecular Biology** Sommerville, J. & Scheer, U. (Eds)
.... Hardback 0-947946-64-0 **£30.00**
.... Paperback 0-947946-54-3 **£19.50**
25. **Teratocarcinomas and Embryonic Stem Cells** Robertson, E.J. (Ed)
.... Hardback 1-85221-005-2 **£19.50**
.... Paperback 1-85221-004-4 **£19.50**
24. **Spectrophotometry and Spectrofluorimetry** Harris, D.A. & Bashford, C.L. (Eds)
.... Hardback 0-947946-69-1 **£30.00**
.... Paperback 0-947946-46-2 **£18.50**
23. **Plasmids** Hardy, K.G. (Ed)
.... Paperback 0-947946-81-0 **£18.50**
22. **Biochemical Toxicology** Snell, K. & Mullock, B. (Eds)
.... Paperback 0-947946-52-7 **£19.50**
19. **Drosophila** Roberts, D.B. (Ed)
.... Hardback 0-947946-66-7 **£32.50**
.... Paperback 0-947946-45-4 **£22.50**
17. **Photosynthesis: Energy Transduction** Hipkins, M.F. & Baker, N.R. (Eds)
.... Hardback 0-947946-63-2 **£30.00**
.... Paperback 0-947946-51-9 **£18.50**
16. **Human Genetic Diseases** Davies, K.E. (Ed)
.... Hardback 0-947946-76-4 **£30.00**
.... Paperback 0-947946-75-6 **£18.50**
14. **Nucleic Acid Hybridisation** Hames, B.D. & Higgins, S.J. (Eds)
.... Hardback 0-947946-61-6 **£30.00**
.... Paperback 0-947946-23-3 **£19.50**
13. **Immobilised Cells and Enzymes** Woodward, J. (Ed)
.... Hardback 0-947946-60-8 **£18.50**
12. **Plant Cell Culture** Dixon, R.A. (Ed)
.... Paperback 0-947946-22-5 **£19.50**
11a. **DNA Cloning: Volume I** Glover, D.M. (Ed)
.... Paperback 0-947946-18-7 **£18.50**
11b. **DNA Cloning: Volume II** Glover, D.M. (Ed)
.... Paperback 0-947946-19-5 **£19.50**
10. **Virology** Mahy, B.W.J. (Ed)
.... Paperback 0-904147-78-9 **£19.50**
9. **Affinity Chromatography** Dean, P.D.G., Johnson, W.S. & others (Eds)
.... Paperback 0-904147-71-1 **£19.50**
7. **Microcomputers in Biology** Ireland, C.R. & Long, S.P. (Eds)
.... Paperback 0-904147-57-6 **£18.00**
6. **Oligonucleotide Synthesis** Gait, M.J. (Ed)
.... Paperback 0-904147-74-6 **£18.50**
5. **Transcription and Translation** Hames, B.D. & Higgins, S.J. (Eds)
.... Paperback 0-904147-52-5 **£22.50**
3. **Iodinated Density Gradient Media** Rickwood, D. (Ed)
.... Paperback 0-904147-51-7 **£19.50**

Sets

Essential Molecular Biology: Volumes I and II as a set Brown, T.A. (Ed)
.... Spiralbound hardback 0-19-963114-X **£58.00**
.... Paperback 0-19-963115-8 **£40.00**
Antibodies: Volumes I and II as a set Catty, D. (Ed)
.... Paperback 0-19-963063-1 **£33.00**
Cellular and Molecular Neurobiology Chad, J. & Wheal, H. (Eds)
.... Spiralbound hardback 0-19-963255-3 **£56.00**
.... Paperback 0-19-963254-5 **£38.00**
Protein Structure and Protein Function: Two-volume set Creighton, T.E. (Ed)
.... Spiralbound hardback 0-19-963064-X **£55.00**
.... Paperback 0-19-963065-8 **£38.00**
DNA Cloning: Volumes I, II, III as a set Glover, D.M. (Ed)
.... Paperback 1-85221-069-9 **£46.00**
Molecular Plant Pathology: Volumes I and II as a set Gurr, S.J., McPherson, M.J. & others (Eds)
.... Spiralbound hardback 0-19-963354-1 **£56.00**
.... Paperback 0-19-963353-3 **£37.00**
Protein Purification Methods, and Protein Purification Applications, two-volume set Harris, E.L.V. & Angal, S. (Eds)
.... Spiralbound hardback 0-19-963048-8 **£48.00**
.... Paperback 0-19-963049-6 **£32.00**
Diagnostic Molecular Pathology: Volumes I and II as a set Herrington, C.S. & McGee, J. O'D. (Eds)
.... Spiralbound hardback 0-19-963241-3 **£54.00**
.... Paperback 0-19-963240-5 **£35.00**
Receptor Biochemistry; Receptor-Effector Coupling; Receptor-Ligand Interactions Hulme, E.C. (Ed)
.... Spiralbound hardback 0-19-963096-8 **£90.00**
.... Paperback 0-19-963097-6 **£62.50**
Signal Transduction Milligan, G. (Ed)
.... Spiralbound hardback 0-19-963296-0 **£30.00**
.... Paperback 0-19-963295-2 **£18.50**
Human Cytogenetics: Volumes I and II as a set (2/e) Rooney, D.E. & Czepulkowski, B.H. (Eds)
.... Hardback 0-19-963314-2 **£58.50**
.... Paperback 0-19-963313-4 **£40.50**
Peptide Hormone Secretion/Peptide Hormone Action Siddle, K. & Hutton, J.C. (Eds)
.... Spiralbound hardback 0-19-963072-0 **£55.00**
.... Paperback 0-19-963073-9 **£38.00**

ORDER FORM for UK, Europe and Rest of World

(Excluding USA and Canada)

Qty	ISBN	Author	Title	Amount
			P&P	
			TOTAL	

Please add postage and packing: £1.75 for UK orders under £20; £2.75 for UK orders over £20; overseas orders add 10% of total.

Name ....

Address ....

....

.... Post code ....

[ ] Please charge £ .... to my credit card

Access/VISA/Eurocard/AMEX/Diners Club (circle appropriate card)

Card No .... Expiry date ....

Signature ....

Credit card account address if different from above:

....

.... Postcode ....

[ ] I enclose a cheque for £.....................

Please return this form to: OUP Distribution Services, Saxon Way West, Corby, Northants NN18 9ES

OR ORDER BY CREDIT CARD HOTLINE: Tel +44-(0)536-741519 or Fax +44-(0)536-746337

ORDER OTHER TITLES OF INTEREST TODAY

Price list for: USA and Canada

123. **Protein Phosphorylation** Hardie, G. (Ed)
.... Spiralbound hardback 0-19-963306-1 **$65.00**
.... Paperback 0-19-963305-3 **$45.00**
121. **Tumour Immunobiology** Gallagher, G., Rees, R.C. & others (Eds)
.... Spiralbound hardback 0-19-963370-3 **$72.00**
.... Paperback 0-19-963369-X **$50.00**
117. **Gene Transcription** Hames, D.B. & Higgins, S.J. (Eds)
.... Spiralbound hardback 0-19-963292-8 **$72.00**
.... Paperback 0-19-963291-X **$50.00**
116. **Electrophysiology** Wallis, D.I. (Ed)
.... Spiralbound hardback 0-19-963348-7 **$66.50**
.... Paperback 0-19-963347-9 **$45.95**
115. **Biological Data Analysis** Fry, J.C. (Ed)
.... Spiralbound hardback 0-19-963340-1 **$80.00**
.... Paperback 0-19-963339-8 **$60.00**
114. **Experimental Neuroanatomy** Bolam, J.P. (Ed)
.... Spiralbound hardback 0-19-963326-6 **$65.00**
.... Paperback 0-19-963325-8 **$40.00**
111. **Haemopoiesis** Testa, N.G. & Molineux, G. (Eds)
.... Spiralbound hardback 0-19-963366-5 **$65.00**
.... Paperback 0-19-963365-7 **$45.00**
113. **Preparative Centrifugation** Rickwood, D. (Ed)
.... Spiralbound hardback 0-19-963208-1 **$90.00**
.... Paperback 0-19-963211-1 **$50.00**
110. **Pollination Ecology** Dafni, A.
.... Spiralbound hardback 0-19-963299-5 **$65.00**
.... Paperback 0-19-963298-7 **$45.00**
109. **In Situ Hybridization** Wilkinson, D.G. (Ed)
.... Spiralbound hardback 0-19-963328-2 **$58.00**
.... Paperback 0-19-963327-4 **$36.00**
108. **Protein Engineering** Rees, A.R., Sternberg, M.J.E. & others (Eds)
.... Spiralbound hardback 0-19-963139-5 **$75.00**
.... Paperback 0-19-963138-7 **$50.00**
107. **Cell-Cell Interactions** Stevenson, B.R., Gallin, W.J. & others (Eds)
.... Spiralbound hardback 0-19-963319-3 **$60.00**
.... Paperback 0-19-963318-5 **$40.00**
106. **Diagnostic Molecular Pathology: Volume I** Herrington, C.S. & McGee, J. O'D. (Eds)
.... Spiralbound hardback 0-19-963237-5 **$58.00**
.... Paperback 0-19-963236-7 **$38.00**
105. **Biomechanics-Materials** Vincent, J.F.V. (Ed)
.... Spiralbound hardback 0-19-963223-5 **$70.00**
.... Paperback 0-19-963222-7 **$50.00**
104. **Animal Cell Culture (2/e)** Freshney, R.I. (Ed)
.... Spiralbound hardback 0-19-963212-X **$60.00**
.... Paperback 0-19-963213-8 **$40.00**
103. **Molecular Plant Pathology: Volume II** Gurr, S.J., McPherson, M.J. & others (Eds)
.... Spiralbound hardback 0-19-963352-5 **$65.00**
.... Paperback 0-19-963351-7 **$45.00**
101. **Protein Targeting** Magee, A.I. & Wileman, T. (Eds)
.... Spiralbound hardback 0-19-963206-5 **$75.00**
.... Paperback 0-19-963210-3 **$50.00**
100. **Diagnostic Molecular Pathology: Volume II: Cell and Tissue Genotyping** Herrington, C.S. & McGee, J.O'D. (Eds)
.... Spiralbound hardback 0-19-963239-1 **$60.00**
.... Paperback 0-19-963238-3 **$39.00**
99. **Neuronal Cell Lines** Wood, J.N. (Ed)
.... Spiralbound hardback 0-19-963346-0 **$68.00**
.... Paperback 0-19-963345-2 **$48.00**
98. **Neural Transplantation** Dunnett, S.B. & Björklund, A. (Eds)
.... Spiralbound hardback 0-19-963286-3 **$69.00**
.... Paperback 0-19-963285-5 **$42.00**
97. **Human Cytogenetics: Volume II: Malignancy and Acquired Abnormalities (2/e)** Rooney, D.E. & Czepulkowski, B.H. (Eds)
.... Spiralbound hardback 0-19-963290-1 **$75.00**
.... Paperback 0-19-963289-8 **$50.00**
96. **Human Cytogenetics: Volume I: Constitutional Analysis (2/e)** Rooney, D.E. & Czepulkowski, B.H. (Eds)
.... Spiralbound hardback 0-19-963288-X **$75.00**
.... Paperback 0-19-963287-1 **$50.00**
95. **Lipid Modification of Proteins** Hooper, N.M. & Turner, A.J. (Eds)
.... Spiralbound hardback 0-19-963274-X **$75.00**
.... Paperback 0-19-963273-1 **$50.00**
94. **Biomechanics-Structures and Systems** Biewener, A.A. (Ed)
.... Spiralbound hardback 0-19-963268-5 **$85.00**
.... Paperback 0-19-963267-7 **$50.00**
93. **Lipoprotein Analysis** Converse, C.A. & Skinner, E.R. (Eds)
.... Spiralbound hardback 0-19-963192-1 **$65.00**
.... Paperback 0-19-963231-6 **$42.00**
92. **Receptor-Ligand Interactions** Hulme, E.C. (Ed)
.... Spiralbound hardback 0-19-963090-9 **$75.00**
.... Paperback 0-19-963091-7 **$50.00**
91. **Molecular Genetic Analysis of Populations** Hoelzel, A.R. (Ed)
.... Spiralbound hardback 0-19-963278-2 **$65.00**
.... Paperback 0-19-963277-4 **$45.00**
90. **Enzyme Assays** Eisenthal, R. & Danson, M.J. (Eds)
.... Spiralbound hardback 0-19-963142-5 **$68.00**
.... Paperback 0-19-963143-3 **$48.00**
89. **Microcomputers in Biochemistry** Bryce, C.F.A. (Ed)
.... Spiralbound hardback 0-19-963253-7 **$60.00**
.... Paperback 0-19-963252-9 **$40.00**
88. **The Cytoskeleton** Carraway, K.L. & Carraway, C.A.C. (Eds)
.... Spiralbound hardback 0-19-963257-X **$60.00**
.... Paperback 0-19-963256-1 **$40.00**
87. **Monitoring Neuronal Activity** Stamford, J.A. (Ed)
.... Spiralbound hardback 0-19-963244-8 **$60.00**
.... Paperback 0-19-963243-X **$40.00**
86. **Crystallization of Nucleic Acids and Proteins** Ducruix, A. & Gieg‹130›, R. (Eds)
.... Spiralbound hardback 0-19-963245-6 **$60.00**
.... Paperback 0-19-963246-4 **$50.00**
85. **Molecular Plant Pathology: Volume I** Gurr, S.J., McPherson, M.J. & others (Eds)
.... Spiralbound hardback 0-19-963103-4 **$60.00**
.... Paperback 0-19-963102-6 **$40.00**
84. **Anaerobic Microbiology** Levett, P.N. (Ed)
.... Spiralbound hardback 0-19-963204-9 **$75.00**
.... Paperback 0-19-963262-6 **$45.00**

83. **Oligonucleotides and Analogues** Eckstein, F. (Ed)
.... Spiralbound hardback 0-19-963280-4 **$65.00**
.... Paperback 0-19-963279-0 **$45.00**
82. **Electron Microscopy in Biology** Harris, R. (Ed)
.... Spiralbound hardback 0-19-963219-7 **$65.00**
.... Paperback 0-19-963215-4 **$45.00**
81. **Essential Molecular Biology: Volume II** Brown, T.A. (Ed)
.... Spiralbound hardback 0-19-963112-3 **$65.00**
.... Paperback 0-19-963113-1 **$45.00**
80. **Cellular Calcium** McCormack, J.G. & Cobbold, P.H. (Eds)
.... Spiralbound hardback 0-19-963131-X **$75.00**
.... Paperback 0-19-963130-1 **$50.00**
79. **Protein Architecture** Lesk, A.M.
.... Spiralbound hardback 0-19-963054-2 **$65.00**
.... Paperback 0-19-963055-0 **$45.00**
78. **Cellular Neurobiology** Chad, J. & Wheal, H. (Eds)
.... Spiralbound hardback 0-19-963106-9 **$73.00**
.... Paperback 0-19-963107-7 **$43.00**
77. **PCR** McPherson, M.J., Quirke, P. & others (Eds)
.... Spiralbound hardback 0-19-963226-X **$55.00**
.... Paperback 0-19-963196-4 **$40.00**
76. **Mammalian Cell Biotechnology** Butler, M. (Ed)
.... Spiralbound hardback 0-19-963207-3 **$60.00**
.... Paperback 0-19-963209-X **$40.00**
75. **Cytokines** Balkwill, F.R. (Ed)
.... Spiralbound hardback 0-19-963218-9 **$64.00**
.... Paperback 0-19-963214-6 **$44.00**
74. **Molecular Neurobiology** Chad, J. & Wheal, H. (Eds)
.... Spiralbound hardback 0-19-963108-5 **$56.00**
.... Paperback 0-19-963109-3 **$36.00**
73. **Directed Mutagenesis** McPherson, M.J. (Ed)
.... Spiralbound hardback 0-19-963141-7 **$55.00**
.... Paperback 0-19-963140-9 **$35.00**
72. **Essential Molecular Biology: Volume I** Brown, T.A. (Ed)
.... Spiralbound hardback 0-19-963110-7 **$65.00**
.... Paperback 0-19-963111-5 **$45.00**
71. **Peptide Hormone Action** Siddle, K. & Hutton, J.C.
.... Spiralbound hardback 0-19-963070-4 **$70.00**
.... Paperback 0-19-963071-2 **$50.00**
70. **Peptide Hormone Secretion** Hutton, J.C. & Siddle, K. (Eds)
.... Spiralbound hardback 0-19-963068-2 **$70.00**
.... Paperback 0-19-963069-0 **$50.00**
69. **Postimplantation Mammalian Embryos** Copp, A.J. & Cockroft, D.L. (Eds)
.... Spiralbound hardback 0-19-963088-7 **$70.00**
.... Paperback 0-19-963089-5 **$50.00**
68. **Receptor-Effector Coupling** Hulme, E.C. (Ed)
.... Spiralbound hardback 0-19-963094-1 **$70.00**
.... Paperback 0-19-963095-X **$45.00**
67. **Gel Electrophoresis of Proteins (2/e)** Hames, B.D. & Rickwood, D. (Eds)
.... Spiralbound hardback 0-19-963074-7 **$75.00**
.... Paperback 0-19-963075-5 **$50.00**
66. **Clinical Immunology** Gooi, H.C. & Chapel, H. (Eds)
.... Spiralbound hardback 0-19-963086-0 **$69.95**
.... Paperback 0-19-963087-9 **$50.00**
65. **Receptor Biochemistry** Hulme, E.C. (Ed)
.... Spiralbound hardback 0-19-963092-5 **$70.00**
.... Paperback 0-19-963093-3 **$50.00**
64. **Gel Electrophoresis of Nucleic Acids (2/e)** Rickwood, D. & Hames, B.D. (Eds)
.... Spiralbound hardback 0-19-963082-8 **$75.00**
.... Paperback 0-19-963083-6 **$50.00**
63. **Animal Virus Pathogenesis** Oldstone, M.B.A. (Ed)
.... Spiralbound hardback 0-19-963100-X **$68.00**
.... Paperback 0-19-963101-8 **$40.00**
62. **Flow Cytometry** Ormerod, M.G. (Ed)
.... Paperback 0-19-963053-4 **$50.00**
61. **Radioisotopes in Biology** Slater, R.J. (Ed)
.... Spiralbound hardback 0-19-963080-1 **$75.00**
.... Paperback 0-19-963081-X **$45.00**
60. **Biosensors** Cass, A.E.G. (Ed)
.... Spiralbound hardback 0-19-963046-1 **$65.00**
.... Paperback 0-19-963047-X **$43.00**
59. **Ribosomes and Protein Synthesis** Spedding, G. (Ed)
.... Spiralbound hardback 0-19-963104-2 **$75.00**
.... Paperback 0-19-963105-0 **$45.00**
58. **Liposomes** New, R.R.C. (Ed)
.... Spiralbound hardback 0-19-963076-3 **$70.00**
.... Paperback 0-19-963077-1 **$45.00**
57. **Fermentation** McNeil, B. & Harvey, L.M. (Eds)
.... Spiralbound hardback 0-19-963044-5 **$65.00**
.... Paperback 0-19-963045-3 **$39.00**
56. **Protein Purification Applications** Harris, E.L.V. & Angal, S. (Eds)
.... Spiralbound hardback 0-19-963022-4 **$54.00**
.... Paperback 0-19-963023-2 **$36.00**
55. **Nucleic Acids Sequencing** Howe, C.J. & Ward, E.S. (Eds)
.... Spiralbound hardback 0-19-963056-9 **$59.00**
.... Paperback 0-19-963057-7 **$38.00**
54. **Protein Purification Methods** Harris, E.L.V. & Angal, S. (Eds)
.... Spiralbound hardback 0-19-963002-X **$60.00**
.... Paperback 0-19-963003-8 **$40.00**
53. **Solid Phase Peptide Synthesis** Atherton, E. & Sheppard, R.C.
.... Spiralbound hardback 0-19-963066-6 **$58.00**
.... Paperback 0-19-963067-4 **$39.95**
52. **Medical Bacteriology** Hawkey, P.M. & Lewis, D.A. (Eds)
.... Spiralbound hardback 0-19-963008-9 **$69.95**
.... Paperback 0-19-963009-7 **$50.00**
51. **Proteolytic Enzymes** Beynon, R.J. & Bond, J.S. (Eds)
.... Spiralbound hardback 0-19-963058-5 **$60.00**
.... Paperback 0-19-963059-3 **$39.00**
50. **Medical Mycology** Evans, E.G.V. & Richardson, M.D. (Eds)
.... Spiralbound hardback 0-19-963010-0 **$69.95**
.... Paperback 0-19-963011-9 **$50.00**
49. **Computers in Microbiology** Bryant, T.N. & Wimpenny, J.W.T. (Eds)
.... Paperback 0-19-963015-1 **$40.00**
48. **Protein Sequencing** Findlay, J.B.C. & Geisow, M.J. (Eds)
.... Spiralbound hardback 0-19-963012-7 **$56.00**
.... Paperback 0-19-963013-5 **$38.00**
47. **Cell Growth and Division** Baserga, R. (Ed)
.... Spiralbound hardback 0-19-963026-7 **$62.00**
.... Paperback 0-19-963027-5 **$38.00**
46. **Protein Function** Creighton, T.E. (Ed)
.... Spiralbound hardback 0-19-963006-2 **$65.00**
.... Paperback 0-19-963007-0 **$45.00**
45. **Protein Structure** Creighton, T.E. (Ed)
.... Spiralbound hardback 0-19-963000-3 **$65.00**
.... Paperback 0-19-963001-1 **$45.00**
44. **Antibodies: Volume II** Catty, D. (Ed)
.... Spiralbound hardback 0-19-963018-6 **$58.00**
.... Paperback 0-19-963019-4 **$39.00**
43. **HPLC of Macromolecules** Oliver, R.W.A. (Ed)
.... Spiralbound hardback 0-19-963020-8 **$54.00**
.... Paperback 0-19-963021-6 **$45.00**
42. **Light Microscopy in Biology** Lacey, A.J. (Ed)
.... Spiralbound hardback 0-19-963036-4 **$62.00**
.... Paperback 0-19-963037-2 **$38.00**
41. **Plant Molecular Biology** Shaw, C.H. (Ed)
.... Paperback 1-85221-056-7 **$38.00**
40. **Microcomputers in Physiology** Fraser, P.J. (Ed)
.... Spiralbound hardback 1-85221-129-6 **$54.00**
.... Paperback 1-85221-130-X **$36.00**
39. **Genome Analysis** Davies, K.E. (Ed)
.... Spiralbound hardback 1-85221-109-1 **$54.00**
.... Paperback 1-85221-110-5 **$36.00**
38. **Antibodies: Volume I** Catty, D. (Ed)
.... Paperback 0-947946-85-3 **$38.00**
37. **Yeast** Campbell, I. & Duffus, J.H. (Eds)
.... Paperback 0-947946-79-9 **$36.00**
36. **Mammalian Development** Monk, M. (Ed)
.... Hardback 1-85221-030-3 **$60.00**
.... Paperback 1-85221-029-X **$45.00**
35. **Lymphocytes** Klaus, G.G.B. (Ed)
.... Hardback 1-85221-018-4 **$54.00**
34. **Lymphokines and Interferons** Clemens, M.J., Morris, A.G. & others (Eds)
.... Paperback 1-85221-035-4 **$44.00**
33. **Mitochondria** Darley-Usmar, V.M., Rickwood, D. & others (Eds)
.... Hardback 1-85221-034-6 **$65.00**
.... Paperback 1-85221-033-8 **$45.00**

32. **Prostaglandins and Related Substances** Benedetto, C., McDonald-Gibson, R.G. & others (Eds)
.... Hardback 1-85221-032-X **$58.00**
.... Paperback 1-85221-031-1 **$38.00**
31. **DNA Cloning: Volume III** Glover, D.M. (Ed)
.... Hardback 1-85221-049-4 **$56.00**
.... Paperback 1-85221-048-6 **$36.00**
30. **Steroid Hormones** Green, B. & Leake, R.E. (Eds)
.... Paperback 0-947946-53-5 **$40.00**
29. **Neurochemistry** Turner, A.J. & Bachelard, H.S. (Eds)
.... Hardback 1-85221-028-1 **$56.00**
.... Paperback 1-85221-027-3 **$36.00**
28. **Biological Membranes** Findlay, J.B.C. & Evans, W.H. (Eds)
.... Hardback 0-947946-84-5 **$54.00**
.... Paperback 0-947946-83-7 **$36.00**
27. **Nucleic Acid and Protein Sequence Analysis** Bishop, M.J. & Rawlings, C.J. (Eds)
.... Hardback 1-85221-007-9 **$66.00**
.... Paperback 1-85221-006-0 **$44.00**
26. **Electron Microscopy in Molecular Biology** Sommerville, J. & Scheer, U. (Eds)
.... Hardback 0-947946-64-0 **$54.00**
.... Paperback 0-947946-54-3 **$40.00**
25. **Teratocarcinomas and Embryonic Stem Cells** Robertson, E.J. (Ed)
.... Hardback 1-85221-005-2 **$62.00**
.... Paperback 1-85221-004-4 **$0.00**
24. **Spectrophotometry and Spectrofluorimetry** Harris, D.A. & Bashford, C.L. (Eds)
.... Hardback 0-947946-69-1 **$56.00**
.... Paperback 0-947946-46-2 **$39.95**
23. **Plasmids** Hardy, K.G. (Ed)
.... Paperback 0-947946-81-0 **$36.00**
22. **Biochemical Toxicology** Snell, K. & Mullock, B. (Eds)
.... Paperback 0-947946-52-7 **$40.00**
19. **Drosophila** Roberts, D.B. (Ed)
.... Hardback 0-947946-66-7 **$67.50**
.... Paperback 0-947946-45-4 **$46.00**
17. **Photosynthesis: Energy Transduction** Hipkins, M.F. & Baker, N.R. (Eds)
.... Hardback 0-947946-63-2 **$54.00**
.... Paperback 0-947946-51-9 **$36.00**
16. **Human Genetic Diseases** Davies, K.E. (Ed)
.... Hardback 0-947946-76-4 **$60.00**
.... Paperback 0-947946-75-6 **$34.00**
14. **Nucleic Acid Hybridisation** Hames, B.D. & Higgins, S.J. (Eds)
.... Hardback 0-947946-61-6 **$60.00**
.... Paperback 0-947946-23-3 **$36.00**
13. **Immobilised Cells and Enzymes** Woodward, J. (Ed)
.... Hardback 0-947946-60-8 **$0.00**
12. **Plant Cell Culture** Dixon, R.A. (Ed)
.... Paperback 0-947946-22-5 **$36.00**
11a. **DNA Cloning: Volume I** Glover, D.M. (Ed)
.... Paperback 0-947946-18-7 **$36.00**
11b. **DNA Cloning: Volume II** Glover, D.M. (Ed)
.... Paperback 0-947946-19-5 **$36.00**
10. **Virology** Mahy, B.W.J. (Ed)
.... Paperback 0-904147-78-9 **$40.00**
9. **Affinity Chromatography** Dean, P.D.G., Johnson, W.S. & others (Eds)
.... Paperback 0-904147-71-1 **$36.00**
7. **Microcomputers in Biology** Ireland, C.R. & Long, S.P. (Eds)
.... Paperback 0-904147-57-6 **$36.00**
6. **Oligonucleotide Synthesis** Gait, M.J. (Ed)
.... Paperback 0-904147-74-6 **$38.00**
5. **Transcription and Translation** Hames, B.D. & Higgins, S.J. (Eds)
.... Paperback 0-904147-52-5 **$38.00**
3. **Iodinated Density Gradient Media** Rickwood, D. (Ed)
.... Paperback 0-904147-51-7 **$36.00**

Sets

Essential Molecular Biology: Volumes I and II as a set Brown, T.A. (Ed)
.... Spiralbound hardback 0-19-963114-X **$118.00**
.... Paperback 0-19-963115-8 **$78.00**
Antibodies: Volumes I and II as a set Catty, D. (Ed)
.... Paperback 0-19-963063-1 **$70.00**
Cellular and Molecular Neurobiology Chad, J. & Wheal, H. (Eds)
.... Spiralbound hardback 0-19-963255-3 **$133.00**
.... Paperback 0-19-963254-5 **$79.00**
Protein Structure and Protein Function: Two-volume set Creighton, T.E. (Ed)
.... Spiralbound hardback 0-19-963064-X **$114.00**
.... Paperback 0-19-963065-8 **$80.00**
DNA Cloning: Volumes I, II, III as a set Glover, D.M. (Ed)
.... Paperback 1-85221-069-9 **$92.00**
Molecular Plant Pathology: Volumes I and II as a set Gurr, S.J., McPherson, M.J. & others (Eds)
.... Spiralbound hardback 0-19-963354-1 **$0.00**
.... Paperback 0-19-963353-3 **$0.00**
Protein Purification Methods, and Protein Purification Applications, two-volume set Harris, E.L.V. & Angal, S. (Eds)
.... Spiralbound hardback 0-19-963048-8 **$98.00**
.... Paperback 0-19-963049-6 **$68.00**
Diagnostic Molecular Pathology: Volumes I and II as a set Herrington, C.S. & McGee, J. O'D. (Eds)
.... Spiralbound hardback 0-19-963241-3 **$0.00**
.... Paperback 0-19-963240-5 **$0.00**
Receptor Biochemistry; Receptor-Effector Coupling; Receptor-Ligand Interactions Hulme, E.C. (Ed)
.... Spiralbound hardback 0-19-963096-8 **$193.00**
.... Paperback 0-19-963097-6 **$125.00**
Signal Transduction Milligan, G. (Ed)
.... Spiralbound hardback 0-19-963296-0 **$60.00**
.... Paperback 0-19-963295-2 **$38.00**
Human Cytogenetics: Volumes I and II as a set (2/e) Rooney, D.E. & Czepulkowski, B.H. (Eds)
.... Hardback 0-19-963314-2 **$130.00**
.... Paperback 0-19-963313-4 **$90.00**
Peptide Hormone Secretion/Peptide Hormone Action Siddle, K. & Hutton, J.C. (Eds)
.... Spiralbound hardback 0-19-963072-0 **$135.00**
.... Paperback 0-19-963073-9 **$90.00**

ORDER FORM for USA and Canada

Qty	ISBN	Author	Title	Amount
			S&H	
			CA and NC residents add appropriate sales tax	
			TOTAL	

Please add shipping and handling: $2.50 for first book, ($1.00 each book thereafter)

Name ....

Address ....

....

.... Zip ....

[ ] Please charge $ .... to my credit card
Mastercard/VISA/American Express (circle appropriate card)

Acct. .... Expiry date ....

Signature ....

Credit card account address if different from above:

....

.... Zip ....

[ ] I enclose a cheque for $...........

Mail orders to: Order Dept. Oxford University Press, 2001 Evans Road, Cary, NC 27513